D1261933

Enzymes

A Practical Introduction to Structure, Mechanism, and Data Analysis

Robert A. Copeland

Withdrawn
IOWA STATE UNIVERSITY
of Science and Technology
Library

WILEY-VCH

NEW YORK • CHICHESTER • WEINHEIM • BRISBANE • SINGAPORE • TORONTO

Robert A. Copeland
Principal Research Scientist
Inflammatory Diseases Research
The DuPont Merck Research Laboratories
P.O. Box 80400
Wilmington, DE 19880-0400

Cover Art: Structure of bacterial dihydrofolate reductase with the competitive inhibitor methotrexate bound in the active site cleft. Based on the x-ray crystal structure reported by Filman, Bolin, Matthews, and Kraut (*J. Biol. Chem.* (1982) **257**, 13650). Figure provided by Dr. James L. Meek, DuPont Merck Research Laboratories.

This book is printed on acid-free paper.

Copyright © 1996 by Wiley-VCH, Inc. All rights reserved.

Originally published as ISBN 1-56081-903-0.

Published simultaneously in Canada.

No part of this publication may be reproduced, stored in a retrieval system or transmitted in any form or by any means, electronic, mechanical, photocopying, recording, scanning or otherwise, except as permitted under Sections 107 or 108 of the 1976 United States Copyright Act, without either the prior written permission of the Publisher, or authorization through payment of the appropriate per-copy fee to the Copyright Clearance Center, 222 Rosewood Drive, Danvers, MA 01923, (508) 750-8400, fax (508) 750-4744. Requests to the Publisher for permission should be addressed to the Permissions Department, John Wiley & Sons, Inc., 605 Third Avenue, New York, NY 10158-0012, (212) 850-6011, fax (212) 850-6008, E-Mail: PERMREQ@WILEY.COM.

Library of Congress Cataloging-in-Publication Data:
Copeland, Robert Allen.
Enzymes : a practical introduction to structure, mechanism, and data analysis / Robert A. Copeland.
p. cm.
Includes bibliographical references and index.
ISBN 0-471-18621-X (alk. paper)
1. Enzymes. 2. Enzymology. I. Title.
QP601.C753 1996
574.19'25--dc20 96-6035
CIP

Printed in the United States of America.

10 9 8 7 6 5 4 3

To Clyde Worthen
for teaching me all the important lessons;
arigato sensei.

And to Theodore (Doc) Janner
for stoking the fire.

Preface

The latter half of this century has seen an unprecedented expansion in our knowledge and use of enzymes in a broad range of basic research and industrial applications. Enzymes are the catalytic cornerstones of metabolism, and as such are the focus of intense research within the biomedical community. Indeed enzymes remain the most common targets for therapeutic intervention within the pharmaceutical industry. Since ancient times enzymes also have played central roles in many manufacturing processes, such as in the production of wine, cheese, and breads. During the 1970s and 1980s much of the focus of the biochemical community shifted to the cloning and expression of proteins through the methods of molecular biology. Recently, some attention has shifted back to physicochemical characterization of these proteins, and their interactions with other macromolecules and small molecular weight ligands (e.g., substrates, activators, and inhibitors). Hence, there has been a resurgence of interest in the study of enzyme structures, kinetics, and mechanisms of catalysis.

The availability of up-to-date, introductory-level textbooks, however, has not kept up with the growing demand. I first became aware of this void while teaching introductory courses at the medical and graduate student level at the University of Chicago. I found that there were a number of excellent advanced texts that covered different aspects of enzymology with heavy emphasis on the theoretical basis for much of the science. The more introductory texts that I found were often quite dated and did not offer the blend of theoretical and practical information that I felt was most appropriate for a broad audience of students. I thus developed my own set of lecture notes for these courses, drawing material from a wide range of textbooks and primary literature.

In 1993, I left Chicago to focus my research on the utilization of basic enzymology and protein science for the development of therapeutic agents to combat human diseases. To pursue this goal I joined the scientific staff of the DuPont Merck Pharmaceutical Company. During my first year with this company, a group of associate scientists expressed to me their frustration at being unable to find a textbook on enzymology that met their needs for guidance in laboratory protocols and data analysis at an appropriate level and at the same time provide them with some relevant background on the scientific basis of their experiments. These dedicated individuals asked if I would prepare and present a course on enzymology at this introductory level.

Using my lecture notes from Chicago as a foundation, I prepared an extensive set of notes and intended to present a year-long course to a small group of associate scientists in an informal, over-brown-bag-lunch fashion. After the lectures had been announced, however, I was shocked and delighted to find that more than 200 people were registered for this course! The makeup of the student body ranged from individuals with associate degrees in medical technology to chemists and molecular biologists who had doctorates. This convinced me that there was indeed a growing interest and need for a new introductory enzymology text that would attempt to balance the theoretical and practical aspects of enzymology in such a way as to fill the needs of graduate and medical students, as well as research scientists and technicians who are actively involved in enzyme studies.

The text that follows is based on the lecture notes for the enzymology course just described. It attempts to fill the practical needs I have articulated, while also giving a reasonable introduction to the theoretical basis for the laboratory methods and data analyses that are covered. I hope that this text will be of use to a broad range of scientists interested in enzymes. The material covered should be of direct use to those actively involved in enzyme research in academic, industrial, and government laboratories. It also should be useful as a primary text for senior undergraduate or first-year graduate course, in introductory enzymology. However, in teaching a subject as broad and dynamic as enzymology, I have never found a single text that would cover all of my students' needs; I doubt that the present text will be an exception. Thus, while I believe this text can serve as a useful foundation, I encourage faculty and students to supplement the material with additional readings from the literature cited at the end of each chapter, and the primary literature that is continuously expanding our view of enzymes and catalysis.

In attempting to provide a balanced introduction to enzymes in a single, readable volume I have had to present some of the material in a rather cursory fashion; it is simply not possible, in a text of this format, to be comprehensive in such an expansive field as enzymology. I hope that the literature citations will at least pave the way for readers who wish to delve more deeply into particular areas. Overall, the intent of this book is to get people *started* in the laboratory and in their thinking about enzymes. It provides sufficient experimental and data handling methodologies to permit one to begin to design and perform experiments with enzymes, while at the same time providing a theoretical framework in which to understand the basis of the experimental work. Beyond this, if the book functions as a stepping-stone for the reader to move on to more comprehensive and in-depth treatments of enzymology, it will have served its purpose.

Robert A. Copeland
Wilmington, Delaware

Acknowledgments

It is a great pleasure for me to thank the many friends and coworkers who have helped me in the preparation of this work. Many of the original lecture notes from which this text has developed were generated while I was teaching a course on biochemistry for first-year medical students at the University of Chicago, along with the late Howard S. Tager. Howard contributed greatly to my development as a teacher and writer. His untimely death was a great loss to many of us in the biomedical community; I dearly miss his guidance and friendship.

As described in the preface, the notes on which this text is based were significantly expanded and reorganized to develop a course of enzymology for employees and students at the DuPont Merck Pharmaceutical Company. I am grateful for the many discussions with students during this course, which helped to refine the final presentation. I especially thank Diana Blessington for the original suggestion of a course of this nature. That a graduate-level course of this type could be presented within the structure of a for-profit pharmaceutical company speaks volumes for the insight and progressiveness of the management of DuPont Merck. I particularly thank James M. Trzaskos, Robert C. Newton, Ronald L. Magolda, and Pieter B. Timmermans for not only tolerating, but embracing this endeavor.

Many colleagues and coworkers contributed suggestions and artwork for this text. I thank June Davis, Petra Marchand, Diane Lombardo, Robert Lombardo, John Giannaras, Jean Williams, Randi Dowling, Drew Van Dyk, Rob Bruckner, Bill Pitts, Carl Decicco, Pieter Stouten, Jim Meek, Bill DeGrado, Steve Betz, Hank George, Jim Wells, and Charles Craik for their contributions.

Finally, and most importantly, I wish to thank my wife, Nancy, and our children, Lindsey and Amanda, for their constant love, support, and encouragement, without which this work could not have been completed.

Contents

"All the mathematics in the world is no substitute for a reasonable amount of common sense."

W. W. Cleland

CHAPTER 1

A Brief History of Enzymology

Life depends on a well-orchestrated series of chemical reactions. Many of these reactions, however, proceed too slowly on their own to sustain life. Hence nature has designed catalysts, which we now refer to as *enzymes*, to greatly accelerate the rates of these chemical reactions. The catalytic power of enzymes facilitates life processes in essentially all life-forms from viruses to man. Many enzymes retain their catalytic potential after extraction from the living organism, and it did not take long for mankind to recognize and exploit the catalytic power of enzyme for commercial purposes. In fact, the earliest known references to enzymes are from ancient texts dealing with the manufacture of cheeses, breads, and alcoholic beverages, and for the tenderizing of meats. Today enzymes continue to play key roles in many food and beverage manufacturing processes and are ingredients in numerous consumer products, such as laundry detergents (which dissolve protein-based stains with the help of proteolytic enzymes). Enzymes are also of fundamental interest in the health sciences, since many disease processes can be linked to the aberrant activities of one or a few enzymes. Hence, much of modern pharmaceutical research is based on the search for potent and specific inhibitors of these enzymes. The study of enzymes and the action of enzymes has thus fascinated scientists since the dawn of history, not only to satisfy erudite interest but also because of the utility of such knowledge for many practical needs of society. This brief chapter sets the stage for our studies of these remarkable catalysts by providing a historic background of the development of enzymology as a science. We shall see that while enzymes are today the focus of basic academic research, much of the early history of enzymology is linked to the practical application of enzyme activity in industry.

1.1 Enzymes in Antiquity

The oldest known reference to the commercial use of enzymes comes from a description of wine making in the Codex of Hammurabi (ancient Babylon, circa 2100 B.C.). The use of microorganisms as enzyme sources for fermentation was widespread among ancient people. References to these processes can be found in writings not only from Babylon but also from the early civilizations of Rome, Greece, Egypt, China, India. Ancient texts also contain a number of references to the related process of vinegar production, which is based on the enzymatic conversion of alcohol to acetic acid. Vinegar, it appears, was a common staple of ancient life, being used not only for food storage and preparation but also for medicinal purposes.

Dairy products were another important food source in ancient societies. Because in those days fresh milk could not be stored for any reasonable length of time, the conversion of milk to cheese became a vital part of food production, making it possible for the farmer to bring his product to distant markets in an acceptable form. Cheese is prepared by curdling milk via the action of any of a number of enzymes. The substances most commonly used for this purpose in ancient times were ficin, obtained as an extract from fig trees, and rennin, as rennet, an extract of the lining of the fourth stomach of a multiple-stomach animal, such as a cow. A reference to the enzymatic activity of ficin can, in fact, be found in Homer's classic, the Iliad:

> As the juice of the fig tree curdles milk, and thickens it in a moment though it be liquid, even so instantly did Paeëon cure fierce Mars.

The philosopher Aristotle likewise wrote several times about the process of milk curdling and offered the following hypothesis for the action of rennet:

> Rennet is a sort of milk; it is formed in the stomach of young animals while still being suckled. Rennet is thus milk which contains fire, which comes from the heat of the animal while the milk is undergoing concoction.

Another food staple throughout the ages is bread. The leavening of bread by yeast, which results from the enzymatic production of carbon dioxide, was well known and widely used in ancient times. The importance of this process to ancient society can hardly be overstated.

Meat tenderizing is another enzyme-based process that has been used since antiquity. Inhabitants of many Pacific islands have known for centuries that the juice of the papaya fruit will soften even the toughest meats. The active enzyme in this plant extract is a protease known as papain, which is used even today in commercial meat tenderizers. When the British Navy began exploring the Pacific islands in the 1700s, they encountered the use of the papaya fruit as a meat tenderizer and as a treatment for ringworm. Reports of these native uses of the papaya sparked a great deal of interest in eighteenth-century Europe, and may, in part, have led to some of the more systematic studies of digestive enzymes that ensued soon after.

1.2 Early Enzymology

While the ancients made much practical use of enzymatic activity, these early applications were based purely on empirical observations and folklore, rather than any systematic studies or appreciation for the chemical basis of the processes being utilized. In the eighteenth and nineteenth centuries scientists began to study the actions of enzymes in a more systematic fashion. The process of digestion seems to have been a popular subject of investigation during the years of the enlightenment. Wondering how predatory birds manage to digest meat without a gizzard, the famous French scientist Réaumur (1683–1757) performed some of the earliest studies on the digestion of buzzards. Réaumur designed a metal tube with a wire mesh at one end that would hold a small piece of meat immobilized, to protect it from the physical action of the stomach tissue. He found that when a tube containing meat was inserted into the stomach of a buzzard, the meat was digested within 24 hours. Thus he concluded that digestion must be a chemical rather than a merely physical process, since the meat in the tube had been digested by contact with the gastric juices (or as he referred to them, "a solvent"). He tried the same experiment with a piece of bone and with a piece of a plant. He found that while meat was digested, and the bone was greatly softened by the action of the gastric juices, the plant material was impervious to the "solvent"; this was probably the first experimental demonstration of enzyme specificity.

Réaumur's work was expanded by Spallanzani (1729–1799), who showed that the digestion of meat encased in a metal tube took place in the stomachs of a wide variety of animals, including humans. Using his own gastric juices, Spallanzani was able to perform digestion experiments on pieces of meat in vitro (in the laboratory). These experiments illustrated some critical features of the active ingredient of gastric juices: by means of a control experiment in which meat treated with an equal volume of water did not undergo digestion Spallanzani demonstrated the presence of a specific active ingredient in gastric juices. He also showed that the process of digestion is temperature dependent, and that the time required for digestion is related to the amount of gastric juices applied to the meat. Finally, he demonstrated that the active ingredient in gastric juices is unstable outside the body; that is, its ability to digest meat wanes with storage time.

Today we recognize all the foregoing properties as common features of enzymatic reactions, but in Spallanzani's day these were novel and exciting findings. The same time period saw the discovery of enzyme activities in a large number of other biological systems. For example, a peroxidase from the horseradish was described, and the action of α-amylase in grain was observed. These early observations all pertained to materials—crude extract from plants or animals—that contained enzymatic activity.

During the latter part of the nineteenth century scientists began to attempt fractionations of these extracts to obtain the active ingredients in pure form.

For example, in 1897 Bertrand partially purified the enzyme laccase from tree sap, and Buchner, using the "pressed juice" from rehydrated dried yeast, demonstrated that alcoholic fermentation could be performed in the absence of living yeast cells. Buchner's report contained the interesting observation that the activity of the pressed juice diminished within 5 days of storage at ice temperatures. However, if the juice was supplemented with cane sugar, the activity remained intact for up to 2 weeks in the ice box. This is probably the first report of a now well-known phenomenon—the stabilization of enzymes by substrate. It was also during this period that Kühne, studying catalysis in yeast extracts, first coined the term "enzyme" (the word derives from the medieval Greek word *enzymos*, which relates to the process of leavening bread).

1.3 The Development of Mechanistic Enzymology

As enzymes became available in pure, or partially pure forms, scientists' attention turned to obtaining a better understanding of the details of the reaction mechanisms catalyzed by enzymes. The concept that enzymes form complexes with their substrate molecules was first articulated in the late nineteenth century. It is during this time period that Emil Fischer proposed the "lock and key" model for the stereochemical relationship between enzymes and their substrates; this model emerged as a result of a large body of experimental data on the stereospecificity of enzyme reactions. In the early twentieth century, experimental evidence for the formation of an enzyme–substrate complex as a reaction intermediate was reported. One of the earliest of these studies, reported by Brown in 1902, focused on the velocity of enzyme-catalyzed reactions. Brown made the insightful observation that unlike simple diffusion-limited chemical reactions, in enzyme-catalyzed reactions "it is quite conceivable... that the time elapsing during molecular union and transformation may be sufficiently prolonged to influence the general course of the action." Brown then went on to summarize the available data that supported the concept of formation of an enzyme–substrate complex:

> There is reason to believe that during inversion of cane sugar by invertase the sugar combines with the enzyme previous to inversion. C. O'Sullivan and Tompson... have shown that the activity of invertase in the presence of cane sugar survives a temperature which completely destroys it if cane sugar is not present, and regard this as indicating the existence of a combination of the enzyme and sugar molecules. Wurtz [1880] has shown that papain appears to form an insoluble compound with fibrin previous to hydrolysis. Moreover, the more recent conception of E. Fischer with regard to enzyme configuration and action, also implies some form of combination of enzyme and reacting substrate.

Observations like these set the stage for the derivation of enzyme rate equations, by mathematically modeling enzyme kinetics with the explicit involvement of an intermediate enzyme–substrate complex. In 1903 Victor

Henri published the first successful mathematical model for describing enzyme kinetics. In 1913, in a much more widely read paper, Michaelis and Menten expanded on the earlier work of Henri and rederived the enzyme rate equation that today bears their names. The Michaelis–Menten equation, or more correctly the Henri–Michaelis–Menten equation, is a cornerstone of much of the modern analysis of enzyme reaction mechanisms.

The question of how enzymes accelerate the rates of chemical reactions puzzled scientists until the development of transition state theory in the first half of the twentieth century. In 1948 the famous physical chemist Linus Pauling suggested that enzymatic rate enhancement was achieved by stabilization of the transition state of the chemical reaction by interaction with the enzyme active site. This hypothesis, which was widely accepted, is supported by the experimental observation that enzymes bind very tightly to molecules designed to mimic the structure of the transition state of the catalyzed reaction.

In the 1950s and 1960s scientists reexamined the question of how enzymes achieve substrate specificity in light of the need for transition state stabilization by the enzyme active site. New hypotheses, such as the "induced fit" model of Koshland emerged at this time to help rationalize the competing needs of substrate binding affinity and reaction rate enhancement by enzymes. During this time period, scientists struggled to understand the observation that metabolic enzyme activities can be regulated by small molecules other than the substrates or direct products of an enzyme. Studies showed that indirect interactions between distinct binding sites within an enzyme molecule could occur, even though these binding sites were quite distant from one another. In 1965 Monod, Wyman, and Changeux developed the theory of allosteric transitions to explain these observations. Thanks in large part to this landmark paper, we now know that many enzymes, and nonenzymatic ligand binding proteins, display allosteric regulation

1.4 Studies of Enzyme Structure

One of the tenets of modern enzymology is that catalysis is intimately related to the molecular interactions that take place between a substrate molecule and components of the enzyme molecule, the exact nature and sequence of these interactions defining per se the catalytic mechanism. Hence, the application of physical methods to elucidate the structures of enzymes has had a rich history and continues to be of paramount importance today. Spectroscopic methods, x-ray crystallography, and more recently, multidimensional NMR methods have all provided a wealth of structural insights on which theories of enzyme mechanisms have been built. In the early part of the twentieth century, x-ray crystallography became the premier method for solving the structures of small molecules. In 1926 James Sumner published the first crystallization of an enzyme, urease (Figure 1.1). Sumner's paper was a landmark contribution, not

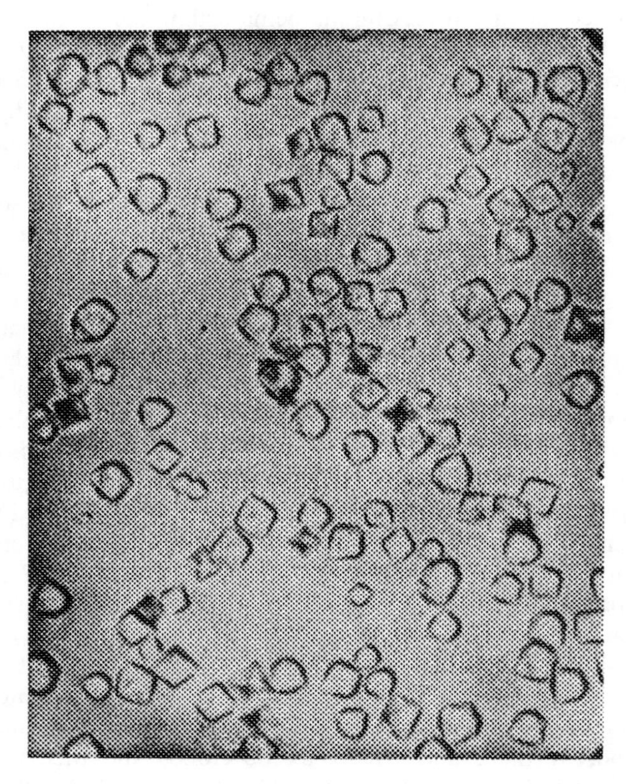

Figure 1.1 Photomicrograph of urease crystals (728 × magnification), the first reported crystals of an enzyme. [From J. B. Sumner, *J. Biol. Chem.* **69**, 435–441 (1926), with permission.]

only because it portended the successful application of x-ray diffraction for solving enzyme structures, but also because a detailed analysis allowed Sumner to show unequivocally that the crystals were composed of protein and that their dissolution in solvent led to enzymatic activity. These observations were very important to the development of the science of enzymology because they firmly established the protein composition of enzymes, a view that had not been widely accepted by Sumner's contemporaries.

Sumner's crystallization of urease opened a floodgate and was quickly followed by reports of numerous other enzyme crystals. Within 20 years of Sumner's first paper more than 130 enzyme crystals had been documented. It was not, however, until the late 1950s that protein structures began to be solved through x-ray crystallography. In 1957 Kendrew became the first to deduce from x-ray diffraction the entire three-dimensional structure of a protein, myoglobin. Soon after, the crystal structures of many proteins, including enzymes, were solved by these methods. Today, the structural insights gained from x-ray crystallography and multidimensional NMR studies are commonly used to elucidate the mechanistic details of enzyme catalysis, and to design new ligands (substrate and inhibitor molecules) to bind at specific sites within the enzyme molecule.

The deduction of three-dimensional structures from x-ray diffraction or NMR methods depends on knowledge of the arrangement of amino acids along the polypeptide chain of the protein; this arrangement is known as the amino acid sequence. To determine the amino acid sequence of a protein, the component amino acids must be hydrolyzed in a sequential fashion from the polypeptide chain and identified by chemical or chromatographic analysis. Edman and coworkers developed a method for the sequential hydrolysis of amino acids from the N-terminus of a polypeptide chain. In 1957 Sanger reported the first complete amino acid sequence of a protein, the hormone insulin, utilizing the chemistry developed by Edman. In 1963 the first amino acid sequence of an enzyme, ribonuclease, was reported.

1.5 Enzymology Today

Fundamental questions still remain regarding the detailed mechanisms of enzyme activity and its relationship to enzyme structure. The two most powerful tools that have been brought to bear on these questions in modern times are the continued development and use of biophysical probes of protein structure, and the application of molecular biological methods to enzymology. X-ray crystallography continues to be used routinely to solve the structures of enzymes and of enzyme–ligand complexes. In addition, new NMR methods and magnetization transfer methods make possible the assessment of the three-dimensional structures of small enzymes in solution, and the structure of ligands bound to enzymes, respectively.

The recent application of Laue diffraction with synchrotron radiation sources holds the promise of allowing scientists to determine the structures of reaction intermediates during enzyme turnover, hence to develop detailed pictures of the individual steps in enzyme catalysis. Other biophysical methods, such as optical (e.g., circular dichroism, UV–visible, fluorescence) and vibrational (e.g., infrared, Raman) spectroscopies, have likewise been applied to questions of enzyme structure and reactivity in solution. Technical advances in many of these spectroscopic methods have made them extremely powerful and accessible tools for the enzymologist. Furthermore, the tools of molecular biology have allowed scientists to clone and express enzymes in foreign host organisms with great efficiency. Enzymes that had never before been isolated have been identified and characterized by molecular cloning. Overexpression of enzymes in prokaryotic hosts has allowed the purification and characterization of enzymes that are available only in minute amounts from their natural sources. This has been a tremendous advance for protein science in general.

Table 1.1 Examples of Enzyme Inhibitors as Potential Drugs

Inhibitor/Drug	Disease/Condition	Enzyme Target
Captopril	Hypertension	Angiotensin-converting enzyme
Aspirin, ibuprofen, DUP697	Inflammation, pain, fever	Prostaglandin synthase
β-Lactam antibiotics	Bacterial infections	D-Ala-D-Ala transpeptidase
Norfloxacin	Urinary tract infections	DNA gyrase
Lovastatin	High cholesterol	HMG CoA reductase
Clavulanate	Bacterial resistance	β-Lactamase
Acyclovir	Herpes	Viral DNA polymerase
Cyclosporin	Organ transplantation	Cyclophilin/calcineurin
Brequinar	Organ transplantation	Dihydroorotate dehydrogenase
Zidovudine	AIDS	HIV reverse transcriptase
Omeprazole	Peptic ulcers	H^+,K^+-ATPase
Allopurinol	Gout	Xanthine oxidase
Trimethoprim	Bacterial infection	Dihydrofolate reductase
Sulfamethoxazole	Bacterial infection, malaria	Dihydropteorate synthase
Methotrexate	Cancer	Dihydrofolate reductase
Fluorouracyl	Cancer	Thymidilate synthase
Phenelzine	Depression	Brain monoamine oxidase
Acetazolamide	Glaucoma	Carbonic anhydrase
Argatroban	Coagulation	Thrombin
Finazteride	Benign prostate hyperplasia	Testosterone-5-α-reductase
Sorbinil	Diabetic retinopathy	Aldose reductase
ICI-200,808	Emphysema	Neutrophil elastase
PD-116124	Metabolism of antineoplastic drugs	Purine nucleoside phosphorylase
Ro 42-5892	Hypertension	Renin
DuP450	AIDS	HIV protease
WIN 51711	Common cold	Rhinovirus coat protein
FK-506	Organ transplantation, autoimmune disease	FK-506 binding protein
Ly-256548	Inflammation	Phospholipase A_2
(2-Furyl)-acryloyl-Gly-Phe-Phe	Lung elastin degradation in cystic fibrosis	*Pseudomonas* elastase
PALA	Cancer	Aspartate transcarbamoylase
Enoximone	Congestive heart failure ischemia	cAMP phosphodiesterase
Zileuton	Allergy	5-Lipoxygenase
3-Fluorovinylglycine	Bacterial infection	Alanine racemase
SQ-29072	Hypertension, congestive heart failure, analgesia	Enkephalinase
Testolactone	Hormone-dependent tumors	Aromatase
Threo-5-fluoro-L-dihydroorotate	Cancer	Dihydroorotase
Nitecapone	Parkinson's disease	Catechol-*O*-methyltransferase
Candoxatril	Hypertension, congestive heart failure	Atriopeptidase

Source: Adapted and expanded from M. A. Navia and M. A. Murcko, *Curr. Opinions Struct. Biol.* **2**, 202–210 (1992).

The tools of molecular biology also allow investigators to manipulate the amino acid sequence of an enzyme at will. The use of site-directed mutagenesis (in which one amino acid residue is substituted for another) and deletional mutagenesis (in which sections of the polypeptide chain of a protein are eliminated) have allowed enzymologists to pinpoint the chemical groups that participate in ligand binding and in specific chemical steps during enzyme catalysis.

The study of enzymes remains of great importance to the scientific community and to society in general. We continue to utilize enzymes in many industrial applications. The traditional roles of enzymes in food and beverage manufacturing are still in use today. In modern times, the role of enzymes in consumer products and in chemical manufacturing has expanded greatly. Enzymes are used today in such varied applications as stereospecific chemical synthesis, laundry detergents, and cleaning kits for contact lenses.

Perhaps one of the most exciting fields of modern enzymology is the application of enzyme inhibitors as drugs in human and veterinary medicine. Many of the drugs that are commonly used today function by inhibiting specific enzymes that are associated with the disease process. Aspirin, for example, one of the most widely used drugs in the world, elicits its anti-inflammatory efficacy by acting as an inhibitor of the enzyme prostaglandin synthase. As illustrated in Table 1.1, enzymes take part in a wide range of human pathophysiologies, and many specific enzyme inhibitors have been developed to combat their activities, thus acting as therapeutic agents. Several of the inhibitors listed in Table 1.1 are the result of the combined use of biophysical methods for assessing enzyme structure and classical pharmacology in what is commonly referred to as rational or structure-based drug design. This approach uses the structural information obtained from x-ray crystallography or NMR spectroscopy to determine the topology of the enzyme active site. Next, model building is performed to design molecules that would fit well into this active site pocket. These molecules are then synthesized and tested as inhibitors. Several iterations of this procedure often lead to extremely potent inhibitors of the target enzyme.

The list in Table 1.1 will continue to grow as our understanding of disease state physiology increases. There remain thousands of enzymes involved in human physiology that have yet to be isolated or characterized. As more and more disease-related enzymes are discovered and characterized, new inhibitors will need to be designed to arrest the actions of these catalysts, in the continuing effort to fulfill unmet human medical needs.

1.6 Summary

We have seen in this chapter that the science of enzymology has a long and rich history. From phenomenological observations, enzymology has grown to a quantitative molecular science. For the rest of this book we shall view enzymes from a chemical prospective, attempting to understand the actions of these proteins in the common language of chemical and physical forces. While the vital importance of enzymes in biology cannot be overstated, the understanding of their structures and functions remains a problem of chemistry.

References and Further Reading

Rather than providing an exhaustive list of primary references for this historical chapter, I refer the reader to a few modern texts that have done an excellent job of presenting a more detailed and comprehensive treatment of the history of enzymology. Not only do these books provide good descriptions of the history of science and the men and women who made that history, but they are also quite entertaining and inspiring reading—enjoy them!

Friedmann, H. C., Ed. (1981) *Enzymes*, Hutchinson Ross, Stroudsburg, PA. [This book is part of the series "Benchmark Papers in Biochemistry." In it, Friedmann has compiled reprints of many of the most influential publications in enzymology from the eighteenth through twentieth centuries, along with insightful commentaries on these papers and their importance in the development of the science.]

Judson, H. F. (1980) *The Eighth Day of Creation*, Simon & Schuster, New York. [This extremely entertaining book chronicles the history of molecular biology, including protein science and enzymology, in the twentieth century.]

Kornberg, A. (1989) *For the Love of Enzymes. The Odyssey of a Biochemist*, Harvard University Press, Cambridge, MA. [An autobiographical look at the career of a Nobel Prize–winning biochemist.]

Werth, B. (1994) *The Billion Dollar Molecule*, Simon & Schuster, New York. [An interesting, if biased, look at the modern science of structure-based drug design.]

CHAPTER

2

Chemical Bonds and Reactions in Biochemistry

The hallmark of enzymes is their remarkable ability to catalyze very specific chemical reactions of biological importance. Some enzymes are so well designed for this purpose that they can accelerate the rate of a chemical reaction by as much as 10^{12}-fold over the spontaneous rate of the uncatalyzed reaction! This incredible rate enhancement results from the juxtaposition of chemically reactive groups within the binding pocket of the enzyme (the enzyme active site) and other groups from the target molecule (substrate), in a way that facilitates the reaction steps required to convert the substrate into the reaction product. In subsequent chapters we shall explore the structural details of these reactive groups and describe how their interactions with the substrate result in the enhanced reaction rates typical of enzymatic catalysis. First, however, we must understand the chemical bonding and chemical reactions that take place both in enzymes and in the simpler molecules on which enzymes act. This chapter is meant as a review of material covered in introductory chemistry courses (basic chemical bonds, some of the reactions associated with these bonds); however, a thorough understanding of the concepts covered here will be essential to understanding the material in Chapters 3–11.

2.1 Atomic and Molecular Orbitals

2.1.1 Atomic Orbitals

Chemical reactions, whether enzyme-catalyzed or not, proceed mainly through the formation and cleavage of chemical bonds. The bonding patterns seen in molecules result from the interactions between electronic orbitals of individual atoms to form molecular orbitals. Here we shall review these orbitals and some properties of the chemical bonds they form.

Recall from your introductory chemical courses that electrons occupy discrete atomic orbitals surrounding the atomic nucleus. The first model of electronic orbitals, proposed by Niels Bohr, viewed these orbitals as a collection of simple concentric circular paths of electron motion orbiting the atomic nucleus. While this was a great intellectual leap in thinking about atomic structure, the Bohr model failed to explain many of the properties of atoms that were known at the time. For instance, the simple Bohr model does not explain many of the spectroscopic features of atoms. In 1926 Erwin Schrödinger applied a quantum mechanical treatment to the problem of describing the energy of a simple atomic system. This resulted in the now-famous Schrödinger wave equation, which can be solved exactly for a simple one-proton, one-electron system (the hydrogen atom).

Without going into great mathematical detail, we can say that the application of the Schrödinger equation to the hydrogen atom indicates that atomic orbitals are quantized; that is, only certain orbitals are possible, and these have well-defined, discrete energies associated with them. Any atomic orbital can be uniquely described by a set of three values associated with the orbital, known as *quantum numbers*. The first or principal quantum number describes the effective volume of the orbital and is given the symbol n. The second quantum number, l, is referred to as the orbital shape quantum number, because this value describes the general probability density over space of electrons occupying that orbital. Together the first two quantum numbers provide a description

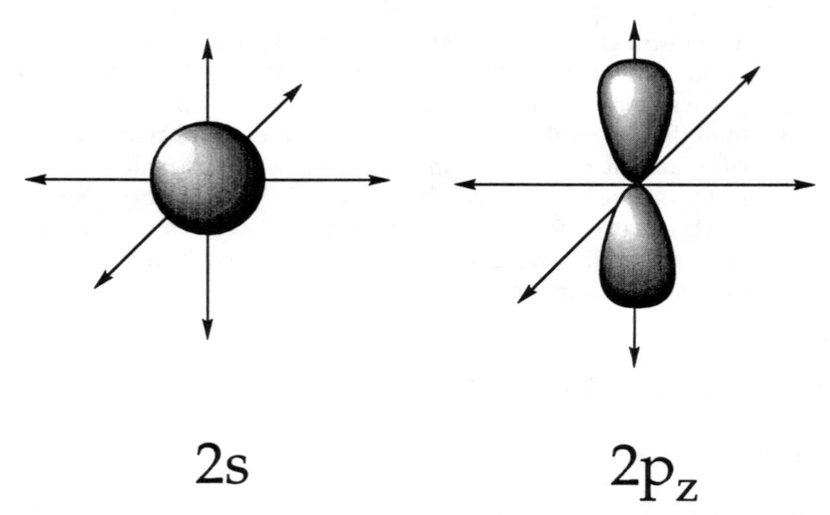

Figure 2.1 Spatial representations of the electron distribution in s and p orbitals.

of the *spatial probability distribution* of electrons within the orbital. These descriptions lead to the familiar pictorial representations of atomic orbitals, as shown in Figure 2.1 for the 1s and 2p orbitals.

The third quantum number, m_l, describes the orbital angular momentum associated with the electronic orbital and can be thought of as describing the orientation of that orbital in space, relative to some arbitrary fixed axis. With these three quantum numbers, one can specify each particular electronic orbital of an atom. Since each of these orbitals is capable of accommodating two electrons, however, we require a fourth quantum number to uniquely identify each individual electron in the atom.

The fourth quantum number, m_s, is referred to as the electron spin quantum number. It describes the direction in which the electron is imagined to spin with respect to an arbitrary fixed axis in a magnetic field (Figure 2.2). Since no

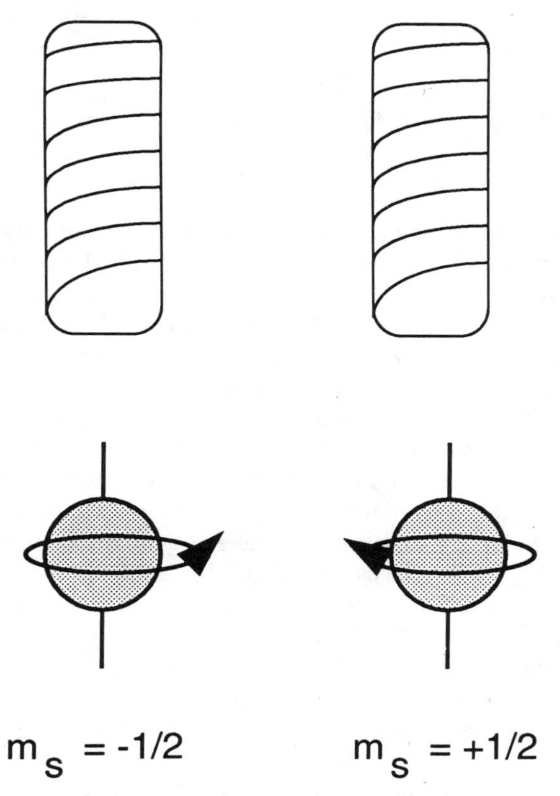

Figure 2.2 Electron spin represented as rotation of a particle in a magnetic field. The two spin "directions" of the electron are represented as clockwise ($m_s = +\frac{1}{2}$) and counterclockwise ($m_s = -\frac{1}{2}$) rotations. The coil-bearing rectangles schematically represent the magnetic fields.

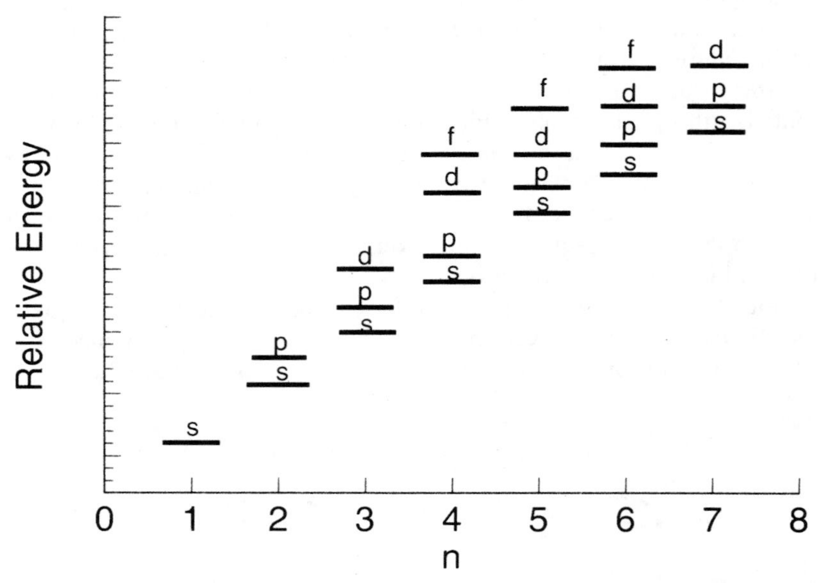

Figure 2.3 The aufbau principle for the order of filling of atomic orbitals, s, p, d, and f.

two electron can have the same values for all four quantum numbers, it follows that two electrons within the same atomic orbital must be spin-paired; that is, if one is spinning clockwise ($m_s = +\frac{1}{2}$), the other must be spinning counter-clockwise ($m_s = -\frac{1}{2}$). This concept, known as the Pauli exclusion principle, is often depicted graphically by representing the spinning electron as an arrow pointing either up or down, within an atomic orbital.

Thus we see that associated with each atomic orbital is a discrete amount of potential energy; that is, the orbitals are quantized. Electrons fill these orbitals according to the potential energy associated with them; low energy orbitals fill first, followed by higher energy orbitals in ascending energetic order (the aufbau principle). By schematizing the energetic order of atomic orbitals, as illustrated in Figure 2.3, we can inventory the electrons in the orbitals of an atom. For example, each atom of the element helium contains two electrons; since both electrons occupy the 1s orbital, we designate this by the shorthand notation $1s^2$. Lithium contains 3 electrons and, according to Figure 2.3, has the configuration $1s^2 2s^1$. When dealing with nonspherical orbitals, such as the p, d, and f orbitals, we must keep in mind that more than one atomic orbital is associated with each orbital set that is designated by a combination of n and l quantum numbers. For example, the 2p orbital set consists of three atomic orbitals: $2p_x$, $2p_y$, and $2p_z$. Hence, the 2p orbital set can accommodate 6 electrons. Likewise, a d orbital set can accommodate a total of 10 electrons (5 orbitals, with 2 electrons per orbital), and an f orbital set can accommodate 14 electrons.

Table 2.1 Electronic Configurations of the Elements Most Commonly Found in Biological Tissues

Element	Number of Electrons	Orbital Configuration
Hydrogen	1	$1s^1$
Carbon	6	$1s^2 2s^2 2p^2$
Nitrogen	7	$1s^2 2s^2 2p^3$
Oxygen	8	$1s^2 2s^2 2p^4$
Phosphorus	15	$1s^2 2s^2 2p^6 3s^2 3p^3$
Sulfur	16	$1s^2 2s^2 2p^6 3s^2 3p^4$

A survey of the biological tissues in which enzyme naturally occur indicates that the elements listed in Table 2.1 are present in highest abundance. Because of their abundance in biological tissue, these are the elements we most often encounter as components of enzyme molecules. For each of these atoms, the highest energy s and p orbital electrons are those that are capable of participating in chemical reactions, and these are referred to as *valence electrons* (the electrons in the lower energy orbitals are chemically inert and are referred to as closed-shell electrons). In the carbon atom, for example, the two 1s electrons are the closed-shell type, while the four electrons in the 2s and 2p orbitals are valence electrons, and are thus available for bond formation.

2.1.2 Molecular Orbitals

If two atoms can approach each other at close enough range, and if their valence orbitals are of appropriate energy and symmetry, the two valence atomic orbitals (one from each atom) can combine to form two molecular orbitals: a *bonding* and an *antibonding* molecular orbital. The bonding orbital occurs at a lower potential energy than the original two atomic orbitals; hence electron occupancy in this orbital promotes bonding between the atoms because of a net stabilization of the system (molecule). The antibonding orbital, in contrast, occurs at a higher energy than the original atomic orbitals; electron occupancy in this molecular orbital would thus be destabilizing to the molecule.

Let us consider the molecule H_2. The two 1s orbitals from each hydrogen atom, each containing a single electron, approach each other until they overlap to the point that the two electrons are shared by both nuclei (i.e., a bond is formed). At this point the individual atomic orbital character is lost and the two electrons are said to occupy a *molecular orbital*, resulting from the mixing of the original two atomic orbitals. Since there were originally two atomic orbitals that mixed, there must result two molecular orbitals. As illustrated in Figure 2.4, one of these molecular orbitals occurs at a lower potential energy than the original atomic orbital, hence stabilizes the molecular bond; this orbital is referred to as a *bonding orbital* (in this case a σ-bonding orbital, as

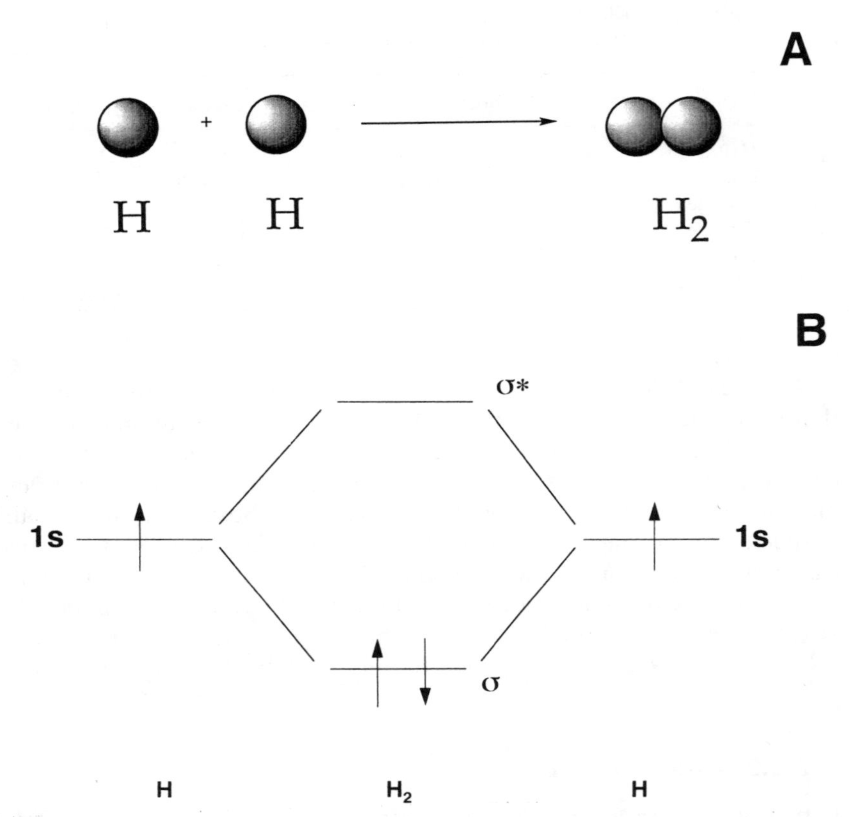

Figure 2.4 (A) Schematic representation of two s orbitals on separate hydrogen atoms combining to form a bonding σ molecular orbital. (B) Energy level diagram for the combination of two hydrogen s orbitals to form a bonding and antibonding molecular orbital in the H_2 molecule.

discussed shortly). The other molecular orbital occurs at higher potential energy (displaced by the same amount as the bonding orbital). Because the higher energy of this orbital makes it destabilizing relative to the atomic orbitals, it is referred to as an *antibonding orbital* (again, in this case a σ-antibonding orbital, σ^*). The electrons fill the molecular orbitals in order of potential energy, each orbital being capable of accommodating two electrons. Thus for H_2 both electrons from the 1s orbitals of the atoms will occupy the σ-bonding molecular orbital in the molecule.

Now let us consider the diatomic molecule F_2. The orbital configuration of the fluorine atom is $1s^2 2s^2 2p^5$. The two s orbitals and two of the three p orbitals are filled and will form equal numbers of bonding and antibonding molecular orbitals, canceling any net stabilization of the molecule. The partially filled p orbitals, one on each atom of fluorine, can come together to form one bonding and one antibonding molecular orbital in the diatomic molecule

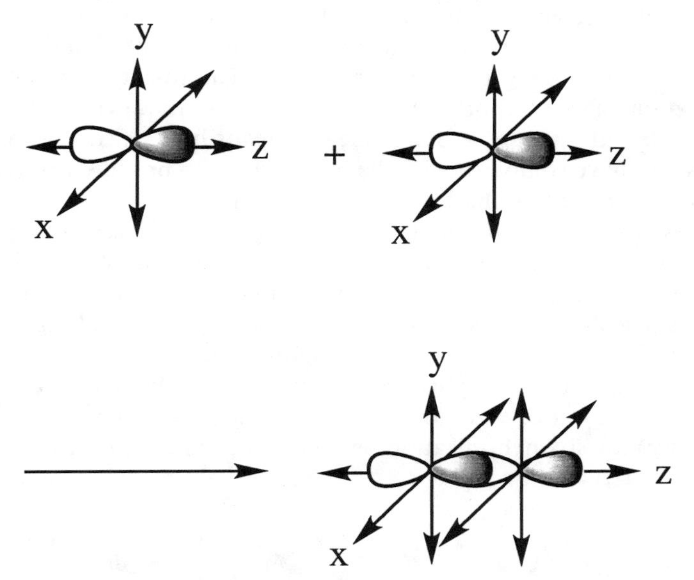

Figure 2.5 Combination of two p_z atomic orbitals by end-to-end overlap to form a σ-type molecular orbital.

F_2. As illustrated in Figure 2.5, the lobes of the two valence p_z orbitals overlap *end to end* in the bonding orbital; molecular orbitals that result from such end-to-end overlaps are referred to as sigma orbitals (σ). The bonding orbital is designated by the symbol σ, and the accompanying antibonding orbital is designated by the symbol σ^*. The bond formed between the two atoms in the F_2 molecules is therefore referred to as a *sigma bond.* Because the σ orbital is lower in energy than the σ^* orbital, both the electrons from the valence atomic orbit will reside in the σ molecular orbital when the molecule is at rest (e.g., when it is in its lowest energy form, referred to as the *ground state* of the molecule).

2.1.3 Hybrid Orbitals

For elements in the second row of the periodic table (Li, Be, C, N, O, F, and Ne), the 2s and 2p orbitals are so close in energy that they can interact to form orbitals with combined, or mixed, s and p orbital character. These *hybrid orbitals* provide a means of maximizing the number of bonds an atom can form, while retaining the greatest distance between bonds, to minimize repulsive forces. The hybrid orbitals formed by carbon are the most highly studied, and the most germane to our discussion of enzymes.

From the orbital configuration of carbon ($1s^2 2s^2 2p^2$), we can see that the similar energies of the 2s and 2p orbital sets in carbon provide four electrons that can act as valence electrons, giving carbon the ability to form four bonds to other atoms. Three types of hybrid orbital are possible, and they result in three different bonding patterns for carbon. The first type results from the combination of one 2s orbital with three 2p orbitals, yielding four hybrid orbitals referred to as sp^3 orbitals (the exponents here reflects the number of p orbitals that have combined with the one s orbital to produce the hybrids). The four sp^3 orbitals allow the carbon atom to form four σ bonds that lie along the apices of a tetrahedron, as shown in Figure 2.6C. The second type of hybrid orbital, sp^2, results from the mixing of one 2s orbital and two 2p orbitals. These hybrid orbitals allow for three trigonal planar bonds to form (Figure 2.6B). When a single 2p orbital combines with a 2s orbital, the resulting single hybrid orbital is referred to as an *sp* orbital (Figure 2.6A).

Let us look at the sp^2 hybrid case in more detail. We have said that the 2p orbital set consists of three p orbitals that can accommodate a total of six electrons. With sp^2 hybridization, we have accounted for two of the three p

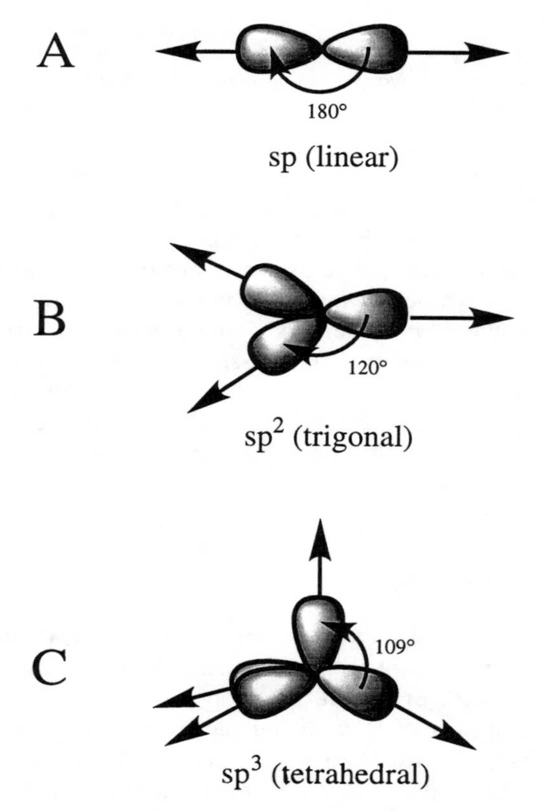

Figure 2.6 Spatial electron distributions of hybrid orbitals: (A) sp hybridization, (B) sp^2 hybridization, and (C) sp^3 hybridization.

orbitals available in forming three trigonal planar σ bonds, as in the case of ethylene (Figure 2.7A). On each carbon atom, this hybridization leaves one orbital, of pure p character, which is available for bond formation. These orbitals can interact with one another to form a bond by edge-to-edge orbital overlap, above and below the plane defined by the sp^2 σ bonds (Figure 2.7B). This type of edge-to-edge orbital overlap results in a different type of molecular orbital, referred to as a π orbital. As illustrated in Figure 2.7B, the overlap of the p orbitals provides for bonding electron density above and below the interatomic axis, resulting in a *pi bond* (π). Of course, as with σ bonds, for every π orbital formed, there must be an accompanying antibonding orbital at higher energy, which is denoted by the symbol π^*. Thus along the interatomic axis of ethylene we find two bonds: one σ bond, and one π bond. This combination is said to form a *double bond* between the carbon atoms. A shorthand notation for this bonding situation is to draw two parallel lines connecting the carbon atoms:

```
H         H
 \       /
  C === C
 /       \
H         H
```

A similar situation arises when we consider sp hybridization. In this case we have two mutually perpendicular p orbitals on each carbon atom available for π orbital formation. Hence, we find one σ bond (from the sp hybrid orbitals) and two π bonds along the interatomic axis. This *triple bond* is denoted by drawing three parallel lines connecting the two carbon atoms, as in acetylene:

H—C≡C—H

Not all the valence electrons of the atoms in a molecule are shared in the form of covalent bonds. In many cases it is energetically advantageous to the molecule to have *unshared* electrons that are essentially localized to a single atom; these electrons are often referred to as non-bonding or *lone pair* electrons. Whereas electrons within bonding orbitals are denoted as lines drawn between atoms of the molecule, lone pair electrons are usually depicted as a pair of dots surrounding a particular atom. (Combinations of atoms and molecules represented by means of these conventions are referred to as Lewis structures.)

2.1.4 Resonance and Aromaticity

Let us consider the ionized form of acetic acid that occurs in aqueous solution at neutral pH (i.e., near physiological conditions). The carbon bound to the oxygen atoms uses sp^2 hybridization; it forms a σ bond to the other carbon, a σ bond to each oxygen atom, and one π bond to one of the oxygen atoms. Thus, one oxygen atom would have a double bond to the carbon atom, while

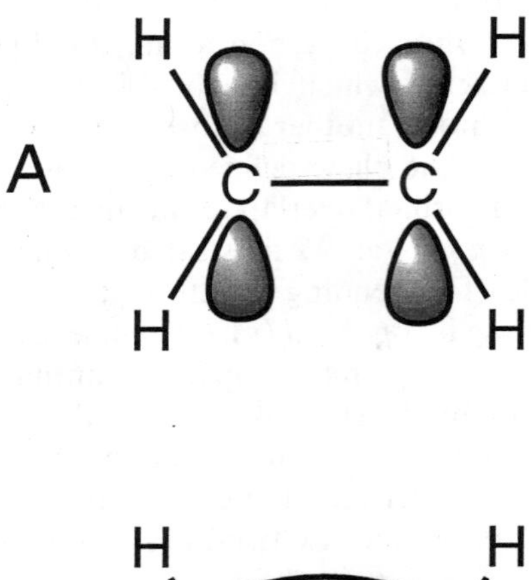

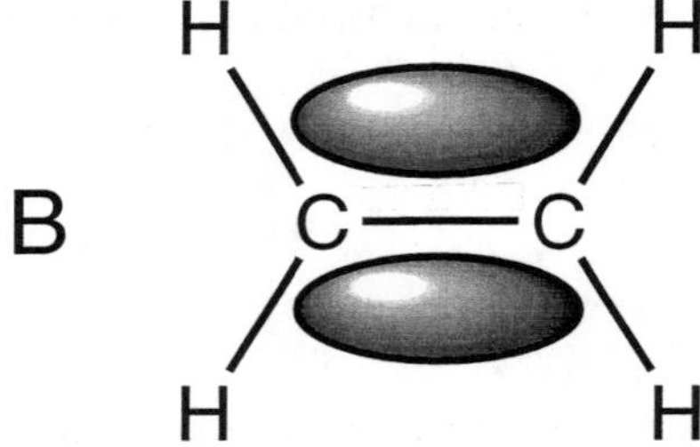

Figure 2.7 Hybrid bond formation in ethylene. (A) The bonds are illustrated as lines, and the remaining p orbitals lobes form edge-to-edge contacts. (B) The p orbital combine to form a π bond with electron density above and below the interatomic bond axis defined by the σ bond between the carbon atoms.

the other has a single bond to the carbon and is negatively charged. Suppose that we could somehow identify the individual oxygen atoms in this molecule—by, for example, using an isotopically labeled oxygen (^{18}O rather than ^{16}O) at one site. Which of the two would form the double bond to carbon, and which would act as the anionic center?

$$H_3C{-}C(-{}^{18}O^-)(={}^{16}O) \quad \text{or} \quad H_3C{-}C(={}^{18}O)(-{}^{16}O^-)$$

Both of these are reasonable electronic forms, and there is no basis on which to choose one over the other. In fact, neither is truly correct, because in reality we find that the π bond (or more correctly, the π electron density) is *delocalized* over both oxygen atoms. In some sense neither forms a single bond nor a double bond to the carbon atom, but rather both behave as if they shared the π bond between them. We refer to these two alternative electronic forms of the molecule as *resonance structures* and sometimes represent this arrangement by drawing a double-headed arrow between the two forms:

$$H_3C{-}C(={}^{16}O){-}{}^{18}O^- \longleftrightarrow H_3C{-}C(={}^{18}O){-}{}^{16}O^-$$

Alternatively, the resonance form is illustrated as follows, to emphasize the delocalization of the π electron density:

$$H_3C{-}C({}^{18}O^{\delta-})({}^{16}O^{\delta-})$$

Now let us consider the organic molecule benzene (C_6H_6). The carbon atoms are arranged in a cyclic pattern, forming a planar hexagon. To account for this, we must assume that there are three double bonds among the carbon–carbon bonds of the molecule. Here are the two resonance structures:

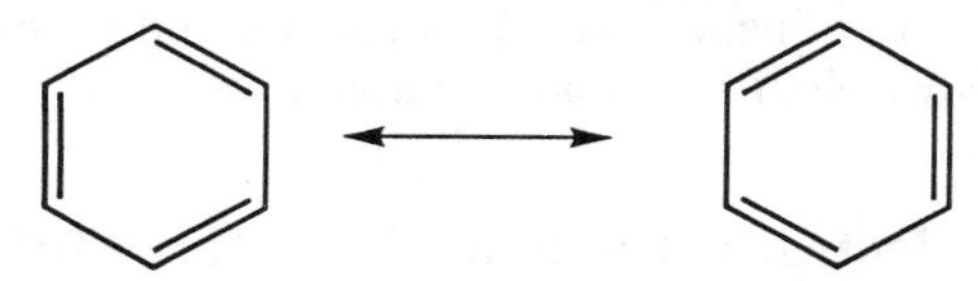

Now a typical carbon–carbon single bond has a bond length of roughly 1.54 Å, while a carbon–carbon double bond is only about 1.35 Å long. When the crystal structure of benzene was determined, it was found that all the carbon–carbon bonds were the same length, 1.45 Å, which is intermediate between the expected lengths for single and double bonds. How can we rationalize this result? The answer is that the π orbitals are not localized to the p_z orbitals of two adjacent carbon atoms (Figure 2.8, left: here the plane defined by the carbon ring system is arbitrarily assigned as the x,y plane); rather, they are delocalized over all six carbon p_z orbitals. To emphasize this π system delocalization, many organic chemists choose to draw benzene as a hexagon enclosing a circle (Figure 2.8, right) rather than a hexagon of carbon with three discrete double bonds.

The delocalization of the π system in molecules like benzene tends to stabilize the molecule relative to what one would predict on the basis of three isolated double bonds. This difference in stability is referred to as the *resonance energy stabilization.* For example, consider the heats of hydrogenation (breaking the carbon–carbon double bond and adding two atoms of hydrogen), using H_2 and platinum catalysis, for the series cyclohexene ($\Delta H = 28.6$ kcal/mol), cyclohexadiene, benzene. If each double bond were energetically equivalent, one would expect the ΔH value for cyclohexadiene hydrogenation to be twice that of cyclohexene (-57.2 kcal/mol), and that is approximately what is observed. Extending this argument further, one would expect the ΔH value for

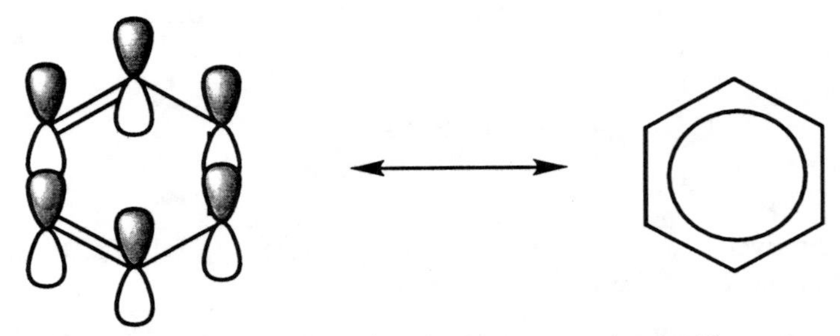

Figure 2.8 Two common representations for the benzene molecule. The representation on the right is meant to emphasize the π-system delocalization in this molecule.

benzene (if it behaved energetically equivalent to cyclohexatriene) to be three times that of cyclohexene, 85.8 kcal/mol. Experimentally, however, the ΔH of hydrogenation of benzene is found to be only -49.8 kcal/mol, a resonance energy stabilization of 36 kcal/mol! This stabilizing effect of π orbital delocalization has an important influence over the structure and chemical reactivities of these molecules, as we shall see in later chapters.

2.1.5 Different Electronic Configurations Have Different Potential Energies

We have seen how electrons distribute themselves among molecular orbitals according to the potential energies of those molecular orbitals. The specific distribution of the electrons within a molecule among the different electronic molecular orbitals defines the electronic configuration or *electronic state* of that molecule. The electronic state that imparts the least potential energy to that molecule will be the most stable form of that molecule under normal conditions. This electronic configuration is referred to as the *ground state* of the molecule. Any alternative electronic configuration of higher potential energy than the ground state is referred to as an *excited state* of the molecule.

Let us consider a simple carbonyl such as formaldehyde (CH_2O):

H
\
C=O
/
H

In the ground state electronic configuration of this molecule, the π-bonding orbital is the highest energy orbital that contains electrons. This orbital is referred to as the Highest Occupied Molecular Orbital or (HOMO). The π^* molecular orbital is the next highest energy molecular orbital and, in the ground state, does not contain any electron density. This orbital is said to be the Lowest Unoccupied Molecular Orbital (LUMO). Suppose that somehow we were able to move an electron from the π to the π^* orbital. The molecule would now have a different electronic configuration that would impart to the overall molecule more potential energy; that is, the molecule would be in an excited electronic state. Now, since in this excited state we have moved an electron from a bonding (π) to an antibonding (π^*) orbital, the overall molecule has acquired more antibonding character. As a consequence, the nuclei will occur at a longer equilibrium interatomic distance, relative to the ground state of the molecule.

In other words, the potential energy minimum (also referred to as the zero-point energy) for the excited state occurs when the atoms are further apart from one another than they are for the potential energy minimum of the ground state. Since the π electrons are localized between the carbon and oxygen atoms in this molecule, it will be the carbon–oxygen bond length that is most affected by the change in electronic configuration; the carbon–hydrogen bond lengths are essentially invariant between the ground and excited states. The nuclei, however, are not fixed in space, but can vibrate in both the ground and excited electronic states of the molecule. Hence, each electronic state of a molecule has built upon it a manifold of vibrational substates.

The foregoing concepts are summarized in Figure 2.9, which shows a potential energy diagram for the ground and one excited state of the molecule. An important point to glean from this figure is that even though the potential

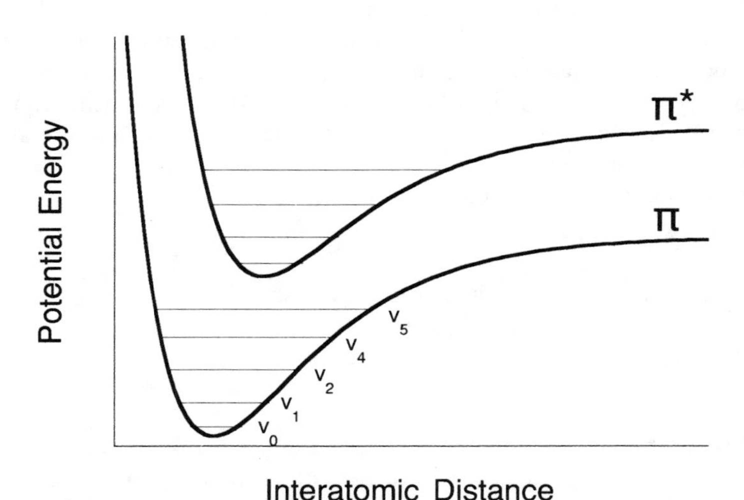

Figure 2.9 Potential energy diagram for the ground and one excited electronic state of a molecule. The potential wells labeled π and π^* represent the potential energy profiles of the ground and excited electronic states, respectively. The sublevels within each of these potential wells, labeled v_n, represent the vibrational substates of the electronic states.

minima of the ground and excited states occur at different equilibrium interatomic distances, vibrational excursions within either electronic state can bring the nuclei into register with their equilibrium positions at the potential minimum of the other electronic state. In other words, a molecule in the ground electronic state can, through vibrational motions, transiently sample the interatomic distances associated with the potential energy minimum of the excited electronic state, and vice versa.

2.2 Thermodynamics of Chemical Reactions

In freshman chemistry we were introduced to the concept of free energy, ΔG, which combined the first and second laws of thermodynamics to yield the familiar formula:

$$\Delta G = \Delta H - T\Delta S \tag{2.1}$$

where ΔG is the change in free energy of the system during a reaction at constant temperature (T) and pressure, ΔH is the change in enthalpy (heat), and ΔS is the change in entropy (a measure of disorder or randomness) associated with the reaction. Some properties of ΔG should be kept in mind. First, ΔG is less than zero (negative) for a spontaneous reaction and greater than zero (positive) for a nonspontaneous reaction. That is, a reaction for which ΔG is negative will proceed spontaneously with the liberation of energy. A reaction for which ΔG is positive will proceed only if energy is supplied to drive the reaction. Second, ΔG is always zero at equilibrium. Third, ΔG is a *path-independent function.* That is, the value of ΔG is dependent on the starting and ending states of the system but not on the path used to go from the starting point to the end point. Finally, while the value of ΔG gives information on the spontaneity of a reaction, it does not tell us anything about the rate at which the reaction will proceed.

Consider the following reaction:

$$A + B \rightleftharpoons C + D$$

Recall that the ΔG for such a reaction is given by:

$$\Delta G = \Delta G^0 + RT \ln\left(\frac{[C][D]}{[A][B]}\right) \tag{2.2}$$

where ΔG^0 is the free energy for the reaction under standard conditions of all reactants and products at a concentration of 1.0 M (1.0 atm for gases). The terms in brackets, such as [C], are the molar concentrations of the reactants and products of the reaction, the symbol "*ln*" is shorthand for the natural, or base *e*, logarithm, and R and T refer to the ideal gas constant (1.98×10^{-3} kcal/mol·degree) and the temperature in degrees Kelvin (298 K for average room temperature, 25°C, and 310 K for physiological temperature, 37°C), respectively. Since, by definition, $\Delta G = 0$ at equilibrium, it follows that under equilibrium conditions:

$$\Delta G^0 = -RT \ln\left(\frac{[C][D]}{[A][B]}\right) \tag{2.3}$$

For many reactions, including many enzyme-catalyzed reactions, the values of ΔG^0 have been tabulated. Thus knowing the value of ΔG^0 one can easily calculate the value of ΔG for the reaction at any displacement from equilibrium. Examples of these types of calculations can be found in any introductory chemistry or biochemistry text.

Because free energy of reaction is a path-independent quantity, it is possible to drive an unfavorable (nonspontaneous) reaction by *coupling* it to a favorable (spontaneous) one. Suppose, for example, that the product of an unfavorable reaction was also a reactant for a thermodynamically favorable reaction. As long as the absolute value of ΔG was greater for the second reaction, the overall reaction would proceed spontaneously. Suppose that the reaction $A \rightleftharpoons B$ had a ΔG^0 of $+5$ kcal/mol, and the reaction $B \rightleftharpoons C$ had a ΔG^0 of -8 kcal/mol. What would be the ΔG^0 value for the net reaction $A \rightleftharpoons C$?

$A \rightleftharpoons B$	$\Delta G^0 = +5$ kcal/mol
$B \rightleftharpoons C$	$\Delta G^0 = -8$ kcal/mol
$A \rightleftharpoons C$	$\Delta G^0 = -3$ kcal/mole

Thus, the overall reaction would proceed spontaneously. In our scheme, B would appear on both sides of the overall reaction and thus could be ignored. Such a species is referred to as a *common intermediate*. This mechanism of providing a thermodynamic driving force for unfavorable reactions is quite common in biological catalysis.

As we shall see in Chapter 3, many enzyme use nonprotein cofactors in the course of their catalytic reactions. In some cases these cofactors participate directly in the chemical transformations of the reactants (referred to as *substrates* by enzymologists) to products of the enzymatic reaction. In many other cases, however, the reactions of the cofactors are used to provide the thermodynamic driving force for catalysis. Oxidation and reduction reactions of metals, flavins, and reduced nicotinamide adenine dinucleotide (NADH) are commonly used for this purpose in enzymes. For example, the enzyme cytochrome *c* oxidase uses the energy derived from reduction of its metal cofactors to drive the transport of protons across the inner mitochondrial membrane, from a region of low proton concentration to an area of high proton concentration. This energetically unfavorable transport of protons could not proceed without coupling to the exothermic electrochemical reactions of the metal centers. Another very common coupling reaction is the hydrolysis of adenosine triphosphate (ATP) to adenosine diphosphate (ADP) and inorganic phosphate (P_i). Numerous enzymes drive their catalytic reactions by coupling to ATP hydrolysis, because of the high energy yield of this reaction.

2.2.1 The Transition State of Chemical Reactions

A chemical reaction proceeds spontaneously when the free energy of the product state is lower than that of the reactant state (i.e., $\Delta G < 0$). As we have stated, the path taken from reactant to product does not influence the free energies of these beginning and ending states, and hence cannot affect the spontaneity of the reaction. The path can, however, greatly influence the rate at which a reaction will proceed, depending on the free energies associated with any intermediate state the molecule must access as it proceeds through the reaction. Most of the chemical transformations observed in enzyme-catalyzed reactions involve the breaking and formation of covalent bonds. If we consider a reaction in which an existing bond between two nuclei is replaced by an alternative bond with a new nucleus, we could envision that at some instant during the reaction a chemical entity would exist that had both the old and new bonds partially formed, that is, a state in which the old and new bonds are simultaneously broken and formed. This molecular form would be extremely unstable, hence would be associated with a very large amount of free energy. For the reactant to be transformed into the product of the chemical reaction, the molecule must transiently access this unstable form, known as the *transition state* of the reaction. Consider, for example, the formation of an alcohol by the nucleophilic attack of a primary alkyl halide by a hydroxide ion:

$$RCH_2Br + OH^- \rightleftharpoons RCH_2OH + Br^-$$

We can consider that the reaction proceeds through a transition state in which the carbon is simultaneously involved in partial bonds between the oxygen and the bromine:

$$RCH_2Br + OH^- \rightarrow [HO\text{---}CH_2R\text{---}Br] \rightarrow RCH_2OH + Br^-$$

where the species in brackets is the transition state of the reaction and partial bonds are indicated by dashes. Figure 2.10 illustrates this reaction scheme in terms of the free energies of the species involved. (Note that for simplicity, the various molecular states are represented as lines designating the position of the potential minimum of each state. Each of these states is more correctly described by the potential wells shown in Figure 2.9, but diagrams constructed according to this convention are less easy to follow.)

In the free energy diagram of Figure 2.10, the x axis is referred to as the reaction coordinate and tracks the progressive steps in going from reactant to product. This figure makes it clear that the transition state represents an energy barrier that the reaction must overcome in order to proceed. The higher the energy of the transition state in relation to the reactant state, the more difficult it will be for the reaction to proceed. Once, however, the system has attained sufficient energy to reach the transition state, the reaction can proceed effortlessly downhill to the final product state (or, alternatively, collapse back to the reactant state). Most of us have experienced a macroscopic analogy of this situation in riding a bicycle. When we encounter a hill we must pedal hard,

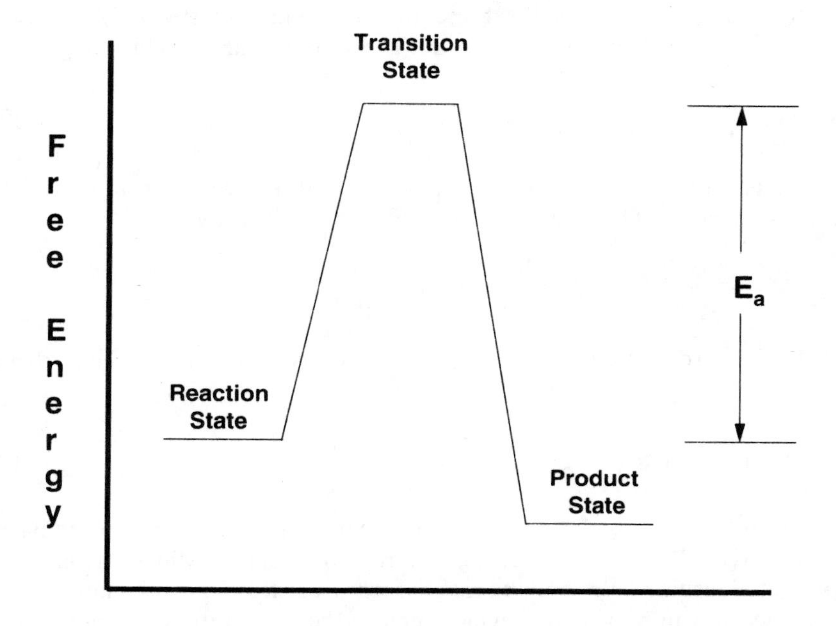

Figure 2.10 Free energy diagram for the reaction profile of a typical chemical reaction, a chemical reaction. The activation energy E_a is the energetic difference between the reactant state and the transition state of the reaction.

exerting energy to ascend the incline. Having reached the crest of the hill, however, we can take our feet off the pedals and coast downhill without further exertion.

The energy required to proceed from the reactant state to the transition state, which is known as the *activation energy* or energy barrier of the reaction, is the difference in free energy between these two states. The activation energy is given the symbol E_a or $\Delta G^{\ddagger}$. This energy barrier is an important concept for our subsequent discussions of enzyme catalysis. This is because the height of the activation energy barrier can be directly related to the rate of a chemical reaction. To illustrate, let us consider a unimolecular reaction in which the reactant A decomposes to B through the transition state $A^{\ddagger}$. The activation energy for this reaction is E_a. The equilibrium constant for A going to $A^{\ddagger}$ will be $[A^{\ddagger}]/[A]$. Using this, and rearranging Equation 2.3 with substitution of E_a for ΔG^0, we obtain:

$$[A^{\ddagger}] = [A]\exp\left(-\frac{E_a}{RT}\right) \tag{2.4}$$

The transition state will decay to product with the same frequency as that of the stretching vibration of the bond that is being ruptured to produce the product molecule. It can be shown that this vibrational frequency is given by:

$$v = \frac{k_B T}{h} \tag{2.5}$$

where v is the vibrational frequency, k_B is the Boltzmann constant, and h is Planck's constant. The rate of loss of [A] is thus given by:

$$\frac{-d[A]}{dt} = v[A^{\ddagger}] = [A]\left(\frac{k_B T}{h}\right)\exp\left(-\frac{E_a}{RT}\right) \tag{2.6}$$

and the first-order rate constant for the reaction is thus given by the Arrhenius equation:

$$k = \left(\frac{k_B T}{h}\right)\exp\left(-\frac{E_a}{RT}\right) \tag{2.7}$$

From Equations 2.6 and 2.7 it is obvious that as the activation energy barrier increases (i.e., E_a becomes larger), the rate of reaction will decrease in an exponential fashion. We shall see in Chapter 4 that this concept relates directly to the mechanism by which enzymes achieve the acceleration of reaction rates characteristic of enzyme-catalyzed reactions.

It is important to recognize that the transition state of a chemical reaction is, under most conditions, an extremely unstable and short-lived species. Some chemical reactions go through intermediate states that are more long-lived and stable than the transition state. In some cases, these intermediate species exist long enough to be kinetically isolated and studied. When present, these intermediate states appear as local free energy minima (dips) in the free energy diagram of the reaction, as illustrated in Figure 2.11. Enzyme-catalyzed reactions go through intermediate states like this, mediated by the specific interactions of the protein and/or enzyme cofactors with the reactants and products of the chemical reaction being catalyzed. We shall have more to say about these intermediate species in Chapter 4.

2.3 Acid–Base Chemistry

In freshman chemistry we were introduced to the common Lewis definition of acids and bases: a *Lewis acid* is any substance that can act as an *electron pair acceptor*, and a *Lewis base* is any substance that can act as an *electron pair donor*. In many enzymatic reactions, protons are transferred from one chemical species to another, hence the alternative *Brønsted–Lowry* definition of acids and bases becomes very useful for dealing with these reactions. In the Brønsted–Lowry classification, an *acid* is any substance that can *donate a proton*, and a *base* is any substance that can *accept a proton* by reacting with

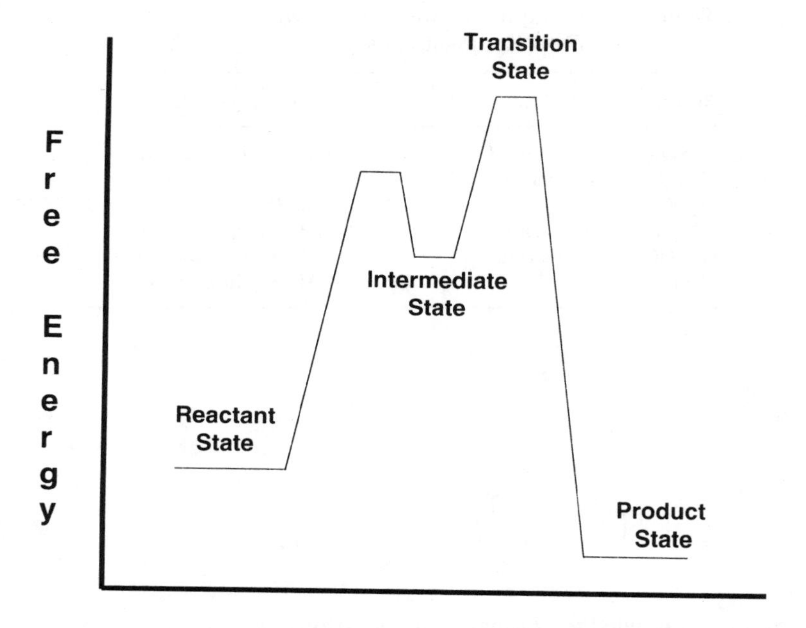

Figure 2.11 Free energy diagram for a chemical reaction that proceeds through the formation of a chemical intermediate.

a Brønsted–Lowry Acid. After donating its proton, a Brønsted–Lowry acid is converted to its *conjugate base*.

Table 2.2 gives some examples of Brønsted–Lowry acids and their conjugate bases. For all these pairs, we are dealing with the transfer of a hydrogen ion (proton) from the acid to some other species (often the solvent) to form the conjugate base. A convenient means of measuring the hydrogen ion concentration in aqueous solutions is the pH scale. The term "pH" is a shorthand notation for the negative base-10 logarithm of the hydrogen ion concentration:

$$\text{pH} = -\log[\text{H}^+] \tag{2.8}$$

Consider the dissociation of a weak Brønsted–Lowry acid (HA) into a proton (H^+) and its conjugate base (A^-) in aqueous solution.

$$\text{HA} \rightleftharpoons \text{H}^+ + \text{A}^-$$

The dissociation constant for the acid, K_a, is given by the ratio $[H^+][A^-]/[HA]$. Let us define the pK_a for this reaction as the negative base-10 logarithm

Table 2.2 Examples of Brønsted–Lowry Acids and Their Conjugate Bases

Brønsted–Lowry Acid	Conjugate Base
H_2SO_4 (sulfuric acid)	HSO_4^- (bisulfate ion)
HCl (hydrochloric acid)	Cl^- (chloride ion)
H_3O^+ (hydronium ion)	H_2O (water)
NH_4^+ (ammonium ion)	NH_3 (ammonia)
CH_3COOH (acetic acid)	CH_3COO^- (acetate ion)
H_2O (water)	OH^- (hydroxide ion)

of K_a:

$$pK_a = -\log\left(\frac{[H^+][A^-]}{[HA]}\right) \tag{2.9}$$

or, using our knowledge of logarithmic relationships, we can write:

$$pK_a = \log(HA) - \log(A^-) - \log(H^+) \tag{2.10}$$

Note that the last term in Equation 2.10 is identical to our definition of pH (Equation 2.8). Using this equality, and again using our knowledge of logarithmic relationships we obtain:

$$pK_a = \log\left(\frac{[HA]}{[A^-]}\right) + pH \tag{2.11}$$

or, rearranging (note the inversion of the logarithmic term):

$$pH = pK_a + \log\left(\frac{[A^-]}{[HA]}\right) \tag{2.12}$$

Equation 2.12 is known as the *Henderson–Hasselbalch* equation, and it provides a convenient means of calculating the pH of a solution from the concentrations of a Brønsted–Lowry acid and its conjugate base. Note that when the concentrations of acid and conjugate base are equal, the value of $[A^-]/[HA]$ is 1.0, and thus the value of $\log([A^-]/[HA])$ is zero. At this point the pH will be exactly equal to the pK_a. This provides a useful working definition of pK_a:

> The pK_a is the pH value at which half the Brønsted–Lowry acid is dissociated to its conjugate base and a proton.

Let us consider a simple example of this concept. Suppose that we dissolve acetic acid into water and begin titrating the acid with hydroxide ion equivalents by addition of NaOH. If we measure the pH of the solution after each addition, we will obtain a titration curve similar to that shown in Figure 2.12. Two points should be drawn from this figure. First, such a titration curve provides a convenient means of graphically determining the pK_a value of the species being titrated. Second, we see that at pH values near the pK_a, it takes a great deal of NaOH to effect a change in the pH value. This resistance to pH change in the vicinity of the pK_a of the acid is referred to as *buffering capacity*, and it is an important property to be considered in the preparation of solutions for enzyme studies. As we shall see in Chapter 6, the pH at which an enzyme reaction is performed can have a dramatic effect on the rate of reaction and on the overall stability of the protein. As a rule, therefore, specific buffering molecules, whose pK_a values match the pH for optimal enzyme activity, are added to enzyme solutions to maintain the solution pH near the pK_a of the buffer.

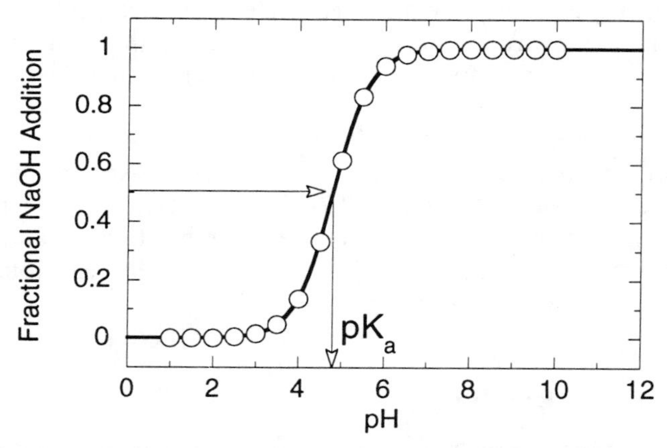

Figure 2.12 Hypothetical titration curve for a weak acid illustrating the graphical determination of the acid's pK_a.

2.4 Noncovalent Interactions

All the properties of molecules we have discussed until now have led us to focus on the formation, stabilization, and breaking of covalent bonds between atoms of the molecule. Molecules can interact with one another by a number of noncovalent forces, as well. These weaker attractive forces are very important in biochemical reactions because they are readily reversible. Three such noncovalent interactions that are particularly important in protein structure and enzyme catalysis are electrostatic interactions, hydrogen bonds, and van der Waals forces. Here we briefly describe the nature of these forces. In subsequent chapters we shall see how each of these can participate in stabilizing the protein structure of an enzyme and may also play important roles in the binding interactions between enzymes and their substrates and inhibitors.

2.4.1 Electrostatic Interactions

When two oppositely charged groups come into close proximity, they are attracted to one another through a Coulombic attractive force that is described by:

$$F = \frac{q_1 q_2}{r^2 D} \tag{2.13}$$

where q_1 and q_2 are the charges on the two atoms involved, r is the distance between them, and D is the dielectric constant of the medium in which the two atoms come together. Since D appears in the denominator of this equation, the attractive force is greatest in low dielectric solvents. Hence electrostatic forces are stronger in the hydrophobic interior of proteins than on the solvent-exposed surface. These attractive interactions are referred to as *ionic bonds, salt bridges*, and *ion pairs*.

Equation 2.13 describes the attractive force only. If two atoms, oppositely charged or not, approach each other too closely, a repulsive force between the outer shell electrons on each atom will come into play. Other factors being constant, it turns out that the balance between these attractive and repulsive forces is such that, on average, the optimal distance between atoms for salt bridge formation is about 2.8 Å (Stryer, 1989).

2.4.2 Hydrogen Bonding

A hydrogen bond (H bond) forms when a hydrogen atom is shared by two electronegative atoms. The atom to which the hydrogen is covalently bonded is referred to as the *hydrogen bond donor*, and the other atom is referred to as the *hydrogen bond acceptor*:

$$N^{\delta^-}—H^{\delta^+}\cdots O^{\delta^-}$$

The donors and acceptors in H bonds are almost exclusively electronegative heteroatoms, and in proteins these are usually oxygen, nitrogen, or sometimes sulfur atoms. Hydrogen bonds are weaker than covalent bonds, varying in bond energy between 2.5 and 8 kcal/mol. Networks of these bonds can occur in proteins, however, collectively adding great stability to certain structural motifs. We shall see examples of this in Chapter 3 when we discuss protein secondary structure. H-bonding also contribute to the binding energy of ligands to enzyme active sites and can play an important role in the catalytic mechanism of the enzyme.

2.4.3 Van der Waals Forces

The distribution of electrons around an atom is not fixed; rather, the character of the so-called electron cloud fluctuates with time. Through these fluctuations, a transient asymmetry of electron distribution, or *dipole moment*, can be established. When atoms are close enough together, this asymmetry on one atom can influence the electronic distribution of neighboring atoms. The result is a similar redistribution of electron density in the neighbors, hence an attractive force between the atoms is developed. This attractive force, referred to as a *van der Waals bond*, is much weaker than either salt bridges or H bonds. Typically a van der Waals bond is worth only about 1 kcal/mol in bond energy. When conditions permit large numbers of van der Waals bonds to simultaneously form, however, their collective attractive forces can provide a significant stabilizing energy to protein–protein and protein–ligand interactions.

As just described, the attractive force between electron clouds increases as the two atoms approach each other but is counterbalanced by a repulsive force at very short distances. The optimal attraction between atoms occurs when they are separated by a critical distance known as the *van der Waal contact distance*. The contact distance for a pair of atoms will be determined by the individual van der Waal contact radius of each atom, which itself depends on the electronic configuration of the atom.

Table 2.3 provides the van der Waal radii for the most abundant atoms found in proteins. Imagine drawing a sphere around each atom with a radius defined by the van der Waal contact radius. These spheres, referred to as *van der Waal surfaces*, would define the closest contacts that atoms in a molecule could make with one another, hence the possibilities for defining atom packing in a molecular structure. Because of the differences in radii, and the interplay between repulsive and attractive forces here, van der Waals bonds and surfaces can play an important role in establishing the specificity of interactions between protein binding pockets and ligands. We shall have more to say about such specificity in Chapter 4 when we discuss enzyme active sites.

Table 2.3 van der Waals Radii for Atoms in Proteins

Atom	Radius (Å)
H	1.2
C	2.0
N	1.5
O	1.4
S	1.9
P	1.9

2.5 Summary

In this chapter we have briefly reviewed atomic and molecular orbitals and the types of bonds formed within molecules as a result of these electronic configurations. We have seen that noncovalent forces also can stabilize interatomic interactions in molecules. Most notably, hydrogen bonds, salt bridges, and van der Waal forces can take on important roles in protein structure and function. In this chapter we have also reviewed some basic thermodynamics and acid–base theories that provide a framework for describing the reactivities of protein components in enzymology. In the chapters to come we shall see how these fundamental forces of chemistry come into play in defining the structures and reactivities of enzymes.

References and Further Reading

Fersht, A. (1985) *Enzyme Structure and Mechanism*, 2nd ed., Freeman, New York.

Gray, H. B. (1973) *Chemical Bonds: An Introduction to Atomic and Molecular Structure*, Benjamin/Cummings, Menlo Park, CA.

Kemp, D. S., and Vellaccio, F. (1980) *Organic Chemistry*, Worth, New York.

Lowry, T. H., and Richardson, K. S. (1981) *Mechanism and Theory in Organic Chemistry*, 2nd ed., Harper & Row, New York.

Palmer, T. (1985) *Understanding Enzymes*, Wiley, New York.

Pauling, L. (1960) *The Nature of the Chemical Bond*, 3rd ed., Cornell University Press, Ithaca, NY.

Stryer, L. (1989) *Molecular Design of Life*, Freeman, New York.

CHAPTER

3

Structural Components of Enzymes

In Chapter 2 we reviewed the forces that come to play in chemical reactions, such as those catalyzed by enzymes. In this chapter we introduce the specific molecular components of enzymes that bring these forces to bear on the reactants and products of the catalyzed reaction. Like all proteins, enzyme are composed mainly of the 20 naturally occurring amino acids. We shall discuss how these amino acids link together to form the polypeptide backbone of proteins, and how these macromolecules fold to form the three-dimensional conformations of enzymes that facilitate catalysis. Individual amino acid side chains supply chemical reactivities of different types that are exploited by the enzyme in catalyzing specific chemical transformations. In addition to the amino acids, many enzymes utilize nonprotein *cofactors* to add additional chemical reactivities to their repertoire. We shall describe some of the more common cofactors found in enzymes, and discuss how they are utilized in catalysis.

3.1 The Amino Acids

An amino acid is any molecule that conforms, at neural pH, to the general formula:

$$H_3N^+\text{—CH(R)—COO}^-$$

The central carbon atom in this structure is referred to as the *alpha carbon* (C_α), and the substituent, R, is known as the *amino acid side chain*. Of all the possible

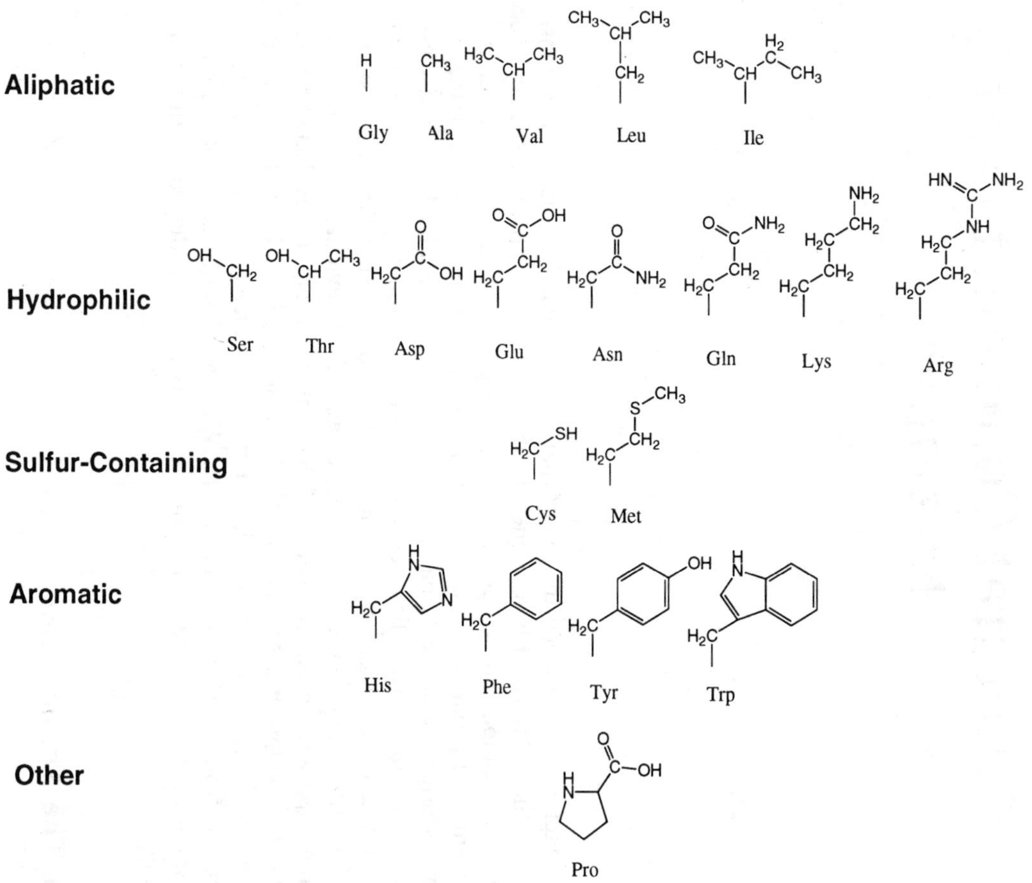

Aliphatic

Gly Ala Val Leu Ile

Hydrophilic

Ser Thr Asp Glu Asn Gln Lys Arg

Sulfur-Containing

Cys Met

Aromatic

His Phe Tyr Trp

Other

Pro

Figure 3.1 Side chain structures of the 20 natural amino acids. The entire proline molecule is shown.

Table 3.1 Physicochemical Properties of the Natural Amino Acids

Amino Acid	Three-Letter Code	One-Letter Code	Mass of Residue in Proteins[a]	Accessible Surface Area ($Å^2$)[b]	Hydro-phobicity[c]	pK_a of Ionizable Side Chain	Occurrence in Proteins (%)[d]	Relative Mutability[e]	Van der Waals Volume ($Å^3$)
Alanine	Ala	A	71.08	115	+1.8		9.0	100	67
Arginine	Arg	R	156.20	225	−4.5	12.5	4.7	65	148
Asparagine	Asn	N	114.11	160	−3.5		4.4	134	96
Aspartate	Asp	D	115.09	150	−3.5	3.9	5.5	106	91
Cysteine	Cys	C	103.14	135	+2.5	8.4	2.8	20	86
Glutamate	Glu	E	128.14	180	−3.5	4.1	3.9	102	109
Glutamine	Gln	Q	129.12	190	−3.5		6.2	93	114
Glycine	Gly	G	57.06	75	−0.4		7.5	49	48
Histidine	His	H	137.15	195	−3.2	6.0	2.1	66	118
Isoleucine	Ile	I	113.17	175	+4.5		4.6	96	124
Leucine	Leu	L	113.17	170	+3.8		7.5	40	124
Lysine	Lys	K	128.18	200	−3.9	10.8	7.0	56	135
Methionine	Met	M	131.21	185	+1.9		1.7	94	124
Phenylalanine	Phe	F	147.18	210	+2.8		3.5	41	135
Proline	Pro	P	97.12	145	−1.6		4.6	56	90
Serine	Ser	S	87.08	115	−0.8		7.1	120	73
Threonine	Thr	T	101.11	140	−0.7		6.0	97	93
Tryptophan	Trp	W	186.21	255	−0.9		1.1	18	163
Tyrosine	Tyr	Y	163.18	230	−1.3	10.1	3.5	41	141
Valine	Val	V	99.14	155	+4.2		6.9	74	105

[a]Values reflect the molecular weights of the amino acids minus that of water.
[b]Accessible surface are for residues as part of a polypeptide chain. Data from Chothia (1976).
[c]Hydrophobicity indices from Kyte and Doolittle (1982).
[d]Based on the frequency of occurrence for each residue in the sequence of 207 unrelated proteins. Data from Klapper (1977).
[e]Likelihood that a residue will mutate within a specified time period during evolution. Data from Dayoff et al. (1978).

chemical entities that could be classified as amino acids, nature has chosen to use 20 as the most common building blocks for constructing proteins and peptides. The structures of the side chains of the 20 naturally occurring amino acids are illustrated in Figure 3.1, and some of the physical properties of these molecules are summarized in Table 3.1. Since the alpha carbon is a chiral center, all the naturally occurring amino acids, except glycine, exist in two enantiomeric forms, L and D. All naturally occurring proteins are composed exclusively from the L enantiomers of the amino acids.

As we shall see later in this chapter, most of the amino acids in a protein or peptide have their charged amino and carboxylate groups neutralized through peptide bond formation (in this situation the amino acid structure that remains is referred to as an amino acid *residue* of the protein or peptide). Hence, what chemically and physically distinguishes one amino acid from another in a protein is the identity of the side chain of the amino acid. As seen in Figure 3.1, these side chains vary in chemical structure from simple substituents, like a proton in the case of glycine, to complex bicyclic ring systems in the case of tryptophan. These different chemical structures of the side chains impart vastly different chemical reactivities to the amino acids of a protein. Let us review here some of the chemical properties of the amino acid side chains that can participate in the interaction of proteins with other molecular and macromolecular species.

3.1.1 Properties of Amino Acid Side Chains

3.1.1.1 Hydrophobicity

Scanning Figure 3.1, we note that several of the amino acids (valine, leucine, alanine, etc.) are composed entirely of hydrocarbons. One would expect that solvation of such nonpolar amino acids in a polar solvent like water would be thermodynamically costly. In general, when hydrophobic molecules are dissolved in a polar solvent, they tend to cluster together to minimize the amount of surface areas exposed to the solvent; this phenomenon is known as *hydrophobic attraction.* The repulsion from water of amino acids in a protein provides a strong driving force for proteins to fold into three-dimensional forms that sequester the nonpolar amino acids within the interior, or *hydrophobic core*, of the protein. Hydrophobic amino acids also help to stabilize the binding of nonpolar substrate molecules in the binding pockets of enzymes.

The hydrophobicity of the amino acids is measured by their tendency to partition into a polar solvent in mixed solvent systems. For example, a molecule can be dissolved in a 50:50 mixture of water and octanol. After mixing, the polar and nonpolar solvents are allowed to separate, and the concentration of the test molecule in each phase is measured. The equilibrium constant for transfer of the molecule from octanol to water is then given by:

$$K_{\text{transfer}} = \frac{[C_{H_2O}]}{[C_{\text{oct.}}]} \tag{3.1}$$

where $[C_{H_2O}]$ and $[C_{oct.}]$ are the molar concentrations of the molecule in the aqueous and octanol phases, respectively. The free energy of transfer can then be calculated from the $K_{transfer}$ value using Equation 2.3. Such thermodynamic studies have been performed for the transfer of the naturally occurring amino acids from a number of nonpolar solvent to water. To make these measurements more representative of the hydrophobicities of amino acids within proteins, workers use analogues of the amino acids in which the amino and carboxylate charges are neutralized (e.g., using *N*-acetyl ethyl esters of the amino acids). Combining this type of thermodynamic information for the different solvent systems, one can develop a rank order of hydrophobicities for the 20 amino acids. A popular rank order used in this regard is that developed by Kyte and Doolittle (1982); the Kyte and Doolittle hydrophobicity indices for the amino acid are listed in Table 3.1. In general, hydrophobic amino acids are found on the interior of folded proteins, where they are shielded from the repulsive forces of the polar solvent, and polar amino acids tend to be found on the solvent-exposed surfaces of folded proteins.

3.1.1.2 Hydrogen Bonding

Associated with the heteroatoms of the side chains of several amino acids are exchangeable protons that can serve as hydrogen donors for H-bonding. Other amino acids can participate as H-bond acceptors through the lone pair electrons on heteroatoms of their side chains. Hydrogen bonding of amino acid side chains and polypeptide backbone groups can greatly stabilize protein structures, as we shall see later in this chapter. Additionally, hydrogen bonds can be formed between amino acid side chains and ligand (substrates, products, inhibitors, etc.) atoms and can contribute to the overall binding energy for the interactions of enzymes with such molecules. Side chains that are capable of acting as H-bond donors include tyrosine (—O—H), serine (—O—H), threonine (—O—H), tryptophan (—N—H), histidine (—N—H), and cysteine (—S—H). At low pH the side chains of glutamic and aspartic acid can also act as H-bond donors (—COO—H). Heteroatoms on the side chains of the following amino acids can serve as H-bond acceptors: tyrosine (—:O:—H), glutamic and aspartic acid (—COO$^-$), serine and threonine (—:O:—H),

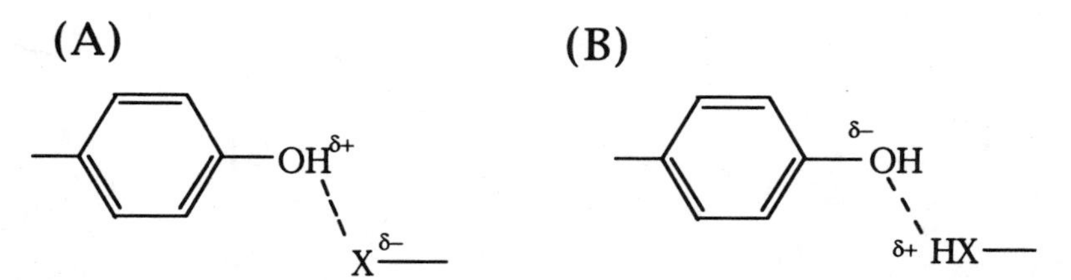

Figure 3.2 Tyrosine participation in hydrogen bonding as (A) a hydrogen donor and (B) a hydrogen acceptor.

histidine (N:), cysteine (—:S:—H), and methionine (:S:). Several of the amino acids can serve as both donors and acceptors, of H bonds, as illustrated for tyrosine in Figure 3.2.

3.1.1.3 Salt Bridge Formation

Noncovalent electrostatic interactions can occur between electonegative and electropositive species within proteins. Figure 3.3 illustrates the formation of such an electrostatic interaction between the side chains of a lysine residue on one polypeptide chain and a glutamic acid residue on another polypeptide. Because these interactions resemble the ionic interactions associated with small molecule salt formation, they are often referred to as *salt bridges.* Salt bridges can occur intramolecularly, between a charged amino acid side chain and other groups within the protein, or intermolecularly, between the amino acid side chain and charged groups on a ligand or other macromolecule. For example, many proteins that bind to nucleic acids derive a significant portion of their binding energy by forming electrostatic interactions betwen positively charged amino acid residues on their surfaces (usually lysine and arginine residues) and the negatively charged phosphate groups of the nucleic acid backbone.

Another example of the importance of these electrostatic interactions comes from the mitochondrial electron transfer cascade. Here electrons flow from the protein cytochrome *c* to the enzyme cytochrome oxidase, where they are used to reduce oxygen to water during cellular respiration. For the electron to jump from one protein to the other, the two must form a tight (dissociation constant $\approx 10^{-9}$ M) complex. When the crystal structure of cytochrome *c* was solved, it became obvious that the surface of this molecule contained an area with an unusually high density of positively charged lysine residues. The putative binding site for cytochrome *c* on the cytochrome oxidase molecule has a corresponding high density of aspartic and glutamic acid residues. It is thus believed that the tight complex formed between these two proteins is facilitated by forming a large number of salt bridges at this interface. This suggestion is supported by the ability of the complex to be dissociated by adding high

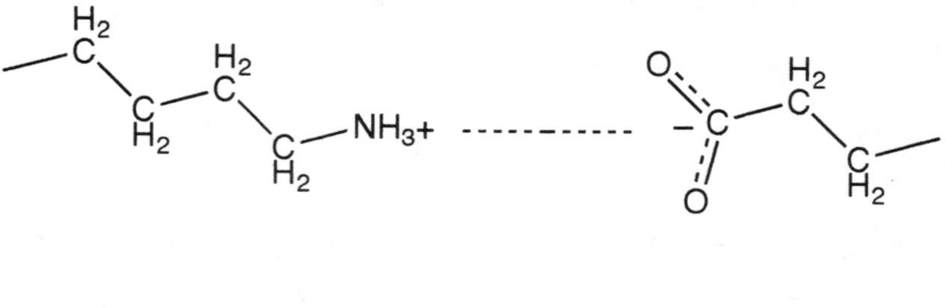

Lysine

Glutamic Acid

Figure 3.3 Salt bridge formation between a lysine and a glutamic acid residue at neutral pH.

concentrations of salt to the solution. As the ionic strength increases, the salt ions compete for the counterions from the amino acid residues that would otherwise participate in salt bridge formation.

3.1.2 Amino Acids as Acids and Bases

Surveying Table 3.1, we see that there are seven amino acid side chains, with titratable protons that can act as Brønsted–Lowry acids and conjugate bases. These are tyrosine, histidine, cysteine, lysine, and arginine, and aspartic and glutamic acids. The ability of these side chains to participate in acid–base chemistry provides enzymes with a mechanism for proton transfer to and from reactant and product molecules. In addition to proton transfer, side chain Lewis acids and bases can participate in nucleophilic and electrophilic reactions with the reactant molecules, leading to bond cleavage and formation. The placement of acid and base groups from amino acid side chains, at critical positions within the active site, is a common mechanism exploited by enzymes to facilitate rapid chemical reactions with the molecules that bind in the active site. For example, hydrolysis of peptide and ester bonds can occur through nucleophilic attack of the peptide by water. This reaction goes through a transition state in which the carbonyl oxygen of the peptide has a partial negative charge, and the oxygen of water has a partial positive charge:

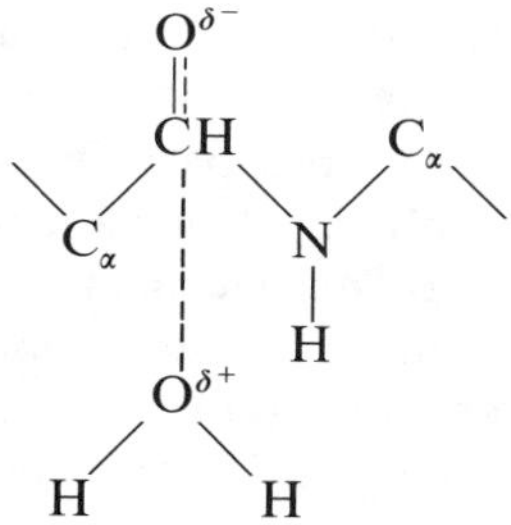

```
        Oδ−
        ||
        CH        Cα
   \   /  \      /   \
    Cα  |   N
        |   |
        |   H
        Oδ+
       /  \
      H    H
```

If one could place a basic group at a fixed position close to the water molecule, it would be possible to stabilize this transition state by partial transfer of one of the water protons to the base. This stabilization of the transition state would allow the reaction to proceed rapidly:

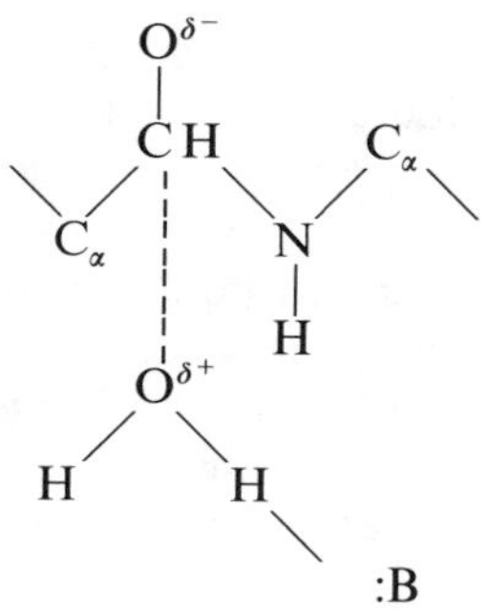

```
        Oδ−
        |
        CH        Cα
   \   /  \      /   \
    Cα  |   N
        |   |
        |   H
        Oδ+
       /  \
      H    H
             \
               :B
```

Alternatively, one could achieve the same stabilization by placing an acidic group at a fixed position in close proximity to the carbonyl oxygen, so that partial transfer of the proton from the acid to the carbonyl would stabilize the partial negative charge at this oxygen in the transition state. When one surveys the active sites of enzymes that catalyze peptide bond cleavage (a family of enzymes known as the proteases), one finds that there are usually acidic or basic amino acid side chains (or both) present at positions that are optimized for this type of transition state stabilization. We shall have more to say about stabilization of the transition states of enzyme reactions in Chapter 4.

In discussing the acid and base character of amino acid side chains, it is important to recognize that the pK_a values listed in Table 3.1 are for the side chain groups in aqueous solution. In proteins, however, these pK_a values can be greatly affected by the local environment that is experienced by the amino acid residue. For example, the pK_a of glutamic acid in aqueous solution is 4.1, but the pK_a of particular glutamic acid residues in some proteins can be as high as 6.5. Thus, while the pK_a values listed in Table 3.1 can provide some insights into the probable roles of certain side chains in chemical reactions, some caution must be exercised to avoid making oversimplifications.

3.1.3 Cation and Metal Binding

Many enzymes incorporate divalent cations (Mg^{2+}, Ca^{2+}, Zn^{2+}) and transition metal ions (Fe, Cu, Ni, Co, etc.) within their structures to stabilize the folded conformation of the protein or to make possible direct participation in the chemical reactions catalyzed by the enzyme. Metals can provide a template for protein folding, as in the zinc finger domain of nucleic acid binding proteins, the calcium ions of calmodulin, and the zinc structural center of insulin. Metal ions can also serve as redox centers for catalysis; examples include heme–iron centers, copper ions, and nonheme irons. Other metal ions can serve as electrophilic reactants in catalysis, as in the case of the active site zinc ions of the metalloproteases. Most commonly metals are bound to the protein portion of the enzyme by formation of coordinate bonds with certain amino acid side chains: histidine, tyrosine, cysteine, and methionine, and aspartic and glutamic acids. Examples of metal coordination by each of these side chains can be found in the protein literature.

The side chain imidazole ring of histidine is a particularly common metal coordinator. Histidine residues are almost always found in association with transition metal binding sites on proteins and are very often associated with divalent metal ion binding as well. Figure 3.4, for example, illustrates the coordination sphere of the active site zinc of the enzyme carbonic anhydrase. Zinc typically forms four coordinate bonds in a tetrahedral arrangement about the metal ion. In carbonic anhydrase, three of the four bonds are formed by coordination to the side chains of histidine residues from the protein. The fourth coordination site is occupied by a water molecule that participates

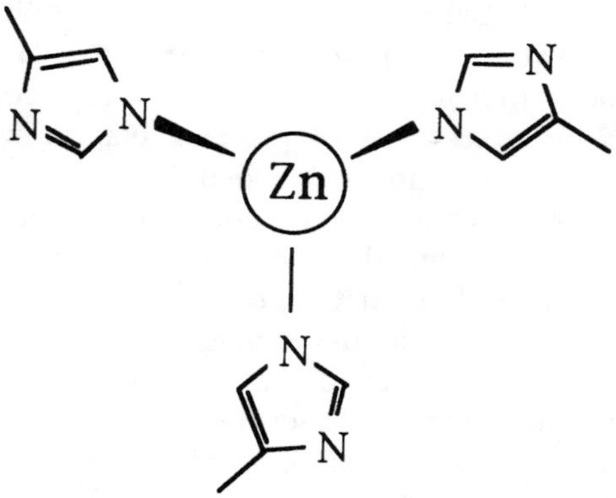

Figure 3.4 The coordination sphere of the active site zinc of carbonic anhydrase.

directly in catalysis. During the course of the enzyme-catalyzed reaction, the zinc–water bond is broken and replaced transiently by a bond between the metal and the carbon dioxide substrate of the reaction.

3.1.4 Anion and Polyanion Binding

The positively charged amino acids lysine and arginine can serve as counter-ions for anion and polyanion binding. Interactions of this type are important in binding of cofactors, reactants, and inhibitors to enzymes. Examples of anionic reactants and cofactors utilized by enzymes include phosphate groups,. nucleotides and their analogues, nucleic acids, and heparin.

3.1.5 Covalent Bond Formation

We have pointed out that the chemical reactivities of amino acid residues within proteins are determined by the structures of their side chains. Several amino acids can undergo posttranslational modification (i.e., alterations that occur after the polypeptide chain has been synthesized at the ribosome) that alter their structure, hence reactivity, by covalent bond formation. In some cases, reversible modification of amino acid side chains is a critical step in the catalytic mechanism of the enzyme. Sections 3.1.5.1–3.1.5.3 give some examples of covalent bonds formed by amino acid side chains.

3.1.5.1 Disulfide Bonds

Two cysteine residues can cross-link, through an oxidative process, to form a sulfur–sulfur bond, referred to as a disulfide bond. These cross-links can occur intramolecularly, between two cysteines within a single polypeptide, or intermolecularly, to join two polypeptides together. Such disulfide bond cross-linking can provide stabilizing energy to the folded conformation of the protein. Numerous examples exist of proteins that utilize both inter- and intramolecular disulfide bonds in their folded forms. Intermolecular disulfide bonds can also occur between a cysteine residue on a protein and a sulfhydryl group on a small molecule ligand or modifying reagent. For example, 4,4′-dithioldipyridine is a reagent used to quantify the number of free cysteines (those not involved in disulfide bonds) in proteins. The reagent reacts with the free sulfhydryls to form intermolecular disulfide bonds, with the liberation of a chromophoric by-product. The formation of the by-product is stoichiometric with reactive cysteines. Thus, one can quantify the number of cysteines that reacted from the absorbance of the by-product.

3.1.5.2 Phosphorylation

Certain amino acid side chains can be covalently modified by addition of a phosphate from inorganic phosphate (P_i). In nature, the phosphorylation of specific residues within proteins is facilitated by a class of enzymes known as the *kinases*. Another class of enzymes, the *phosphatases*, will selectively remove phosphate groups from these amino acids. This reversible phosphorylation/dephosphorylation can greatly affect the biological activity of enzymes, receptors, and proteins involved in protein–protein and protein–nucleic acid complex formation.

The most common sites for phosphorylation on proteins are the hydroxyl groups of threonine and serine residues; however, the side chains of tyrosine, histidine, and lysine can also be modified in this way (Figure 3.5). Tyrosine

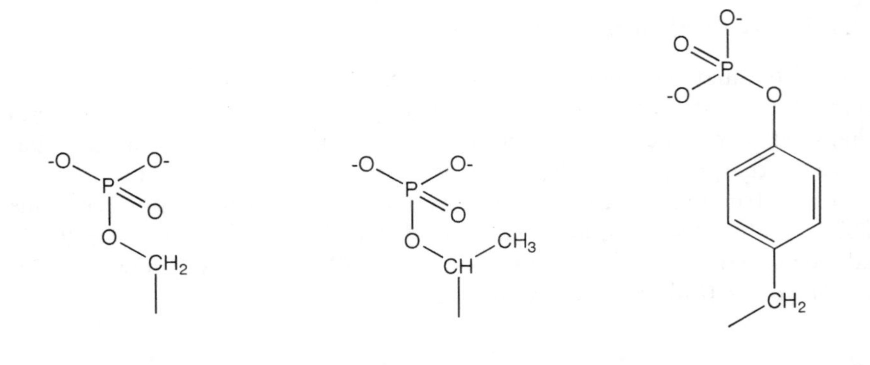

Figure 3.5 The structures of the phosphorylated forms of serine, threonine, and tyrosine.

kinases, enzymes that specifically phosphorylate tyrosine residues within certain proteins, are of great current interest in biochemistry and cell biology. This is because it has recently been recognized that tyrosine phosphorylation and dephosphorylation are critical in the transmission of chemical signals within cells (signal transduction).

Enzymes can also transiently form covalent bonds to phosphate groups during the course of catalytic turnover. In these cases, a *phosphoryl–enzyme intermediate* is formed by the transfer of an phosphate from substrate molecule or inorganic phosphate to specific amino acid side chains within the enzyme active site. Several examples of phosphoryl–enzyme intermediates are now known, which involve phosphoserine, phosphohistidine, and even phosphoaspartate formation. For example, the ATPases are enzymes that catalyze the hydrolysis of ATP (adenosine triphosphate) to form ADP (adenosine diphosphate) and P_i. In a subgroup of these enzymes, the Na^+, K^+, and the Ca^{2+} ATPases, the γ-phosphate of ATP is transferred to the β-carboxylate of an aspartic acid residue of the enzyme during the reaction. Since the phosphoaspartate is thermodynamically unstable, it very quickly dissociates to liberate inorganic phosphate.

3.1.5.3 Glycosylation

In eukaryotic cells, sugars can attach to proteins by covalent bond formation at the hydroxyl groups of serine and threonine residues (O-linked glycosylation) or at the nitrogen of asparagine side chains (N-linked glycosylation). The resulting protein–sugar complex is referred to as a *glycoprotein*. The sugars used for this purpose are composed of monomeric units of galactose, glucose, manose, *N*-acetylglucosamine, *N*-acetylgalactosamine, sialic acid, fructose, and xylose. The presence of these sugar moieties can significantly affect the solubility, folding, and biological reactivity of proteins.

3.1.6 Steric Bulk

Aside from the chemical reactivities already discussed, the stereochemistry of the amino acid side chains plays an important role in protein folding and intermolecular interactions. The size and shape of the side chain determines the type of packing interactions that can occur with neighboring groups, according to their van der Waals radii. It is the packing of amino acid side chains within the active site of an enzyme molecule that give overall size and shape to the binding cavity (pocket), which accommodates the substrate molecule; hence these packing interactions help determine the specificity for binding of substrate and inhibitor molecules at these sites. This is a critical aspect of enzyme catalysis; in Chapter 4 we shall discuss further the relationship between the size and shape of the enzyme binding pocket and the structure of ligands.

For the aliphatic amino acids, side chain surface area also influences the overall hydrophobicity of the residue. The hydrophobicity of aliphatic molecules, in general, has been correlated with their exposed surface area. Hansch and Coats (1970) have made the generalization that the $\Delta G_{transfer}$ from a nonpolar solvent, like *n*-octanol, to water increases by about 0.68 kcal/mol for every methylene group added to an aliphatic structure. While this is an oversimplification, it serves as a useful rule of thumb for predicting the relative hydrophobicities of structurally related molecules. This relationship between surface area and hydrophobicity holds not only for the amino acids that line the binding pocket of an enzyme, but also for the substrate and inhibitor molecules that might bind in that pocket.

3.2 The Peptide Bond

The peptide bond is the primary structural unit of the polypeptide chains of proteins. Peptide bonds result from the condensation of two amino acids, as follows:

$$ {}^{+}H_3N{-}C(H)(R_1){-}COO^- \; + \; {}^{+}H_3N{-}C(H)(R_2){-}COO^- \longrightarrow {}^{+}H_3N{-}C(H)(R_1){-}C(=O){-}N(H){-}C(H)(R_2){-}COO^- \; + \; H_2O $$

The product of such a condensation is referred to as a dipeptide, because it is composed of two amino acids. A third amino acid could condense with this dipeptide to form a tripeptide, a fourth to form a tetrapeptide, and so on. In this way chains of amino acids can be linked together to form *polypeptides* or *proteins*.

Until now we have drawn the peptide bond as an amide with double-bond character between the peptide carbon and the oxygen, and single-bond character between the peptide carbon and nitrogen atoms. Table 3.2 provides typical bond lengths for carbon–oxygen and carbon–nitrogen double and single bonds. Based on these data, one would expect the peptide carbon–oxygen bond length to be 1.22 Å and the carbon–nitogen bond length to be

Table 3.2 Typical Bond Lengths for Carbon–Oxygen and Carbon–Nitrogen Bonds

Bond Type	Bond Length (Å)
C—O	1.27
C=O	1.22
C—N	1.45
C=N	1.25

1.45 Å. In fact, however, when x-ray crystallography was first applied to small peptides and other amide containing molecules, it was found that the carbon–oxygen bond length was longer than expected, 1.24 Å, and the carbon–nitrogen bond length was shorter than expected, 1.32 Å (Figure 3.6). These values are intermediate between those expected for double and single bonds. These results were rationalized by the chemist Linus Pauling by invoking two resonance structures for the peptide bond; one as we have drawn it at the beginning of this section, with all the double-bond character on the carbon–oxygen bond, and another in which the double bond is between the carbon and the nitrogen, and the oxygen is negatively charged. Thus the π system is actually delocalized over all three atoms, O—C—N. Based on the observed bond lengths, it was concluded that a peptide bond has about 60% C=O character and about 40% C=N character.

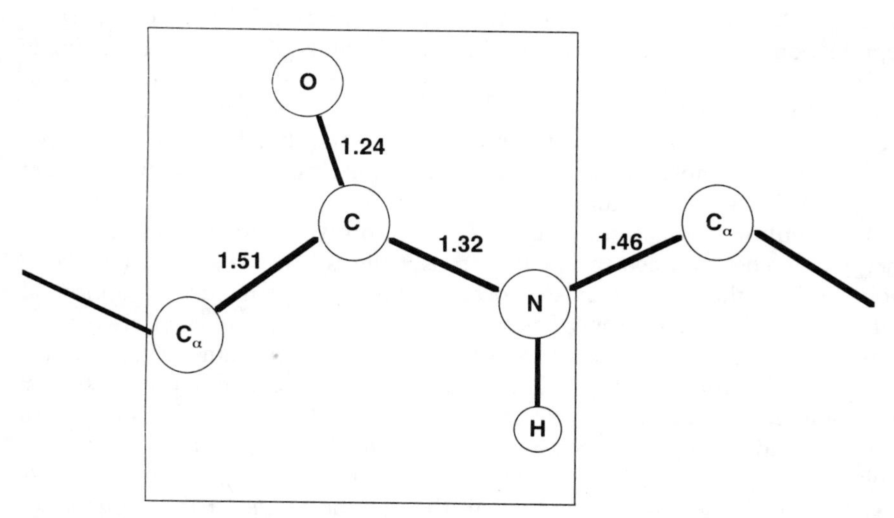

Figure 3.6 Schematic diagram of a typical peptide bond; numbers are typical bond lengths in angstrom units.

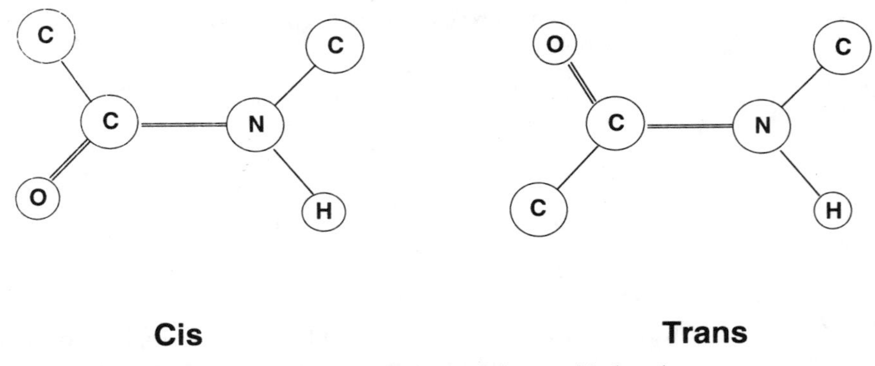

Figure 3.7 The cis and trans configurations of the peptide bond.

The 40% double-bond character along the C–N axis results in about a 20 kcal/mol resonance energy stabilization of the peptide. It also imposes a severe barrier to rotation about this axis. Hence, the six atoms associated with the peptide unit of polypeptides occur in a *planar* arrangement. The planarity of this peptide unit limits the possible configurations the polypeptide chain can adopt. It also allows for two stereoisomers of the peptide bond to occur: a *trans* configuration, in which the carbonyl oxygen and the nitrogenous proton are on opposite sides of the axis defined by the C—N bond, and a *cis* configuration, in which these groups are on the same side of this axis (Figure 3.7). When proteins are produced on the ribosomes of cells, they are synthesized stereospecifically. They could be synthesized with either all cis or all trans peptide bonds. However, because of the steric bulk of the amino acid side chains, polypeptides composed of cis peptide bonds are greatly restricted in terms of the conformational space they can survey. Thus, there is an significant thermodynamic advantage to utilizing trans peptide bonds for proteins and, unsurprisingly, almost all the peptide bonds in naturally occurring proteins are present in the trans configuration.

An exception to this general rule is found in prolyl–peptide bonds. Here the energy difference between the cis and trans isomers is much smaller ($\approx$2 kcal/mol), and so the cis isomer can occur without a significant disruption in stability of the protein. Nevertheless, only a very few examples of cis prolyl–peptide bonds have ever been observed in nature. Three examples are known of cis prolyl–peptide bonds within enzymes from x-ray crystallographic studies: in ribonuclease-S (before Pro 93 and before Pro 114), in a subtilisin (before Pro 168), and in staphylococcal nuclease (before Pro 116). Thus, while the cis prolyl–peptide bond is energetically feasible, it is extremely rare. For our purposes, then, we can assume that all the peptide bonds in the proteins we shall be discussing are present in the trans configuration.

3.3 Amino Acid Sequence or Primary Structure

The structure and reactivity of a protein are defined by the identity of the amino acids that make up its polypeptide chain, and the order in which those amino acids occur in the chain. This information constitutes the *amino acid sequence* or *primary structure* of the protein. Recall that we can link amino acids together through condensation reactions to form polypeptide chains. For most of the amino acids in such a chain, we have seen that the condensation results in loss of the charged amino and carboxylate moieties. No matter how many times we perfom this condensation reaction, however, the final polypeptide will always retain a charged amino group at one end of the chain and a charged carboxylate at the other end. The terminal amino acid that retains the positively charged amino group is referred to as the *N-terminus* or *amino terminus.* The other terminal amino acid, retaining the negatively charges carboxylate group, is referred to as the *C-terminus* or *carboxy terminus.*

The individul amino acids in a protein are identified numerically in sequential order, starting with the N-terminus. The N-terminal amino acid is labeled number 1, and the numbering continues in ascending numerical order, ending with the residue at the carboxy terminus. Thus, when we read in the literaure about "active site residue Ser 530," it means that the 530th amino acid from the N-terminus of this enzyme is a serine, and it occurs within the active site of the folded protein.

With the advent of recombinant DNA technologies, it has become commonplace to substitute amino acid residues within proteins (see Davis and Copeland, 1996, for recent review of these methods). One may read, for example, about a protein in which a His 323-Asn mutation was induced by means of site-directed mutagenesis. This means that in the natural, or *wild-type* protein, a histidine residue occupies position 323, but through mutagenesis this residue has been replaced by an asparagine to create a mutant (or altered) protein. Nowadays it is very common to read about studies in which mutant proteins have been purposely created through the methods of molecular biology. It is important to remember, however, that mutations in protein sequences occur naturally as well. For the most part, point mutations of protein sequences occur with little effect on the biological activity of the protein, but in some cases the result is devastating. Consider, for example, the disease sickle cell anemia.

Patients with sickle cell anemia have a point mutation in their hemoglobin molecules. Hemoglobin is a tetrameric protein composed of four polypeptide chains: two identical α chains, and two identical β chains. Together, the four polypeptides of a hemoglobin molecule contain about 600 amino acids. The β chains of normal hemoglobin have a glutamic acid residue at position 6. The crystal structure of hemoglobin reveals that residue 6 of the β chains is at the solvent-exposed surface of the protein molecule, and it is thus not surprising to find a highly polar side chain, like glutamic acid, at this position. In the hemoglobin from sickle cell anemia patients, this glutamic acid is replaced by a valine, a very nonpolar amino acid. When the hemoglobin molecule is devoid of bound oxygen (the deoxy form, which occurs after hemoglobin has released its oxygen supply to the muscles), these valine residues on different molecules of hemoglobin will come together to shield themselves from the polar solvent through protein aggregation. This aggregation leads to long fibers of hemoglobin in the red blood cells, causing the cells to adopt the narrow elongated "sickle" shape that is characteristic of this disease. Thus with only two amino acid changes out of 600 (one residue per β chain), the entire biological activity of the protein is severely altered. (For a very clear and interesting account of the biochemistry of sickle cell anemia, see Stryer, 1989.) Sickle cell anemia was the first human disease that was shown to be caused by mutation of a specific protein (Pauling et al. 1949). We now know that there are many examples of such genetic-based human diseases. Current efforts in "gene therapy" are aimed at correcting mutation-based diseases of these types.

3.4 Secondary Structure

We mentioned earlier that the delocalization of the peptide π system restricts rotation about the C—N bond axis. While this is true, it should be noted that free rotation is possible about the N—C_α and the C_α—C′ bond axes (where C′ represents the carbonyl carbon of the amino acid residue). The dihedral angles defined by these two rotations are represented by the symbols ϕ and ψ, respectively, and are illustrated for a dipeptide in Figure 3.8.

Upon surveying the observed values for ϕ and ψ for amino acid residues in the crystal structures of proteins, one finds that certain values occur with high frequency. For any amino acid except glycine, a plot of the observed ϕ and ψ angles looks like Figure 3.9. This type of graph is known as a Ramachandran plot, after the scientist who first measured these angles and constructed such plots. The most obvious feature of Figure 3.9 is that the values of ϕ and ψ tend to cluster around two pairs of angles. The two regions of high frequency correspond to the angles associated with two commonly found regular and repeating structural motifs within proteins, the *right-handed* α *helix* and the β*-pleated sheet*. Both of these structural motifs are examples of protein *secondary structure*, an important aspect of the overall conformation of any protein.

3.4.1 The Right-Handed α Helix

The right-handed α helix is one of the most commonly found secondary structures within proteins (Figure 3.10). This structure was first predicted by Linus Pauling, on the basis of the stereochemical properties of polypeptides. The structure is stabilized by a network of hydrogen bonds between the

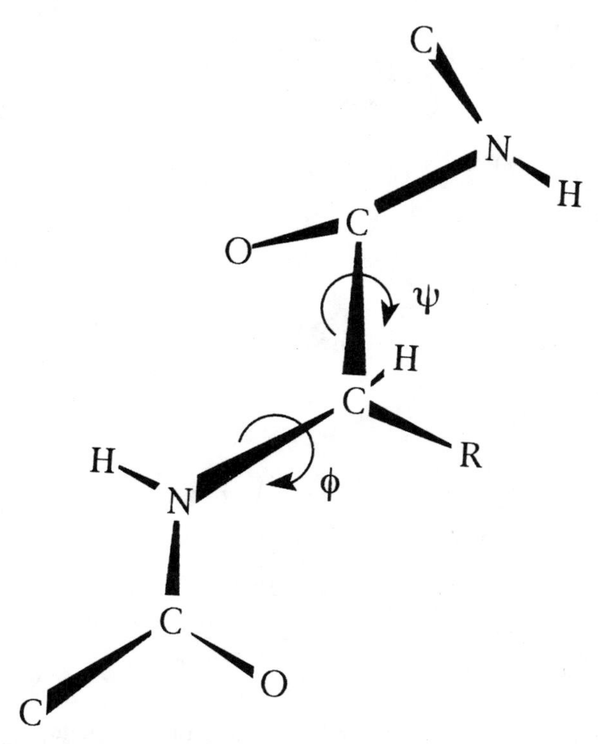

Figure 3.8 The dihedral angles of rotation for an amino acid in a peptide chain.

carbonyl oxygen of one residue (i) and the nitrogenous proton of residue i + 4. For most of the residues in the helix, there are thus two hydrogen bonds formed with neighboring peptide bonds, each contributing to the overall stability of the helix. As seen in Figure 3.10, this hydrogen-bonding network is possible because of the arrangement of C=O and N—H groups along the helical axis. The side chains of the amino acid residues all point away from the axis in this structure, thus minimizing steric crowding. The individual peptide bonds are aligned within the α-helical structure, producing, in addition, an overall dipole moment associated with the helix; this too adds some stabilization to the structure. The amino acid residues in an α helix conform to a very precise stereochemical arrangement. Each turn of an α helix requires 3.6 amino acid residues, with a translation along the helical axis of 1.5 Å per residue, or 5.4 Å per turn.

The network of hydrogen bonds formed in an α helix eliminates the possibility of those groups' participating in hydrogen bond formation with solvent water molecules. For small peptides, the removal of these competing solvent hydrogen bonds, by dissolving the peptide in aprotic solvents, tends to

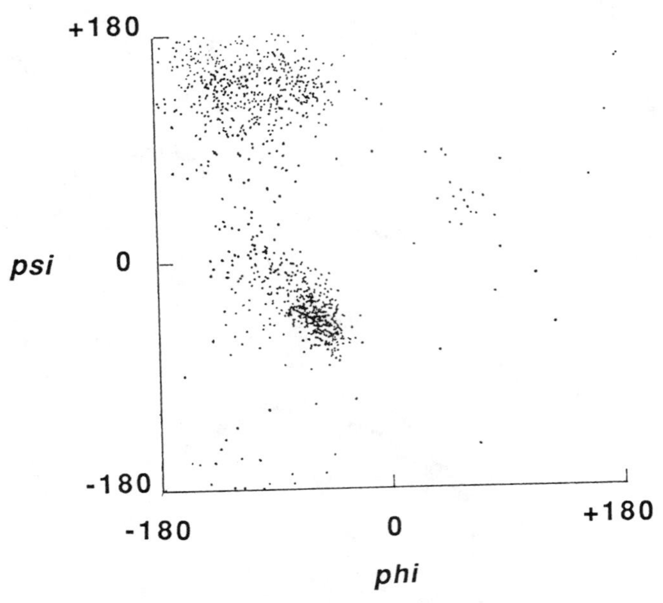

Figure 3.9 A Ramachandran plot for the amino acid alanine.

promote the formation of α helices. The same tendency is observed when regions of a polypeptide are embedded in the hydrophobic interior of a cell membrane. Within the membrane bilayer, where hydrogen bonding with solvent is not possible, peptides and polypeptides tend to form α-helical structures. The hydrocarbon core cross section of a typical biological membrane is about 30 Å. Since an α helix translates 1.5 Å per residue, one can calculate that it takes, on average, about 20 amino acid residues, arranged in an α-helical structure, to traverse a membrane bilayer. For many proteins that are embedded in cell membranes (known as integral membrane proteins), one or more segments of 20 hydrophobic amino acids are threaded through the membrane bilayer as an α helix. These structures, often referred to as *transmembrane α helices*, are one of the main mechanisms by which proteins associate with cell membranes.

3.4.2 The β-Pleated Sheet

The β-pleated sheet, another very abundant secondary structure found in proteins, is composed of fully extended polypeptide chains linked together through interamide hydrogen bonding between adjacent strands of the sheet (Figure 3.11). Figure 3.11A illustrates a β-pleated sheet composed of two

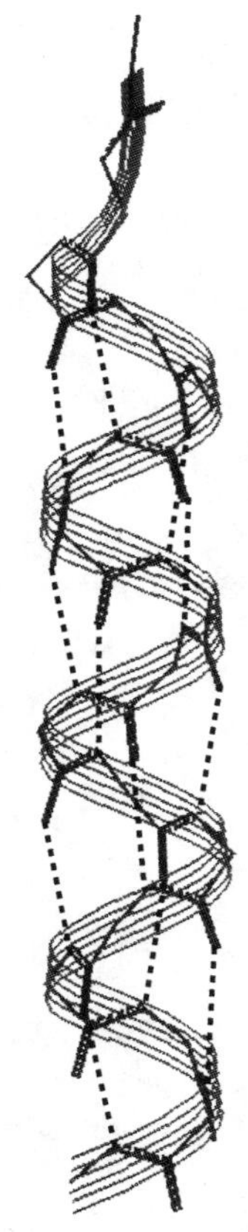

Figure 3.10 The right-handed α helix. (Figure provided by Dr. Steve Betz.)

polypeptide chains. Note that the arrangement of peptide C=O and N—H groups permits this structure to be extended in either direction (i.e., to the left or right in the plane of the page as shown in Figure 3.11A) through the same type of interamide hydrogen bonding (Figure 3.11B).

For the moment, let us focus on a two-stranded β sheet, as in Figure 3.11A: the two component polypeptide chains could come from two distinct polypeptides (an *intermolecular β sheet*) or from two regions of the same contiguous polypeptide (an *intramolecular β sheet*); both types are found in natural proteins. If we imagine a β sheet within the plane of this page, we could have both chains running in the same direction, say from C-terminus at the top of the page to N-terminus at the botton. Alternatively, we could have the two chains running in opposite directions with respect to the placement of their N- and C-termini. These two situations describe structures referred to as *parallel*

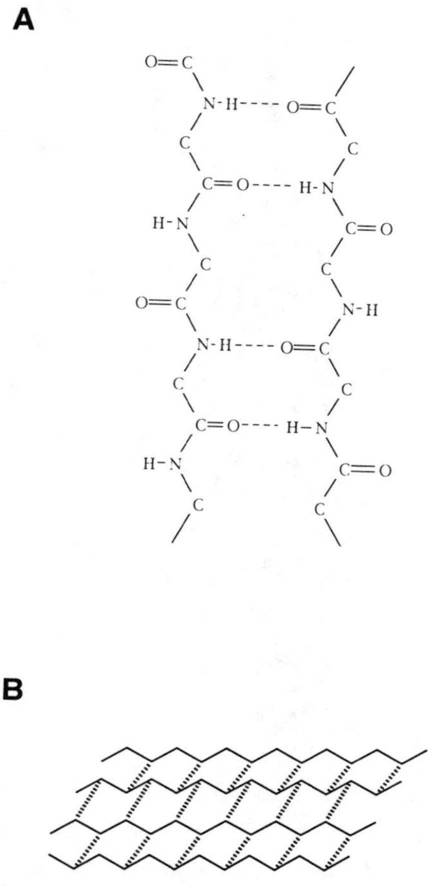

Figure 3.11 (A) A β-pleated sheet composed of two segments of polypeptide held together by interchain hydrogen bonding. (B) An extended β-pleated sheet composed of four segments of polypeptide.

and *antiparallel* β-pleated sheets, respectively. Again, one finds both types in nature.

3.4.3 β Turns

A third common secondary structure found in natural proteins is the β turn (also known as a reverse turn, hairpin turn, or β bend). The β turns are short segments of the polypeptide chain that allow it to change direction—that is, to *turn* upon itself. Turns are composed of four amino acid residues in a compact configuration in which an interamide hydrogen bond is formed

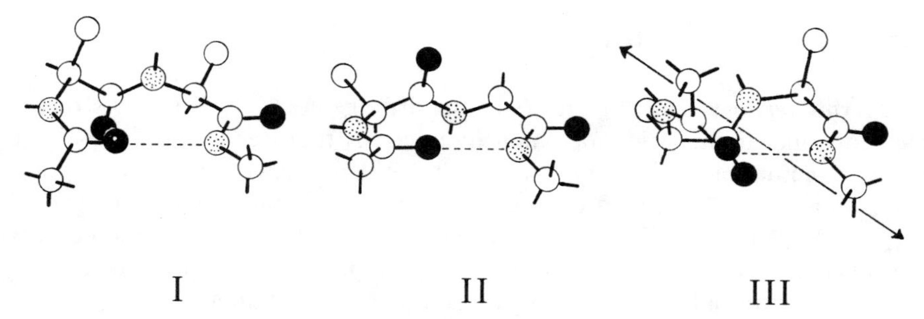

Figure 3.12 Three common forms of β turn.

between the first and fourth residue to stabilize the structure. Three types of β turn are commonly found in proteins: types I, II, and III (Figure 3.12). Although turns represent small segments of the polypeptide chain, they occur often in a protein, allowing the molecule to adopt a compact three-dimensional structure. Consider, for example, an intramolecular antiparallel β sheet within a contiguous segment of a protein. To bring the two strands of the sheet into register for the correct hydrogen bonds to form, the polypeptide chain would have to change direction by 180°. This can be accomplished only by incorporating a type I or type II β turn into the polypeptide chain, between the two segments making up the β sheet. Thus β turns play a very important role in establishing the overall three-dimensional structure of a protein.

3.4.5 Other Secondary Structures

One can imagine other regular repeating structural motifs that are stereochemically possible for polypeptides. In a series of adjacent type III β turns, for example, the polypeptide chain would adopt a helical structure, different from the α helix, that is known as a 3_{10} helix. This structure is indeed found in proteins, but it is rare. Some proteins, composed of high percentages of a single amino acid type, can adopt specialized helical structures, such as the polyproline helices and polyglycine helices. Again, these are special cases, not commonly found in the vast majority of proteins.

Most proteins contain regions of well-defined secondary structures interspersed with segments of nonrepeating, unordered structure in a conformation commonly referred to as *random coil* structure. These regions provide dynamic flexibility to the protein, allowing it to change shape, or conformation. These structural fluctuations can play an important role in facilitating the biological activities of proteins in general. They have particular significance in the cycle of substrate binding, catalytic transformations, and product release that is required for enzymes to function.

3.5 Tertiary Structure

The term "*tertiary structure*" refers to the arrangement of secondary structure elements and amino acid side chain interactions that define the three-dimensional structure of the folded protein. Imagine that a newly synthesized protein exists in as a fully extended polypeptide chain—it is said then to be *unfolded* (Figure 3.13A) [Actually there is debate as to how fully extended the polypeptide chain really is in the unfolded state of a protein; some data suggest that even in the unfolded state, proteins retain a certain amount of structure. However, this is not an important point for our present discussion.] Now suppose that this protein is placed under the set of conditions that will lead to the formation of elements of secondary structure at appropriate locations along the polypeptide chain (Figure 3.13B). Next, the individual elements of secondary structure arrange themselves in three-dimensional space, so that specific contacts are made between amino acid side chains and between backbone groups (Figure 3.13C). The resulting *folded* structure of the protein is referred to as its *tertiary structure*.

What we have just described is the process of protein folding, which occurs naturally in cells as new proteins are synthesized at the ribosomes. The process is remarkable because under the right set of conditions it will also proceed spontaneously outside the cell in a test tube (in vitro). For example, at high concentrations chemicals like urea and guanidine hydrochloride will cause most proteins to adopt an unfolded conformation. In many cases, the subsequent removal of these chemicals (by dialysis, gel filtration chromatography, or dilution) will cause the protein to refold spontaneously into its correct native conformation (i.e., the folded state that occurs naturally and best facilitates the biological activity of the protein). The very ability to perform

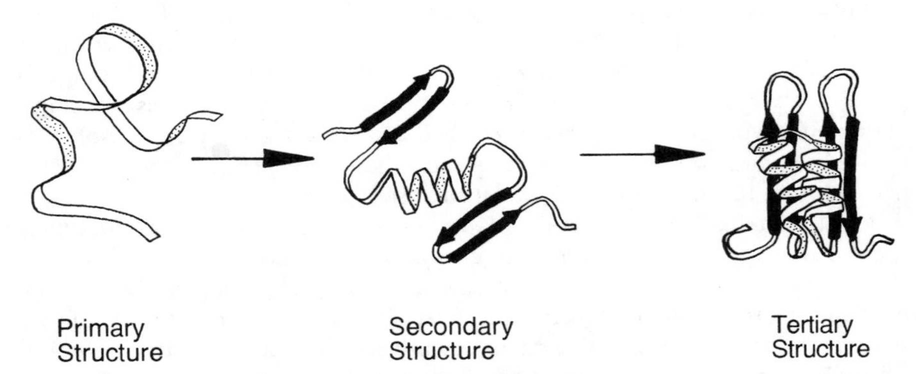

Figure 3.13 The folding of a polypeptide chain illustrating the hierarchy of protein structure from primary structure or amino acid sequence through secondary structure and tertiary structure. [Adapted from Dill et al., *Protein Sci.* **4**, 561 (1992).]

such experiments in the laboratory indicates that all the information required for the folding of a protein into its proper secondary and tertiary structures is encoded within the amino acid sequence of that protein.

Why is it that proteins fold into these tertiary structures? There are several important advantages to proper folding for a protein. First, folding provides a means of burying hydrophobic residues away from the polar solvent and exposing polar residues to solvent for favorable interactions. In fact, many scientists believe that the shielding of hydrophoic residues from the solvent is one of the strongest thermodynamic driving forces for protein folding. Second, through folding the protein can bring together amino acid side chains that are distant from one another along the polypeptide chain. By bringing such groups into close proximity, the protein can form chemically reactive centers, such as the active sites of enzymes. An excellent example is provided by the serine protease chymotrypsin.

Serine proteases are a family of enzymes that cleave peptide bonds in proteins at specific amino acid residues (see Chapter 4 for more details). All these enzymes must have a serine residue within their active sites which functions as the primary nucleophile—that is, to attack the substrate peptide, thereby initiating catalysis. To enhance the nucleophilicity of this residue, the hydroxyl group of the serine side chain participates in hydrogen bonding with an active site histidine residue, which in turn may hydrogen-bond to an active site aspartate as shown in Figure 3.14. This "active site triad" of amino acids is a structural feature common to all serine proteases. In chymotrypsin this triad is composed of Asp 102, His 57, and Ser 195. As the numbering indicates, these three residues would be quite distant from one another along the fully extended polypeptide chain of chymotrypsin. However, the tertiary structure of chymotrysin is such that when the protein is properly folded, these three residues come together to form the necessary interactions for effective catalysis.

The tertiary structure of a protein will often provide folds or pockets within the protein structure that can accommodate small molecules. We have already

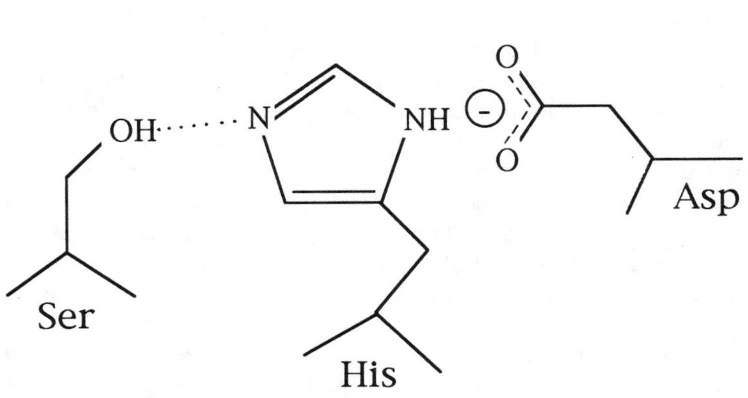

Figure 3.14 The active site triad of serine proteases.

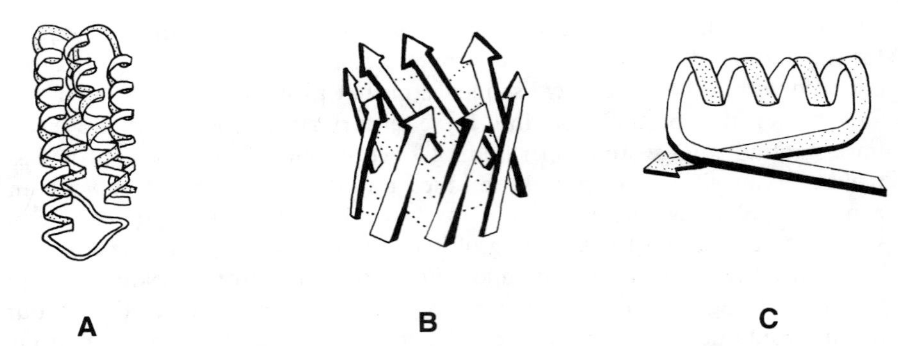

Figure 3.15 Examples of super secondary structures: (A) a helical bundle, (B) a β barrel, and (C) a β-α-β loop.

used the term "active site" several times, referring, collectively, to the chemically reactive groups of the enzyme that facilitate catalysis. The active site of an enzyme is also defined by a cavity or pocket into which the substrate molecule binds to initiate the enzymatic reaction; the interior of this binding pocket is lined with the chemically reactive groups from the protein. As we shall see in Chapter 4, there is a precise stereochemical relationship between the structure of the molecules that bind to the enzyme and that of the active site pocket. The same is generally true for the binding of agonists and antagonists to the binding pockets of protein receptors. In all these cases, the structure of the binding pocket is dictated by the tertiary structure of the protein.

While no two proteins have completely identical three-dimensional structures, enzymes that carry out similar functions often adopt similar active site structures, and sometimes similar overall folding patterns. Some arrangements of secondary structure elements, which occur commonly in folded proteins, are referred to by some workers as *supersecondary structure.* Three examples of supersecondary structures are the helical bundle, the β barrels, and the β–α–β loop, illustrated in Figure 3.15.

In some proteins one finds discrete regions of compact tertiary structure that are separated by stretches of the polypeptide chain in a more flexible arrangement. These discrete folded units are known as *domains*, and often they define functional units of the protein. For example, many cell membrane receptors play a role in signal transduction by binding extracellular ligands at the cell surface. In response to ligand binding, the receptor undergoes a structural change that results in macromolecular interactions between the receptor and other proteins within the cell cytosol. These interactions in turn set off a cascade of biochemical events that ultimately lead to some form of cellular response to ligand binding. To function in this capacity, such a receptor requires a minimum of three separate domains: an extracellular ligand binding domain, a transmembrane domain that anchors the protein within the cell

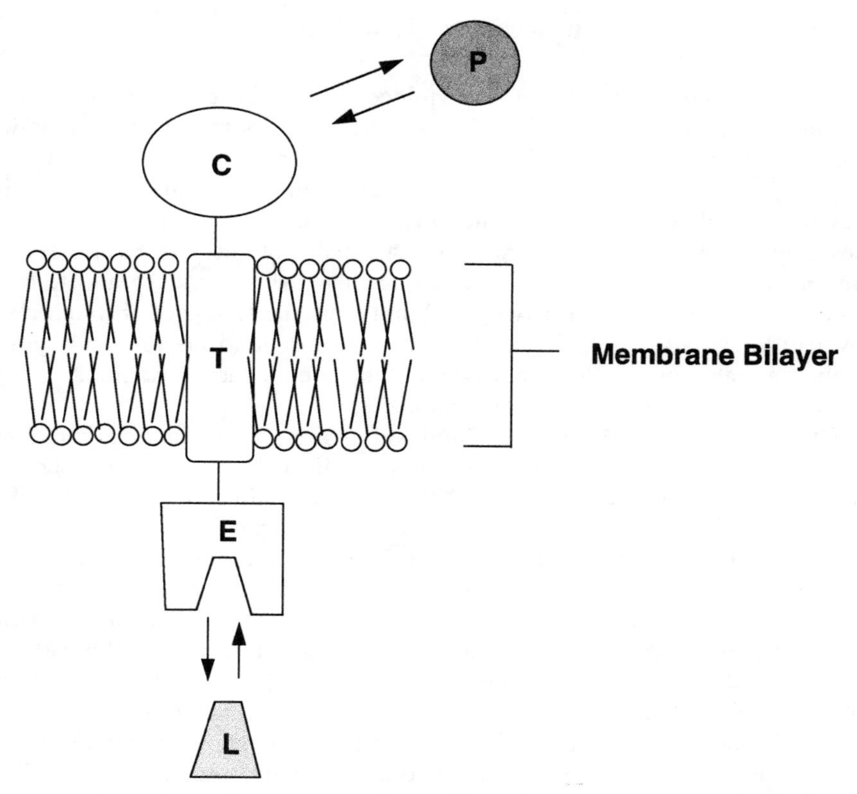

Figure 3.16 Cartoon illustration of the domains of a typical transmembrane receptor. The protein consists of three domains. The extracellular domain (E) forms the center for interaction with the receptor ligand (L). The transmembrane domain (T) anchors the receptor within the phospholipid bilayer of the cellular membrane. The cytosolic domain (C) extends into the intracellular space and forms a locus for interactions with other cytosolic proteins (P), which can then go on to transduce signals within the cell.

membrane, and an intracellular domain that forms the locus for protein–protein interactions. These concepts are schematically illustrated in Figure 3.16.

Many enzymes are composed of discrete domains as well. For example, the crystal structure of the integral membrane enzyme prostaglandin synthase was recently solved by Garavito and his coworkers (Picot et al., 1994). The structure reveals three separate domains of the folded enzyme monomer: a β-sheet domain that functions as an interface for dimerization with another molecule of the enzyme, a membrane-incorporated α-helical domain that anchors the enzyme to the biological membrane, and a extramembrane globular (i.e., compact folded region) domain that contains the enzymatic active site and is thus the catalytic unit of the enzyme.

3.6 Subunits and Quaternary Structure

Not every protein functions as a single folded polypeptide chain. In many cases the biological activity of a protein requires two or more folded polypeptide chains to associate to form a functional molecule. In such cases the individual polypeptides of the active molecule are referred to as *subunits*. The subunits may be multiple copies of the same polypeptide chain (a homomultimer), or they may represent distinct polypeptides (a heteromultimer). In both cases the subunits fold as individual units, acquiring their own secondary and tertiary structures. The association between subunits may be stabilized through non-covalent forces, such as hydrogen bonding, salt bridge formation, and hydrophobic interactions, and may additionally include covalent disulfide bonding between cysteines on the different subunits.

There are numerous examples of multisubunit enzymes in nature, and a few are listed in Table 3.3. In some cases, the subunits act at as quasi-independent catalytic units. For example, the enzyme prostaglandin synthase exists as a homodimer, with each subunit containing an independent active site that processes substrate molecules to product. In other cases, the active site of the enzyme is contained within a single subunit, and the other subunits serve to stabilize the structure, or modify the reactivity of that active subunit. In the cytochrome oxidases, for example, all the active sites are contained in subunit I, and the other 3–12 subunits (depending of species) modify the stability and specific activity of subunit I. In still other cases the active site of the enzyme is formed by the coming together of the individual subunits. A good illustration of this comes from the aspartyl protease of the human immunodeficiency virus, HIV (the causal agent of AIDS). The active sites of all aspartyl proteases require a pair of aspartic acid residues for catalysis. The HIV protease is synthesized as a 99-residue polypeptide chain that dimerizes to form the active enzyme (a homodimer). Residue 25 of each HIV protease monomer is an aspartic acid residue. When the monomers combine to form the active homodimer, the two Asp 25 residues (designated Asp 25 and Asp 25′ to denote their locations on separate polypeptide chains) come together to form the

Table 3.3 Examples of Multisubunit Enzymes

Enzyme	Number of Subunits
HIV protease	2
Hexokinase	2
Bacterial cytochrome oxidase	3
Lactate dehydrogenase	4
Aspartate carbamoyl transferase	12
Human cytochrome oxidase	13

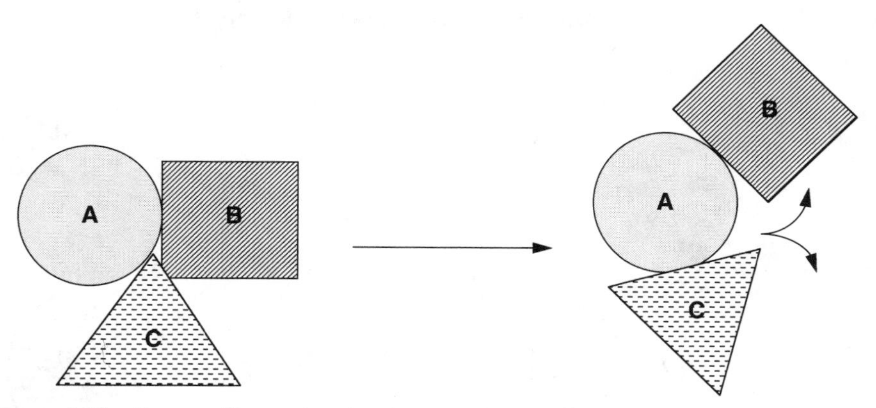

Figure 3.17 Cartoon illustrating the changes in subunit arrangements for a hypothetical heterotrimer that might result from a modification in quarternary structure.

active site structure. Without this subunit association, the enzyme could not perform its catalytic duties.

The arrangement of subunits of a protein relative to one another defines the *quaternary structure* of the protein. Consider a heterotrimeric protein composed of subunits A, B, and C. Each subunit folds into its own discrete tertiary structure. As illustrated schematically in Figure 3.17, these three subunits could take up a number of different arrangements with respect to one another in three-dimensional space. This cartoon depicts two particular arrangements, or quaternary structures, that exist in equilibrium with each other. Changes in quaternary structure of this type can occur as part of the activity of many proteins, and these changes can have dramatic consequences.

An example of the importance of protein quaternary structure comes from examination of the biological activity of hemoglobin. Hemoglobin is the protein in blood that is responsible for transporting oxygen from the lungs to the muscles (as well as transporting carbon dioxide in the opposite direction). The active unit of hemoglobin is a heterotetramer, composed of two α subunits and two β subunits. Each of these four subunits contains a heme cofactor (see Section 3.7) that is capable of binding a molecule of oxygen. The affinity of the heme for oxygen depends on the quaternary structure of the protein and on the state of oxygen binding of the heme groups in the other three subunits (a property known as cooperativity). Because of the cooperativity of oxygen binding to the hemes, hemoglobin molecules almost always have all four heme sites bound to oxygen (the oxy form) or all four heme sites free of oxygen (the deoxy form); intermediate form with one, two, or three oxygen molecules bound are almost never observed.

When the crystal structures of oxy- and deoxyhemoglobin were solved, it was discovered that the two forms differed significantly in quaternary structure. If we label the four subunits of hemoglobin α_1, α_2, β_1 and β_2, we find that at

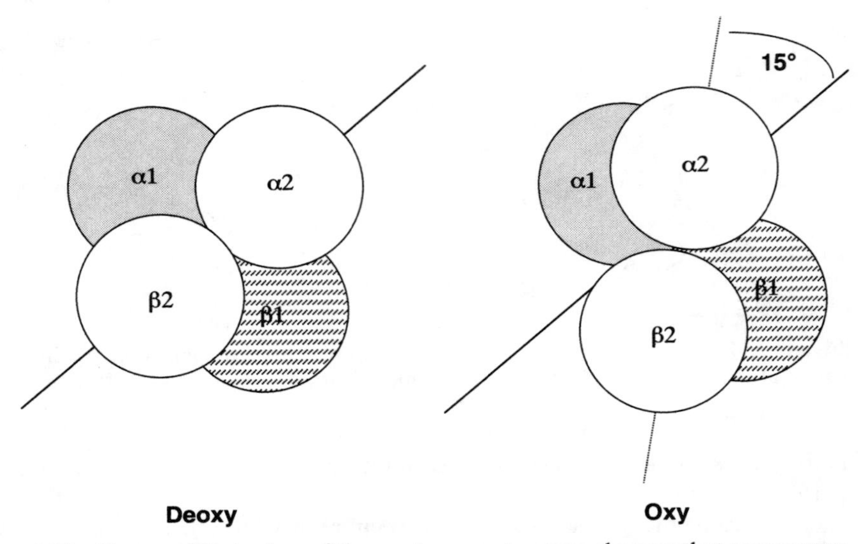

Figure 3.18 Cartoon illustration of the quaternary structure changes that accompany the binding of oxygen to hemoglobin.

the interface between the α_1 and β_2 subunits, oxygen binding causes changes in hydrogen bonding and salt bridges that lead to a compression of the overall size of the molecule, and a rotation of 15° for the $\alpha_1\beta_1$ pair of subunits relative to the $\alpha_2\beta_2$ pair (Figure 3.18). These changes in quarternary structure in part affect the relative affinity of the four heme groups for oxygen, providing a means of reversible oxygen binding by the protein. It is the reversibility of the oxygen binding of hemoglobin that allows it to function as a biological transporter of this important energy source; hemoglobin can bind oxygen tightly in the lungs and then release it in the muscles, thus facilitating cellular respiration in higher organisms. (For a very clear description of all of the factors leading to reversible oxygen binding and structural transitions in hemoglobin, see Stryer, 1989.)

3.7 Cofactors in Enzymes

As we have seen, the structures of the 20 amino acid side chains can confer on enzymes a vast array of chemical reactivities. Often, however, the reactions catalyzed by enzymes require the incorporation of additional chemical groups to facilitate rapid reaction. Thus to fulfill reactivity needs that cannot be achieved with the amino acids alone, many enzymes incorporate nonprotein chemical groups into the structures of their active sites. These nonprotein chemical groups are collectively referred to as enzyme *cofactors* or *coenzymes*. In some cases the cofactors are covalently bonded to the polypeptide of the enzyme. Examples of covalently associated cofactors include the heme group of the electron transfer protein cytochrome *c*, which is bound to the protein through thioether bonding with two modified cysteine residues, and the pyridoxal phosphate cofactor of aspartate aminotransferase, which forms a

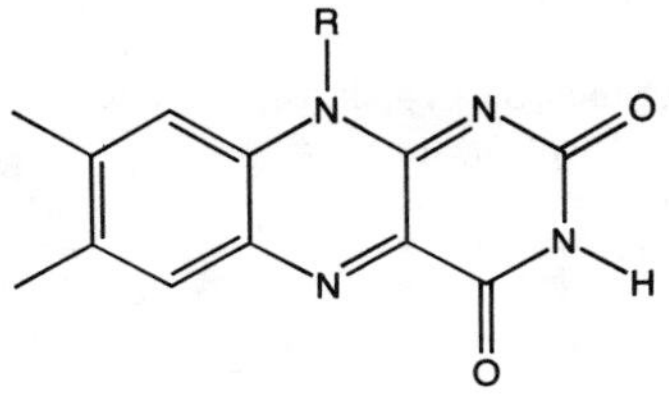

Flavin

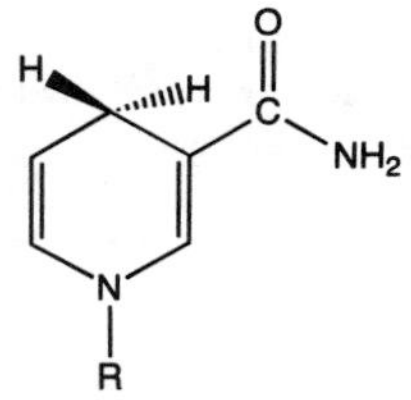

Nicotinamide

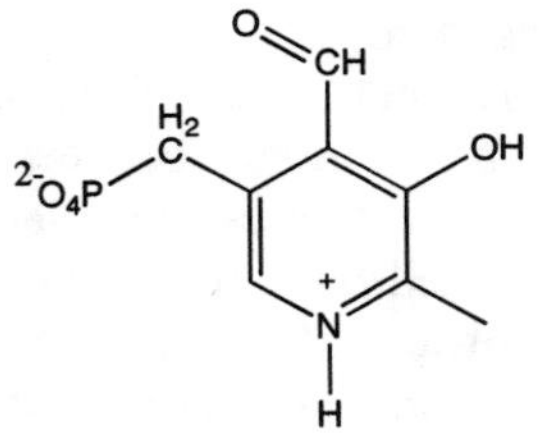

Pyridoxal phosphate

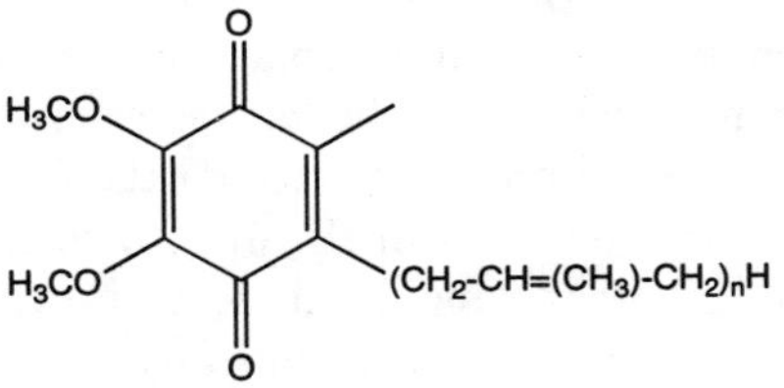

Ubiquinone (n)

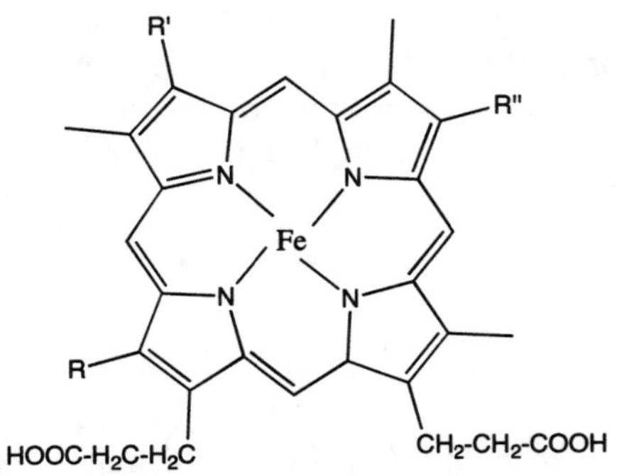

Heme

Figure 3.19 Examples of some common cofactors found in enzymes.

Table 3.4 Some Examples of Cofactors Found in Enzymes

Cofactor	Enzymatic Use	Examples of Enzyme
Copper ion	Redox center–ligand binding	Cytochrome oxidase, superoxide
Magnesium ion	Active site electrophile–phosphate binding	dismutase phosphodiesterases, ATP synthases
Zinc ion	Active site electrophile	Matrix metalloproteases, carboxypeptidase A
Flavins	Redox center–proton transfer	Glucose oxidase, succinate dehydrogenase
Hemes	Redox center–ligand binding	Cytochrome oxidase, cytochrome P450s
NAD and NADP	Redox center–proton transfer	Alcohol dehydrogenase, ornithine cyclase
Pyridoxal phosphate	Amino group transfer–stabilizer of intermediate carbanions	Aspartate transaminase, arginine racemase
Quinones	Redox center–hydrogen transfer	Cytochrome *bo*, dihydroorotate dehydrogenase
Coenzyme A	Acyl group transfer	Pyruvate dehydrogenase,

Schiff base with a lysine residue in the active site of the enzyme. In most cases, however, the protein and cofactors associated noncovalently.

In enzymes requiring a cofactor for activity, the protein portion of the active unit is referred to as the *apoenzyme*, and the active complex between the protein and the cofactor is called the *holoenzyme*. In some cases the cofactors can be removed to form the apoenzyme and later added back to reconstitute the active holoenzyme. In these cases, to facilitate structural and mechanistic studies of the enzyme, chemically or isotopically modified versions of the cofactor are sometimes incorporated into the apoenzyme.

Cofactors fulfill a broad range of reactions in enzymes and are composed of a wide variety of chemical groups. Figure 3.19 illustrates the structure of some common cofactors found in enzyme catalysis, and Table 3.4 lists the enzymatic functions and examples of enzymes utilizing these and other cofactors. The list is by no means comprehensive, but it gives a flavor of the breadth of structures and reactivities that nature has chosen to utilize in catalysis. (See Dixon and Webb, 1979, for a more complete description of cofactor utilization in enzymes.)

3.8 Summary

In this chapter we have seen the diversity of chemical reactivities that are imparted to enzymes by the structures of the amino acid side chains. We have described how these amino acids can be linked together to form a polypeptide chain, and how these chains fold into regular patterns of secondary and tertiary structure. The folding of an enzyme into its correct tertiary structure provides a means of establishing the binding pockets for substrate ligands and presents, within these binding pockets, the chemically reactive groups required for catalysis. The active site of the enzyme is defined by these reactive groups, and by the overall topology of the binding pocket. We have seen that the chemically reactive groups used to convert substrate to product molecules are recruited by enzymes, not only from the amino acids that make up the protein, but from cofactor molecules as well; these cofactors are critical components of the biologically active enzyme molecule. In the next chapter, we shall see how the structural details of the enzyme active site facilitate substrate binding and the acceleration of reaction rates, which are the hallmarks of enzymatic catalysis.

References and Further Reading

Branden, C., and Tooze, J. (1991) *Introduction to Protein Structure*, Garland, New York.

Copeland, R. A. (1994) *Methods for Protein Analysis: A Practical Guide to Laboratory Protocols*, Chapman & Hall, New York.

Creighton, T. E. (1984) *Proteins, Structure and Molecular Properties*, Freeman, New York.

Davis, J. P., and Copeland, R. A. (1996) *Protein Engineering*, in *Kirk-Othmer Encyclopedia of Chemical Technology*, 4th ed., Wiley, New York.

Dixon, M., and Webb, E. C. (1979) *Enzymes*, 3rd ed., Academic Press, New York.

Hansch, C., and Coats, E. (1970) *J. Pharm. Sci.* **59**, 731–743.

Kyte, J., and Doolittle, R. F. (1982) *J. Mol. Biol.* **157**, 105–132.

Pauling, L., Itano, H. A., Singer, S. J., and Wells, I. C. (1949) *Science*, **110**, 543–548.

Picot, D., Loll, P., and Garavito, R. M. (1994) *Nature* **367**, 243–249.

Stryer, L. (1989) *Molecular Design of Life*, Freeman, New York.

CHAPTER

4

Chemical Mechanisms in Enzyme Catalysis

In this chapter we discuss how the elements of protein structure, described in Chapter 3, come together to form the active site of an enzyme. The active site serves the dual purpose of binding the substrate molecule and bringing into close proximity the chemically reactive groups that facilitate the transformation of the substrate(s) to the product(s) of the catalyzed reaction. We shall see that the three-dimensional structure of the active site is, in some ways, stereochemically complementary to the structure of the substrate molecule. This fitting together of the substrate and active site will provide a means of stabilizing the transition state of the reaction, leading to a significant acceleration of the reaction rate. Nature has evolved a broad range of enzyme molecules, each catalyzing specific reactions in accordance with the complementarity of its active site structure and that of its target substrate. Scientists, in turn, want to keep track of the myriad reactions catalyzed by enzymes. We shall introduce the system of nomenclature devised for the categorization of enzymes on the basis of the reactions they promote.

4.1 Substrate Specificity and Rate Enhancement in Enzyme Catalysis

In Chapter 1 we said that chemical reactions, such as molecule S (for substrate) going to molecule P (for product), will proceed through the formation of a high energy, short-lived state known as the transition state (here we shall give the transition state the symbol $S^{\ddagger}$). Thus at a minimum we shall have three molecular species involved in a chemical reaction: the reactant molecule S, the transition state $S^{\ddagger}$, and the product P. This is likewise true for enzyme-catalyzed reactions. Here, however, the transition state occurs while the substrate molecule is bound at the enzyme active site. For an enzyme-catalyzed reaction, we also must include molecular forms in which the enzyme and substrate are bound together to form a tight complex (known as the enzyme–substrate complex and given the symbol ES), and one in which the product molecule and the enzyme are likewise bound together as the enzyme–product complex, EP. Consider a simple enzyme-catalyzed reaction in which one substrate is transfomed into one product. The reaction path would be described as follows:

$$E + S \rightleftharpoons ES \rightleftharpoons ES^{\ddagger} \rightleftharpoons EP \rightleftharpoons E + P$$

In this scheme the free enzyme E and free substrate S must encounter each other (typically through molecular collisions in solution) to form the enzyme–substrate complex ES. This process is a reversible equilibrium, but under typical laboratory conditions the formation of ES is favored; the ΔG of binding for a typical ES pair is about -3 to -12 kcal/mol. With the substrate bound, the enzyme active site structure facilitates the formation of the bound transition state by mechanisms we shall discuss shortly. Once formed, the $ES^{\ddagger}$ complex could decay back to ES or proceed forward to form the enzyme-bound product complex EP. As with uncatalyzed reactions, however, once the energy barrier to the transition state has been overcome, the reaction is much more likely to proceed energetically downhill to the formation of the product state. Thus we can consider the reaction of $ES^{\ddagger}$ going to EP as essentially a one-way irreversible reaction. Once formed, the EP complex must finally dissociate, leading to the liberation of the free product molecule and our starting point, the free enzyme molecule. Since the free enzyme molecule appears as both reactant and as final product, it can be ignored when considering the thermodynamics of the overall reaction (see Chapter 1). Hence, the free energy of reaction here will depend only on the relative concentrations of S and P:

$$\Delta G = -RT \ln\left(\frac{[P]}{[S]}\right) \tag{4.1}$$

This is exactly the same way we would calculate the ΔG for the uncatalyzed reaction $S \rightarrow P$, and it reflects the *path independence* of the function ΔG. In other words, ΔG depends only on the initial and final states of the reaction, not on the various intermediates (i.e., ES, $ES^{\ddagger}$, and EP) formed during the reaction. This leads to the important realization that *enzymes cannot alter the equilibrium between products and substrates.*

What then is the value of using an enzyme to catalyze a chemical reaction? The answer is that enzymes, and in fact all catalysts, speed up the rate at which equilibrium is established in a chemical system: *enzymes accelerate the rate of chemical reactions.* Hence, with an ample supply of substrate, one can form much greater amounts of product *per unit time* in the presence of an enzyme than in its absence. Thus the great value of enzymes, both for biological systems and in commercial use, is that they provide a means of making more product at a faster rate than can be achieved without catalysis.

How is it that enzymes achieve this acceleration in reaction rate? The answer lies in a consideration of the activation energy of the chemical reaction. Many uncatalyzed reactions proceed slowly because they must overcome a substantial energy barrier to reach the transition state, as we saw in Chapter 2. If we could somehow lower that energetic barrier, by stabilizing the transition state, the reaction would proceed faster. Recall from Chapter 2 that the rate of loss of substrate, also referred to as the *velocity* of substrate loss v, is related to the activation energy of the reaction as follows:

$$v = \frac{-d[\mathrm{S}]}{dt} = \left(\frac{k_{\mathrm{B}}T}{h}\right)[\mathrm{S}]\exp\left(-\frac{E_{\mathrm{a}}}{RT}\right) \tag{4.2}$$

Now, for simplicity, let us fix the reaction temperature at 25°C and fix [S] at a value of 1 in some arbitrary units. At 25°C, $RT = 0.59$, and $k_{\mathrm{B}}T/h = 6.2 \times 10^{12}\,\mathrm{s}^{-1}$. Suppose that the activation energy of a chemical reaction at 25°C is 7 kcal/mol. The velocity of the reaction will thus be:

$$v = 6.2 \times 10^{12}\,\mathrm{s}^{-1} \cdot 1 \cdot e^{-7/0.59} = 4 \times 10^{7}\,\mathrm{s}^{-1} \tag{4.3}$$

If we somehow reduce the activation energy to 3 kcal/mol, a difference of only 4 kcal/mol, the velocity now becomes:

$$v = 6.2 \times 10^{12}\,\mathrm{s}^{-1} \cdot 1 \cdot e^{-3/0.59} = 4 \times 10^{10}\,\mathrm{s}^{-1} \tag{4.4}$$

Thus by lowering the activation energy by 4 kcal/mol we have achieved a thousand fold increase in reaction velocity! This is exactly how enzymes function. They accelerate the velocity of chemical reactions by stabilizing the transition state of the reaction, hence lowering the energetic barrier that must be overcome.

Let us look at the energetics of a chemical reaction in the presence and absence of an enzyme, using the energy level diagrams introduced in Chapter 2. These are illustrated in Figure 4.1 for the situation of a substrate present at a concentration less than the equilibrium constant for formation of the ES complex, conditions under which there is no substantial buildup of the ES complex. In the absence of enzyme, the reaction proceeds from substrate to product by overcoming the sizable energy barrier required to reach the transition state $\mathrm{S}^{\ddagger}$. In the presence of enzyme, on the other hand, the reaction

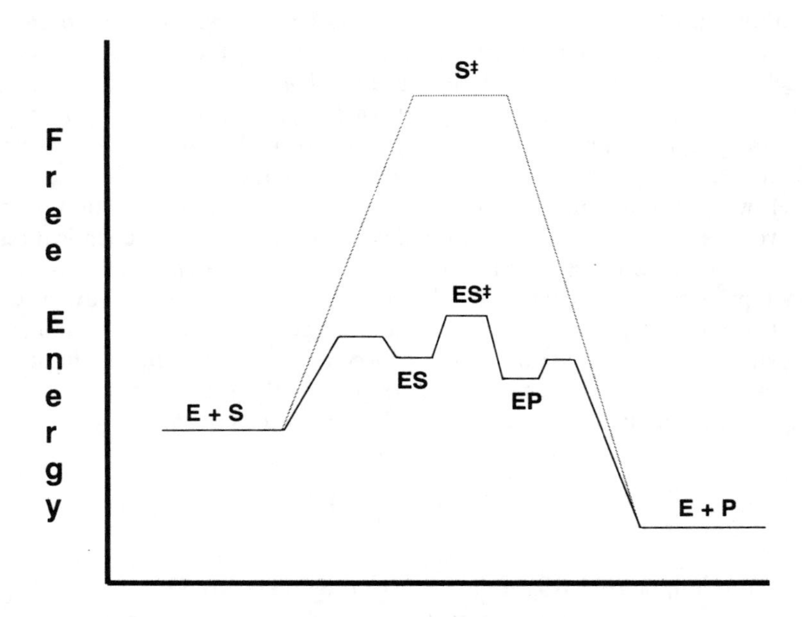

Figure 4.1 Energy level diagram of an enzyme-catalyzed reaction (solid line) and the corresponding uncatalyzed chemical reaction (dashed line). The symbols E, S, ES, ES‡, EP, and P represent the free enzyme, the free substrate, the enzyme–substrate complex, the enzyme–transition state complex, the enzyme–product complex, and the free product states, respectively. At the energy levels depicted, the substrate is present in concentrations far below the dissociation constant for the ES complex, with the result that this complex does not build up to any significant concentration.

first proceeds through the formation of the enzyme–substrate complex ES. The ES complex represents an intermediate along the reaction pathway that forms when the free enzyme and free substrate molecules meet and combine. Once binding has occurred, molecular strain in the bound molecule and other factors (discussed shortly) have the effect of simultaneously destabilizing the ground state configuration of the substrate molecule, in the ES complex, and energetically favoring the transition state. The complex ES‡ thus occurs at a lower energy than the free S‡ state, as shown in Figure 4.1.

The reaction next proceeds through formation of another intermediate state, the enzyme–product complex EP, before finally reaching the free product plus free enzyme state. Again, the initial and final states are energetically identical in the catalyzed and uncatalyzed reactions. However, the overall activation energy barrier has been greatly reduced in the enzyme-catalyzed case. This reduction in activation barrier results in a significant acceleration of reaction velocity in the presence of enzyme, as we have seen (Equations 4.2–4.4). This is the common strategy for rate acceleration utilized by all enzymes:

> Enzymes accelerate the rate of chemical reactions by stabilizing the transition state of the reaction, hence lowering the activation energy barrier to product formation.

Several mechanisms in the active site of the enzyme are responsible for the stabilization of the transition state, as well as the destabilization of the substrate ground state. The reactions catalyzed in enzyme active sites follow general organic reaction mechanisms, involving nucleophilic and electrophilic attacks, as well as acid–base chemistry. As we have seen, the reactive groups required for these mechanisms are provided by the enzyme active site through a combination of amino acid side chains and cofactor groups. By binding the substrate, the enzyme brings together the reactive groups required for reactions within close proximity. This has a rate enhancing effect by itself.

For a reaction to occur in solution, the reactants must encounter one another by random collisions. We could marginally increase the rate of these collisional encounters by elevating the temperature of the solution (but this would be energetically costly), or by increasing the concentration of reactants (but there are practical limits on this, too). When substrates are sequestered with the active site of the enzyme, however, the *effective* concentration of reactants is greatly increased. The reactants also are placed in close contact and in an orientation that permits the reaction to proceed rapidly. This property, which has been termed the *proximity effect* or *propinquity effect*, is thought to play a key role in enzyme rate enhancement.

In addition to simply bringing reactants together, forces within the enzyme active site can align the substrate and enzyme reactive groups into proper orientation for reaction. Hydrogen bonding, electrostatic, hydrophobic, and van der Waals interactions in the active site can distort the substrate molecule toward the transition state. These forces must have a strong directionality, to ensure that the bond distortions are of the correct orientation to promote the transition state. Put another way, the interactions in the active site can perturb the molecular orbitals of the substrate by aligning them with key reactive groups of the enzyme active site, so that they resemble the orbital configurations of the transition state; this concept of *orbital steering* was first proposed by Daniel Koshland and his coworkers (Storm and Koshland, 1970).

After the substrate molecule binds at the enzyme active site, the further chemical steps leading to the transition state and beyond proceed with less loss of overall entropy than would be experienced by the corresponding bimolecular reaction in solution; this is another energetic advantage that contributes to enzymatic rate enhancement. Recall that the entropy of a system is a measure of the randomness or disorder of that system. In molecular terms, entropy is determined mainly by the degrees of rotational and translational freedom of the molecule. In a biomolecular reaction, where two molecule combine in solution to form a third,

$$X + Y \rightleftharpoons [XY^{\ddagger}] \rightleftharpoons X{-}Y$$

the formation of the transition state occurs with loss of translational and rotational freedom for both X and Y individually, hence is costly in entropic terms. If, however, these groups already have been brought together in the correct orientation for reaction by formation of the enzyme–substrate complex, the entropic loss that accompanies attainment of the transition state will be much less. The rotational and translational restrictions have largely been effected during binding to form the ES complex, and this rearrangement is, in part, compensated for by the substrate binding energy. Hence the entropic cost in the enzyme-catalyzed case has been shifted from the chemical events to the binding event.

To fully understand the rate enhancement and specificity of the chemical reactions performed by enzyme, we must consider the structure of the enzyme active site and its relationship to the structures of the substrate in its ground and transition states. The active site structure differs from enzyme to enzyme, but some generalizations can be made.

1. The active sites of enzymes are small relative to the total volume of the enzyme molecule.
2. The active site is three-dimensional—that is, amino acids and cofactors in the active site are held in a precise arrangement with respect to one another and with respect to the structure of the substrate molecule. This three-dimensional structure is formed as a result of the overall tertiary structure of the protein.
3. In most cases, the initial interactions between the enzyme and substrate molecules (i.e., the binding events) are noncovalent, making use of hydrogen bonding, electrostatic and hydrophobic interactions, and van der Waals forces to effect binding.
4. The active sites of enzymes usually occur in clefts and crevices in the protein. This design has the effect of excluding bulk solvent, which would reduce the catalytic activity of the enzyme. In other words, the substrate molecule is *shielded* from the bulk solvent in the enzyme active site.
5. The specificity of substrate binding depends on well-defined arrangements of atoms in the enzyme active site that in some way complement the structure of the substrate molecule.

As we have said, one of the hallmarks of enzyme catalysis is the substrate specificity that can be achieved by these proteins. The degree of substrate specificity, however, varies from enzyme to enzyme. Some enzymes recognize and bind a broad range of molecules in a specific structural class, while others bind to one molecule exclusively. This is illustrated by the class of hydrolytic enzymes known as the serine proteases (Table 4.1). As seen in Table 4.1, certain

Table 4.1 Specificity for Peptide Bond Hydrolysis in the Serine Proteases: Three Examples That Span a Range of Specificities

Enzyme	Substrate[a]	Specificity
Subtilisin	Many peptide bonds	Low
Trypsin	Lys-X and Arg-X bonds, where X $\neq$ Pro	Moderate
Thrombin	Arg-Gly	High

[a]Refers to the peptide linkage, in peptides and proteins, that is most effectively recognized and hydrolyzed by the enzyme.

of the proteases, which hydrolyze peptide (and sometimes ester) bonds, will catalyze the hydrolysis of a broad range of peptide bonds, as in the case of subtilisin. Others, like thrombin, are highly specific with regard to the peptide linkages they recognize. A high degree of specificity can be vital to the physiological roles played by these enzymes.

Early in the history of enzymology, scientists noted that enzymes recognize specific substrates based on the structures of these molecules. Often, an enzyme will recognize and catalyze most efficiently the reaction of a particular substrate molecule. Molecules that are structurally related to the substrate will also bind to the enzyme active site, the strength of binding being directly related to how closely they resemble the true substrate for catalysis. This property is often exploited to make inhibitors of enzymes by preparing structurally related, but unreactive, substrate analogues (see Chapters 8 and 9). To explain these observations, Emil Fisher (1894) proposed the *lock-and-key* model of enzyme–substrate interactions. In this famous model, summarized in Figure 4.2, the enzyme active site is viewed as a static structure that stereochemically complements the structure of the substrate to provide for tight binding. The insertion of the substrate into the static enzyme active site is analogous to a key fitting into a lock; the specificities of the lock for the correct key, and the enzyme active site for the correct substrate, are based on the structural complementarity of the two components.

While the lock-and-key model is conceptually useful, it does not take into consideration the conformational flexibility of enzyme active sites, nor the need for structural distortions of the substrate to achieve transition state stabilization. An alternative model was proposed by Koshland (1958) that has been termed *the induced-fit* model (Figure 4.3). In this model, the enzyme active site is conformationally flexible. It may encounter the substrate molecule in a form that is not optimal for binding but does allow for some affinity between the enzyme and substrate. The enzyme then "molds" the active site structure around the substrate to engage specific intermolecular interactions and adopt a three-dimensional structure that best complements that of the substrate, hence leading to much greater binding affinity (i.e., tighter binding between the

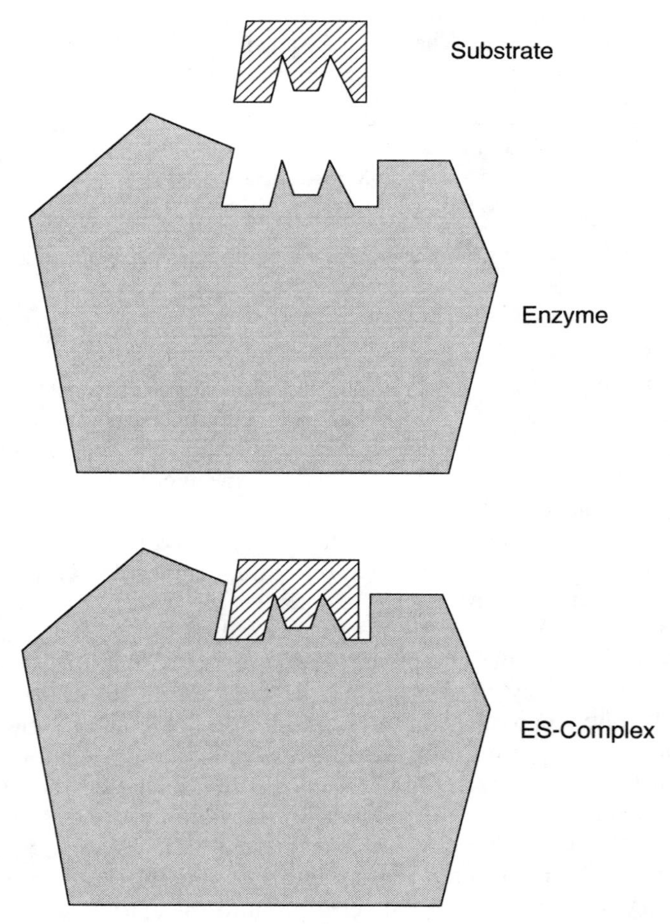

Figure 4.2 Schematic illustration of the lock-and-key model of enzyme–substrate interactions.

enzyme and substrate). Several examples are now known of protein conformational transitions attending substrate binding that support the induced-fit model (see, e.g., Goldsmith and Kuo, 1993).

In discussing the lock-and-key and induced-fit models, we have illustrated the enzyme active site and substrate in two-dimensional drawings, but we must not lose sight of the three-dimensionality of these objects. The overall tertiary structure of the enzyme imparts a precise three-dimensional structure to the enzyme active site. Because of this, enzyme can also display exquisite stereoselectivity for substrate. Consider the example (first used by Segel, 1975) of the alcohol dehydrogenases that catalyze the transfer of a methylene

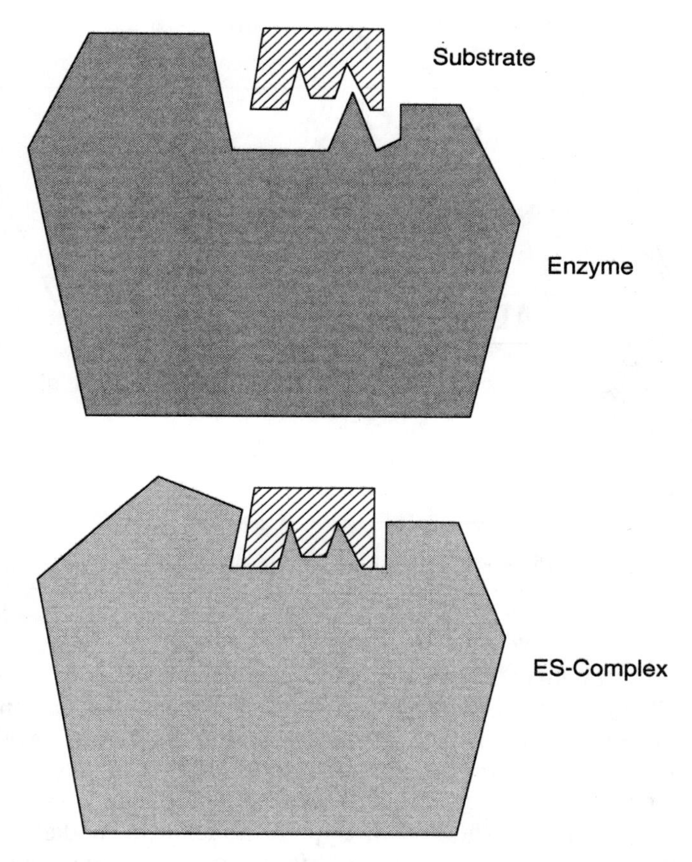

Figure 4.3 Schematic illustration of the induced-fit model of enzyme–substrate interactions.

hydrogen of an alcohol to the cofactor NAD^+:

$$CH_3CH_2OH + NAD^+ \rightleftharpoons CH_3CHO + NADH + H^+$$

A particular enzyme of this family will stereospecifically transfer the same methylene hydrogen to NAD^+ with every catalytic cycle. This stereospecificity is best explained by assuming that the alcohol binds to the active site of the enzyme through specific interactions of its methyl group, its hydroxyl group, and one of its hydrogens to form a *three-point attachment* with the enzyme

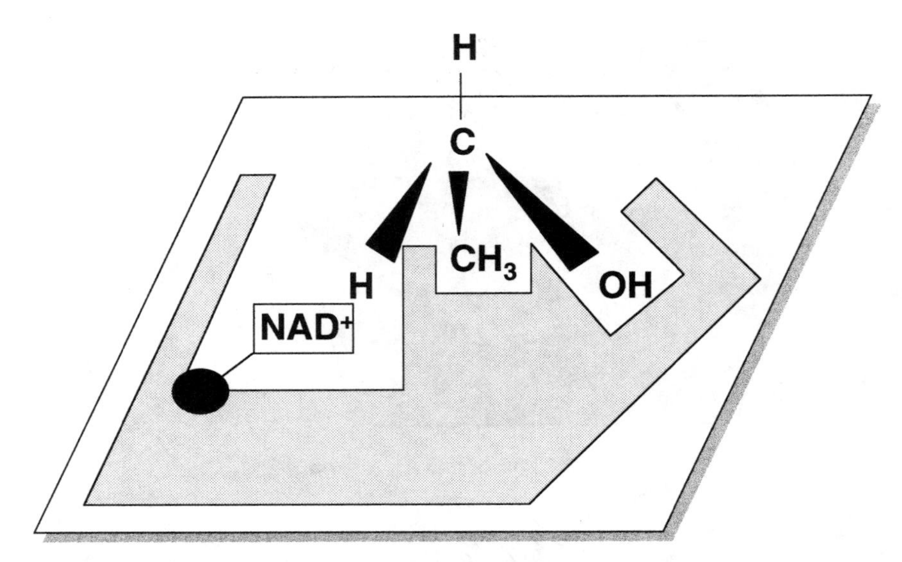

Figure 4.4 Illustration of three-point attachment in enzyme–substrate interactions.

active site, as illustrated in Figure 4.4. In fact, having anchored down the methyl and hydroxyl groups as depicted in Figure 4.4, the enzyme is committed to the transfer of the specific hydrogen atom because of its relative proximity to the NAD^+ cofactor. This three-point attachment hypothesis is often invoked to explain the stereospecificity commonly displayed by enzymes.

The lock-and-key, induced-fit, and three-point attachment models help to explain the specificity displayed by enzymes for unique substrates, but these models do not, by themselves, provide an explanation for enzymatic rate enhancement. There is, in fact, considerable controversy among scientists about how to best reconcile the need for specificity of substrate binding with the need for transition state stabilization in enzyme catalysis. Most scientists agree that the structure of the enzyme active site is specifically designed to bind not the ground state, but the transition state of the substrate molecule. The three models just described are still applicable in this view, but they pertain to the transition state of the substrate rather than the ground state. In this view, the enzyme might "pick" out of solution a subpopulation of substrate molecules that, through vibrational motions, have momentarily achieved a transition state-like structure (Pauling, 1948; Schowen, 1978). This view predicts that molecules whose ground state structure mimics that of the substrate transition state would bind to the enzyme active site with much greater affinity than the ground state substrate molecule.

Experimentally, one finds that transition state analogues (i.e., stable molecules designed to structurally mimic the transition state of the reaction) indeed not only bind to enzyme molecules, but in most cases bind with much greater affinity than the corresponding substrate molecule. Recently, this view has gained further support from the experimental work on catalytic antibodies (see, e.g., Lerner et al., 1991). Normally, the immune system produces antibodies that recognize and bind tightly to proteins or peptides from an infecting organism; but antibodies that recognize and bind smaller molecules, referred to as haptens, can be produced as well. It has been found that if antibodies are raised against a transition state analogue of an enzymatic reaction, the resulting antibody not only recognizes and binds tightly to this hapten, but also displays modest catalytic activity! This is a clear indication of the importance of the transition state in protein-based catalysis. However, it is important to point out that to date none of these catalytic antibodies have achieved anywhere near the level of reaction rate enhancement that is seen for the corresponding natural enzyme.

Schowen (1978) has taken the preceding view to its extreme, arguing that "the entire and sole source of catalytic power is the stabilization of the transition state; reactant-state interactions are by nature inhibitory and only waste catalytic power." The reactant state that Schowen refers to is the enzyme–substrate ground state complex.Other scientists have taken issue with this "fundamentalist" position. They point out that all enzyme reactions proceed through the formation of an initial noncovalent complex between the enzyme and the substrate. In fact, whenever the substrate concentration exceeds the equilibrium constant for complex formation (typical laboratory conditions), the ES complex is found to be stabilized relative to the free enzyme plus substrate state (in other words, under these conditions a free energy plot as in Figure 4.1 would have the ES state at *lower* energy than the E + S state). These scientists question why the ES state would be formed if it were, by its very nature, inhibitory to catalysis. Recent work using site-directed mutagenesis studies has shown that in at least some enzymes, substrate specificity can be altered without affecting catalytic efficiency (Wilson and Agard, 1991). These results suggest that the structural determinants of substrate specificity can at least in part be distinguished from the mechanism of transition state stabilization.

To reconcile the need for substrate specificity and reasonable enzyme affinity for the substrate ground state with the need for substrate ground state destabilization and transition state stabilization, several hypotheses have been put forth. In the *split-site* model proposed by Menger (1992), the enzyme active site is subdivided into a region of binding and a region of reaction. Certain portions of the active site are most responsible for the initial binding of the substrate in its ground state configuration (i.e., specific contacts between amino acids or cofactors of the enzyme are made with the substrate molecule in the ground state). Other components of the enzyme active site engage the substrate molecule for catalysis and are thus responsible for the bond distortions leading to the formation of transition and product states. In this model, the ES complex is energetically subdivided into two quasi-separate states: a binding state ES_B and a reactive state ES_R. As a result, the free energy of the ES complex is given by:

$$\Delta G(\mathrm{ES}) = \Delta G(\mathrm{ES_B}) + \Delta G(\mathrm{ES_R})$$

The substrate binding and reactive sites of the enzyme do not necessarily need to be structurally distinct. Let us go a step further with the concept of active site structural flexibility, as described for the induced-fit model. We recognize that once the substrate has bound to the active site, additional conformational changes of the enzyme can occur. These structural transitions of the active site can be directed along the vibrational coordinates of the substrate that bring it from its ground to transition state configuration. Such structural changes would induce a great deal of molecular strain in the ground state substrate molecule, hence destabilizing it in favor of the distorted transition state configuration. This idea of induced molecular strain may also play a major role in enzyme catalysis, providing a structural basis for the transition state stabilization we have thus far discussed only in energetic terms.

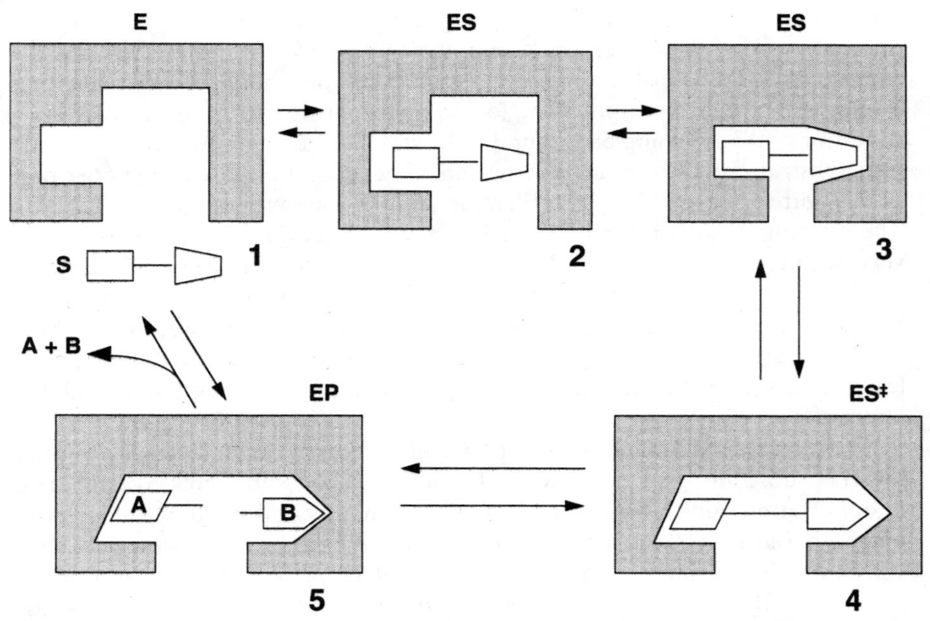

Figure 4.5 The rack model of enzyme–substrate interaction and catalysis: **1**, the free enzyme is in solution with the free substrate; **2**, initial binding occurs to form the ES complex without structural distortion of the substrate; **3**, the enzyme undergoes a conformational change to induce better binding with the substrate (induced fit); **4**, the enzyme undergoes further conformational adjustments that induce strain into the substrate molecule, distorting it from its ground state structure toward its transition state structure; **5**, the strain induced in **4** leads to bond rupture and release of the two hydrolytic products A and B.

Figure 4.5 illustrates schematically the concept of induced molecular strain. In this example the binding of the substrate to the enzyme is followed by conformational transitions that first mold the active site structure around the substrate molecule for tighter binding and subsequently strain the bound substrate along a coordinate that stabilizes the transition state configuration of the molecule, leading to catalysis (in this schematic example, bond cleavage). As illustrated here, the distortions of the substrate within the active site are akin to the medieval torture mode in which a prisoner was stretched on a device known as a rack. For this reason, the term *rack mechanism* is sometimes used (Segel, 1975).

The concepts just discussed emphasize the importance of conformational flexibility in enzyme catalysis. The binding of substrates to enzymes may effect protein structural changes, at least in the vicinity of the enzyme active site. It is important to realize that enzymes in solution access these conformational variants continuously over time, even when substrate is not bound. The binding of substrates or other ligands to the active site may stabilize or "lock" the enzyme in a particular conformation, but at any moment in time, some subpopulation of our enzyme molecules in solution may exist in this conformation even without ligands present.

While the details of how enzymes engage their substrates and achieve transition state lowering are still being debated, there is general agreement that enzyme active sites are structurally designed to destabilize the bound substrate, release the bound product, and maximally stabilize the bound transition state of the reactant molecule, relative to the transition state of the uncatalyzed reaction. These are the essential features that dictate the efficiencies of enzyme-catalyzed reactions.

4.2 The Serine Proteases: An Illustrative Example

The serine proteases are a family of enzymes that catalyze the cleavage of specific peptide bonds in proteins and peptides. As we briefly mentioned earlier, the serine proteases have a common mechanism of catalysis that requires a triad of amino acids at the active sites of these enzymes: a serine residue (hence the family name) acts as the primary nucleophile for attack of the peptide bond, and the nucleophilicity of this group is enhanced by specific interactions with a histidine side chain, which in turn interacts with an aspartate side chain. The catalytic importance of the active site serine and histidine residues has been demonstrated by site-directed mutagenesis studies in which replacement of the serine or the histidine or both reduced the rate of reaction by the enzyme by as much as 10^6-fold (Carter and Wells, 1988).

The serine proteases were among the first enzymes studied, because of their ease of isolation and availability in large quantities from the gastric juices of large animals. While the first members of this family to be studied were all digestive enzymes, we now know that the serine proteases perform a wide variety of catalytic functions in most organisms from bacteria to higher mammals. In man, for example, serine proteases take part not only in digestive processes, but also in the blood clotting cascade, inflammation, the general immune response, and other physiologically important events. These enzymes are among the most well-studied proteins in biochemistry. A great deal of structural and mechanistic information on this class of enzymes is available from crystallographic, classical biochemical, mechanistic, and mutagenesis studies (see Perona and Craik, 1995, for a recent review). Because of this wealth of information, the serine proteases provide a model for discussing in concrete chemical and structural terms some of the concepts of substrate binding specificity and transition state stabilization covered thus far in this chapter.

To cleave a peptide bond within a polypeptide or protein, a protease must recognize and bind a region of the polypeptide chain that brackets the scissile peptide bond (i.e., the bond that is to be cleaved). Proteases vary in the length of polypeptide that forms their respective recognition sequences, but most bind several amino acid residues in their active sites. A nomenclature system has been proposed by Schechter and Berger (1967) to keep track of the substrate amino acid residues involved in binding and catalysis, and the corresponding sites in the enzyme active site where these residues make contact. In this system, the bond that is to be hydrolyzed is formed between residue P1 and P1′ of the substrate; P1 is the residue on the N-terminal side of the scissile bond, and P1′ is the C-terminal-hydrolyzed residue. (The P stands for "peptide" to designate these residues as belonging to the substrate of the reaction.) The residue adjacent to P1 on the N-terminal side of the scissile bond is designated P2, and the residue adjacent to the P1′ residue on the C-terminal side is P2′. The "subsite" in the enzyme active site into which residue P1 fits is designated S1, and the "subsite" into which residue P1′ fits is designated S1′. The numbering continues in this manner, as illustrated in Figure 4.6 for a six-residue peptide substrate. We shall use this nomenclature system from now on when discussing proteolytic enzymes.

On the basis of their structural properties, the serine proteases have been divided into three classes, called the chymotrypsin-like, the subtilisin-like, and the serine carboxypeptidase II–like families. The secondary and tertiary structures of the proteins vary considerably from one family to another, yet in all three the active site serine, histidine, and aspartate are conserved and a common mechanism of catalysis is used. All these enzymes catalyze the hydrolysis of ester and peptide bonds through the same acyl transfer mechanism (Figure 4.7), with a rate acceleration of 10^9 or more relative to the uncatalyzed reaction. After formation of the ES complex, the carbonyl carbon of the scissile peptide bond (i.e., that on P1) is attacked by the active site serine, forming a tetrahedral intermediate with an oxyanionic center on the carbonyl carbon that is highly reminiscent of the transition state of the reaction. This transition state is stabilized by specific hydrogen-bonding interactions between

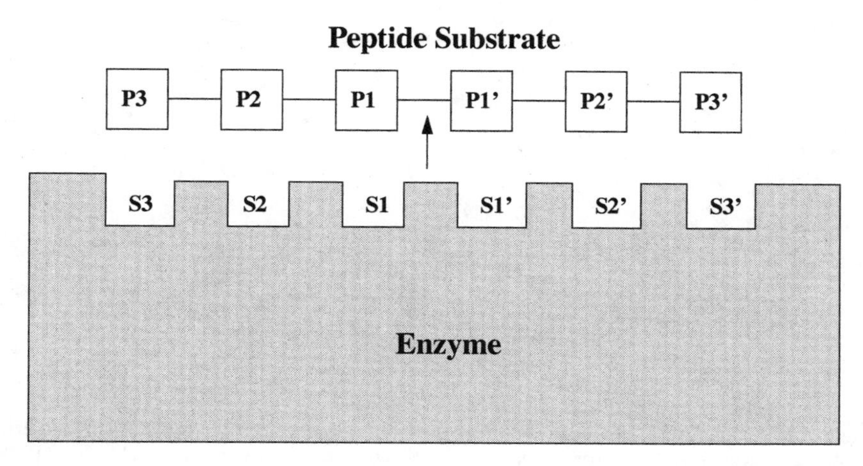

Figure 4.6 The protease subsite nomenclature of Schechter and Berger (1967): residues on the peptide substrate are labeled P1–P*n* on the N-terminal side of the scissile bond, and P1′–P*n*′ on the C-terminal side; the scissile bond is thus between residues P1 and P1′. The corresponding subsites into which these residues fit in the enzyme active site are labeled S1–S*n* and S1′–S*n*′.

residues in the active site pocket and the oxyanion center of the substrate. In subtilisin this hydrogen bonding is provided by the backbone nitrogen of Ser 195 (the active site nucleophile) and the side chain of Asn 155. In the chymotrypsin-like enzymes, these H bonds are provided by the backbone nitrogens of the nucleophilic serine residue and an active site glycine, while in the serine carboxypeptidases these bonds are formed by the backbone nitrogens of a tyrosine and a glycine in the active site. Crystallographic data from studies of subtilisin indicate that a weak hydrogen bond exists between Asp 155 and the substrate in the ES complex, and that this H-bonding is significantly strengthened in the transition state (an example of conformational alterations in the enzyme active site that facilitate transition state stabilization). Mutation of Asp 155 to any of a variety of non-hydrogen-bonding amino acids significantly decreases the rate of the enzymatic reaction, as would be expected from our discussion of enzymatic rate enhancement by transition state stabilization.

The transition state then decays as a proton is donated from the active site histidine to the amine group of P1′, followed by dissociation of the first product of the reaction, the peptide starting at P1′, and simultaneous formation of a covalent intermediate with an acyl group (from P1) bound to the active site serine. The enzyme is then deacylated by nucleophilic attack by a water molecule that enters the enzyme active site from the cavity resulting from the leaving of the first product peptide. The deacylation reaction proceeds with

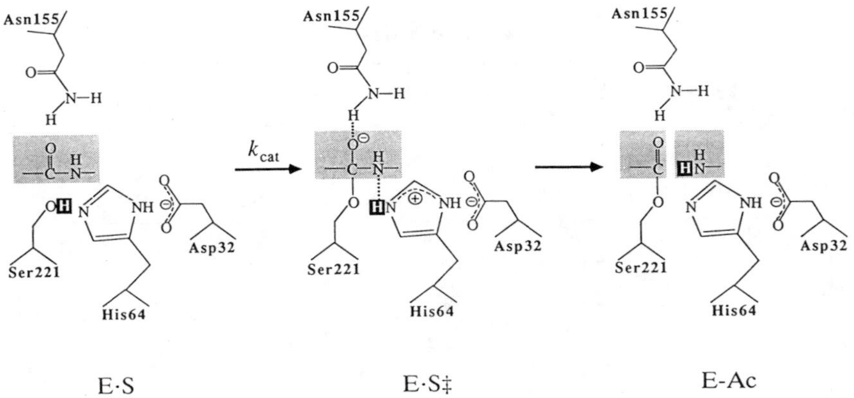

Figure 4.7 Schematic representation of the general acyl transfer mechanism of serine proteases. [Reprinted with permission from Carter and Wells, *Nature*, **332**, 564 (1988), copyright 1988, Macmillan Magazines Limited.]

formation of another tetrahedral transition state, very similar to that formed during the acylation reaction, and engaging the same stabilizing H bonds with the enzyme. This transition state decays with proton transfer to the active site histidine and release of the second peptide product having the P1′ group at its amino terminus.

The active site aspartate residue is a common feature of the serine proteases and has been shown, through mutagenesis studies, to be critical for catalysis. The role of this residue in catalysis is not completely clear. Early studies suggested that together with the serine and histidine residues, this aspartate forms a catalytic triad that acts as a proton shuttle. Such a specific interaction requires a precise geometric relationship between the side chains of the aspartate and histidine residues to ensure strong H-bonding. However, the orientation of the aspartate side chain relative to the Ser-His active site residues varies considerably among the three classes of serine proteases, making it unlikely that direct proton transfer occurs between the histidine and aspartate side chains in all these enzymes. In fact, some workers have suggested that the catalytic machinery of the serine proteases is most correctly viewed as two distinct catalytic dyads—one comprising the serine and histidine residues and the other comprising the histidine and aspartate residues—rather than a single catalytic triad (Liao et al., 1992). Regardless of the molecular details, it is clear that the presence of the carboxylate anion of the aspartate influences the reactivity of the histidine residues in a way that is critical for catalysis.

The foregoing discussion provides a good example of the interplay between substrate and enzyme active site that must accompany transition state stabil-

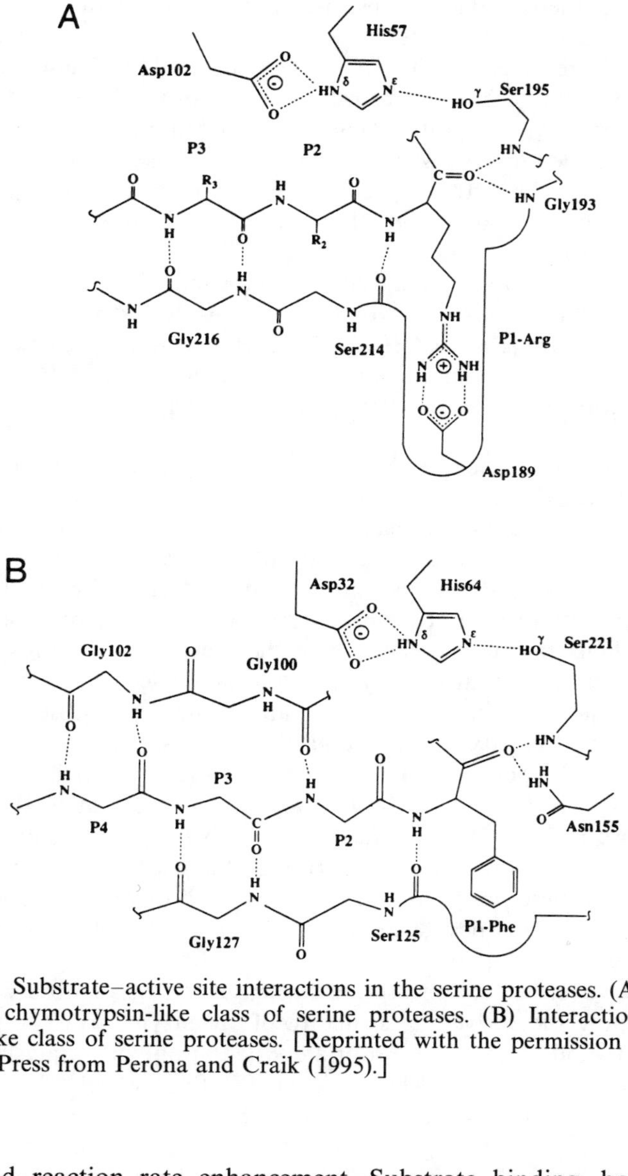

Figure 4.8 Substrate–active site interactions in the serine proteases. (A) Interactions within the chymotrypsin-like class of serine proteases. (B) Interactions within the subtilisin-like class of serine proteases. [Reprinted with the permission of Cambridge University Press from Perona and Craik (1995).]

ization and reaction rate enhancement. Substrate binding, however, must precede these catalytic steps, and formation of the enzyme–substrate complex is also governed by the stereochemical relationships between groups on the substrate and their counterpart subsites in the enzyme active site. For example, the differences in substrate specificity within the chymotrypsin-like and subtilisin-like serine proteases can be explained on the basis of their active site structures. In all these enzymes, substrate binding is facilitated by H-bonding to form β-pleated sheet structures between residues in the enzyme active site and the P1–P4 residues of the substrate (Figure 4.8). These interactions provide binding affinity for the substrates but do not significantly differentiate one peptide substrate from another. The P1–S1 site interactions appear to play a major role in defining substrate specificity in these enzymes. Subtilisins generally show broad substrate specificity, with a preference for large, hydrophobic groups at the substrate P1 position. The order of preference is approximately Tyr, Phe > Leu, Met, Lys > His, Ala, Gln, Ser ≫ Glu, Gly (Perona and Craik, 1995). The S1 site for the subtilisins occurs as a broad, shallow cleft that is formed by two strands of β-sheet structure and a loop region of variable size (Figure 4.8B). In a subclass of these enzymes, in which a P1 Lys residue is accommodated, the loop contains a glutamate residue positioned to form a salt bridge with the substrate lysine, hence neutralizing the charge.

In the chymotrypsin-like enzymes, the S1 pocket consists of a deep cleft into which the substrate P1 residue must fit (Figure 4.8A). The identity of the amino acid residues in this cleft will influence the types of substrate residue that are tolerated at this site. The chymotrypsin-like enzymes can be further subdivided into three subclasses on this basis. Enzymes in the trypsinlike subclass have a conserved aspartate residue at position 189, located at the bottom of the S1 well; this explains the high preference of these enzymes for substrates with arginine or lysine residues at P1. This aspartate is replaced by a serine or small hydrophobic residue in the chymotrypsin and elastase subclasses; hence both subclasses show specificity for nonpolar P1 residues. The other amino acid residues lining the S1 pocket further influence substrate specificity. The elastase subclass contains in this pocket large nonpolar groups that tend to exclude bulky substrate residues. Hence, the elastase subclass favors substrates with small hydrophobic residues at P1. In contrast, the chymotrypsin subclass has small residues in these positions (e.g., glycines), and these enzymes thus favor larger hydrophobic residues, such as tyrosine and phenylalanine at the P1 position of their substrates.

Within this overview of the serine proteases, we have observed specific examples of how the active site structure of an enzyme can (1) engage the substrate and bind it in an appropriate orientation for catalysis (e.g., the H-bonding network developed between the P1–P4 residues of the substrate and the active site residues), (2) lead to stabilization of the transition state to accelerate the reaction rate (e.g., the stabilization of the tetrahedral oxyanionic intermediate through H-bonding interactions), and (3) differentiate between potential substrates on the basis of stereochemical relationships between the substrate and active site subsites. While the molecular details differ from one enzyme to another, the general types of interaction illustrated here with the serine proteases also govern substrate binding and chemical transformations in all the enzymes nature has devised.

4.3 Types of Reaction Catalyzed by Enzymes

The hydrolytic activity illustrated by the serine proteases is but one of a wide variety of bond cleavage and bond formation reactions catalyzed by enzymes. From the earliest studies of these proteins, scientists have attempted to categorize them by the nature of the reactions they promote. Group names have been assigned to enzymes that share common reactivities. For example, "protease" and "proteinase" are used to collectively refer to enzymes that hydrolyze peptide bonds. Common names for particular enzymes are not always universally used, however, and their application in individual cases can lead to confusion. For example, there is a metalloproteinase known by the common names stromelysin, MMP-3 (for matrix metalloproteinase number 3), transin, and proteoglycanase. Some workers refer to this enzyme as stromelysin, others call it MMP-3, and still others use the terms transin or proteoglycanase. A newcomer to the metalloproteinase field could be quite frustrated by this confusing nomenclature. For this reason, the International Union of Pure and Applied Chemistry (IUPAC) formed the Enzyme Commission (EC) to develop a systematic numerical nomenclature for enzymes. While most workers still use common names for the enzymes they are working with, literature references should always include the IUPAC EC designations, which have been universally accepted, to let readers know precisely what enzymes are being discussed. The EC classifications are based on the reactions that enzymes catalyze. Six general categories have been defined, as summarized in Table 4.2. Within each of these broad categories, the enzymes are further differentiated by a second number that more specifically defines the substrates on which they act. For example, 11 types of hydrolase (category 3) can be defined, as summarized in Table 4.3.

Individual enzymes in each subclass are further defined by a third and a fourth number. In this way any particular enzyme can be uniquely identified. Examples of the common names for some enzymes and their EC designations

Table 4.2 The IUPAC EC Classification of Enzymes into Six General Categories According to the Reactions They Catalyze

First EC Number	Enzyme Class	Reaction
1	Oxidoreductases	Oxidation–reduction
2	Transferases	Chemical group transfers
3	Hydrolases	Hydrolytic bond cleavages
4	Lyases	Nonhydrolytic bond cleavages
5	Isomerases	Changes in arrangements of atoms in molecules
6	Ligases	Joining together of two or more molecules

Table 4.3 The IUPAC EC Subclassifications of the Hydrolases

First Two EC Numbers	Substrates[a]
3.1	Esters, —C(O)—O---R, or with S or P replacing C, or —C(O)—S—R
3.2	Glycosyl, sugar—C—O---R, or with N or S replacing O
3.3	Ether, R—O---R, or with S replacing O
3.4	Peptides, C---N
3.5	Nonpeptide C---N
3.6	Acid anhydrides, R—C(O)—O---C(O)—R′
3.7	C---C
3.8	Halides (X), C---X, or with P replacing C
3.9	P---N
3.10	S---N
3.11	C---P

[a]Hydrolyzed bonds shown as dashed lines.

are given in Table 4.4. (These enzyme selected have served as illustrative examples in my laboratory.)

The detailed rules for assigning an EC number to a newly discovered enzyme have been set forth in Volume 13 of the series *Comprehensive Biochemistry* (Florkin and Stoz, 1973); an updated version of the nomenclature system has recently been published by Academic Press (Webb, 1992). Most of the enzymes the reader is likely to encounter or work with already have EC numbers. One can often obtain the EC designation directly from the literature pertaining to the enzyme of interest. Another useful source for this information is the Medical Subject Headings Supplementary Chemical Records, published by the National Library of Medicine (U.S. Department of Health and Human Services, Bethesda, MD). This volume lists the common names of chemicals and reagents (including enzymes) that are referred to in the medical literature that is covered by the *Index Medicus* (a sourcebook for literature searching of medically related subjects). Enzymes are listed here under their common names (with cross-references where an enzyme has more than one common name),

Table 4.4 Some Examples of Enzyme Common Names and Their EC Designations

Common Name(s)	EC Designation
Cytochrome oxidases (cytochrome *c* oxidase)	EC 1.9.3.1
Prostaglandin G/H synthase (cyclooxygenase)	EC 1.14.99.1
Stromelysin (MMP-3, proteoglycanase)	EC 3.4.24.17
Dihydroorotate dehydrogenase	EC 1.3.99.11
Rhodopsin kinase	EC 2.7.1.125

and the EC designation is provided for each. Most college and university libraries carry the *Index Medicus* and will have this supplement available, or one can purchase the supplement directly from the National Library of Medicine. Yet another resource for determining the EC designation of an enzyme is the Enzyme Data Bank, which can be accessed on the Internet: http://192.239.77.6/Dan/proteins/ec-enzyme.html. This data bank provides EC numbers, recommended names, alternative names, catalytic activities, and information on cofactor utilization and associated diseases for a very large collection of enzymes. A complete description of the data bank and its uses can be found in Bairoch (1993).

As we have said, enzymes catalyze a staggering number of specific chemical transformations. Nevertheless, all these varied reactivities fall within the six general categories of bond breaking, bond forming, and group transfer reactions summarized in Table 4.2. The mechanisms by which these enzymes work also are varied, but some common strategies can be identified. These are based on the chemical reactivities of the amino acid side chain and cofactors of the enzymes, as described in Chapter 2. Enzymes utilize general acid–base chemistry, nucleophilic catalysis, and electrophilic catalysis to facilitate the myriad of reactions they catalyze. We have already encountered examples of some of these strategies in discussing the serine proteases. We conclude this section with a brief description of each of these strategies in terms of general utility in enzyme catalysis.

4.3.1 Acid–Base Catalysis

Just about every enzyme-catalyzed reaction involves some type of proton transfer that requires acid and base group participation. In some cases protons (H^+) and hydroxide ions (OH^-) act directly as the acid and base groups in activities referred to as *specific acid* and *specific base catalysis*, respectively. More commonly, however, organic substrates, cofactors, or amino acid side chains from the enzyme fulfill this role by acting as Brønsted–Lowry acids and bases in *general acid* and *general base catalysis*. In solution, specific versus general acid or base catalysis can be distinguished by the effects of pH and acid or base group concentration on the rate of chemical reaction.

General acids and bases will be functional only below or above their pK_a values, respectively. Hence, a plot of reaction rate as a function of solution pH will display the type of sigmoidal curves we are used to seeing for acid–base titrations (Figure 4.9A, B). These reactions will show a linear dependence of reaction rate on the overall concentration of the acid or base group; in contrast, the rate of a specific acid- or base-catalyzed reaction will show no dependence on buffer concentration. Figure 4.9C shows both cases.

Plots such as those of Figure 4.9 not only can be used to identify general versus specific acid or base catalysis, they also can provide a quantitative estimate of the relevant pK_a for the acid or base group. This information can

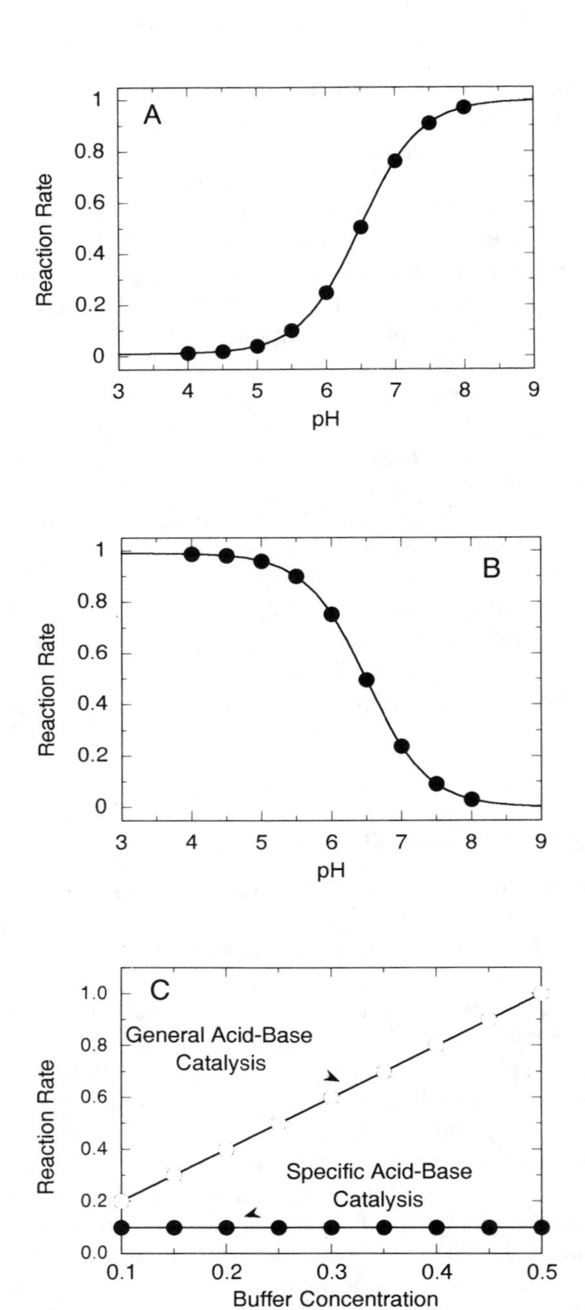

Figure 4.9 (A) Effect of pH on the rate of reaction for a general base catalyzed reaction. (B) Effect of pH on the rate of reaction for a general acid catalyzed reaction. (C) Effect of buffer (general acid or base) concentration on the rate of reaction for a general and specific acid or base catalyzed reaction.

be used to identify potential participants in enzymatic reactions. For example, if the rate of an enzyme-catalyzed reaction displayed a pH profile similar to that shown in Figure 4.9A, with a transition pK_a of 6.5, one would know that the reaction depended on the conjugate base form of some molecule with this pK_a. Surveying the pK_a values of the amino acid side chain, one might suspect that a histidine residue was fulfilling this role. This type of information can be quite useful in divining the mechanistic details of an enzyme-catalyzed reaction. However, some caution must be exercised here, inasmuch as side chain pK_a values can be perturbed significantly by the local protein environment in which they are found. Two extreme examples of this are histidine 159 of papain, which has a side chain pK_a value of 3.4, and glutamic acid 35 of lysozyme, with a greatly elevated pK_a value of 6.5!

The fundamental feature of acid and base catalysis is that the acid or base group participates in proton transfers that stabilize the transition state of the chemical reaction. A good example of this comes from the hydrolysis of ester bonds in water, a reaction carried out by many hydrolytic enzymes. The mechanism of ester hydrolysis requires formation of a transition state involving partial charge transfer between the ester and a water molecule, as shown in Figure 4.10. This transition state can be stabilized by a base acting as a partial proton acceptor from the water molecule, thus enhancing the stability of the

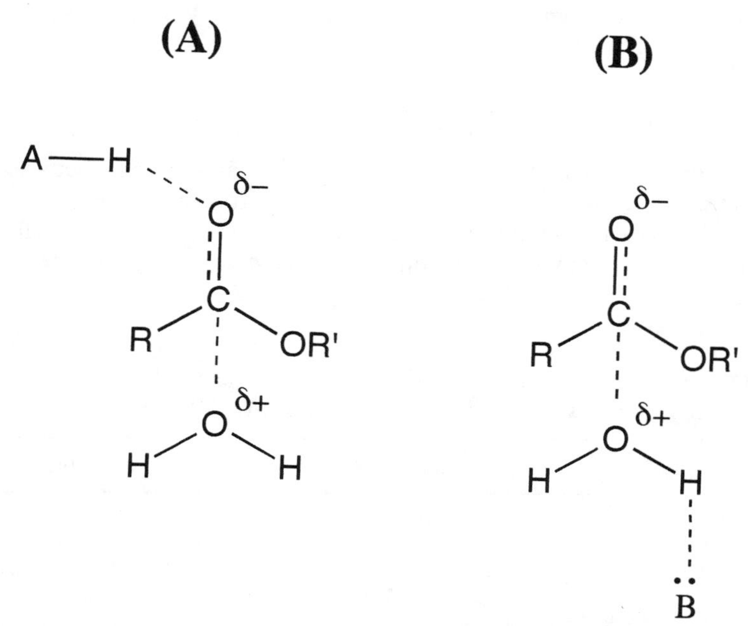

Figure 4.10 Transition state stabilization by a general acid (A) or general base (B) in ester bond hydrolysis by water.

partial positive charge on the water oxygen (Figure 4.10B). Alternatively, the transition state can be stabilized by an acidic group that acts as a partial proton donor to the carbonyl oxygen of the ester (Figure 4.10A). This type of acid or base group transition state stabilization is extremely common in enzyme catalysis. The active site histidine residue of the serine protease, for example, acts as a general acid–base in stabilizing the transition states during peptide hydrolysis, as described earlier.

4.3.2 Nucleophilic Catalysis

In nucleophilic catalysis a covalent bond is transiently formed between the enzyme nucleophile and a group on the substrate molecule in the transition state of the reaction mechanism. The formation of this covalent linkage distinguishes nucleophilic catalysis from general base catalysis and is also the factor that most greatly increases the reactivity of the transition state relative to the ground state of the substrate. We have already seen a specific example of nucleophilic catalysis in the attack of the active site serine residues on the substrate P1 carbonyl in the serine proteases, with formation of an acyl–enzyme intermediate. Other amino acid residues that can act as nucleophiles, and examples of the covalent intermediates they form are shown in Table 4.5. A more comprehensive description of nucleophilic catalysis and examples of its role in enzyme mechanisms can be found in the text by Walsh (1979).

4.3.3 Electrophilic Catalysis

In electrophilic catalysis covalent intermediates are also formed between the cationic electrophile of the enzyme and an electron-rich portion of the substrate molecule. The amino acid side chains do not provide very effective electrophiles. Hence enzyme electrophiles are most commonly electron-defi-

Table 4.5 Some Examples of Enzyme Nucleophiles and the Covalent Intermediates Formed in Their Reactions with Substrates

Nucleophilic Group	Example Enzyme	Covalent Intermediate
Serine (—OH)	Serine proteases	Acyl enzyme
Cysteine (—SH)	Thiol proteases	Acyl enzyme
Aspartate ($—COO^-$)	ATPases	Phosphoryl enzyme
Lysine ($—NH_2$)	Pyridoxal-containing enzymes	Schiff bases
Histidine	Phosphoglycerate mutase	Phosphoryl enzyme
Tyrosine (—OH)	Glutamine synthase	Adenyl enzyme

Source: Adapted from Hammes (1982).

Table 4.6 Some Examples of Electrophilic Catalysis in Enzymatic Reactions

Enzyme	Electrophile
Acetoacetate decarboxylase	Lysine–substrate Schiff base
Aldolase	Lysine–substrate Schiff base
Aspartate aminotransferase	Pyridoxal phosphate
Carbonic anhydrase	Zn^{2+}
L-Malate dehydrogenase	Mn^{2+}
Pyruvate decarboxylase	Thiamine pyrophosphate

cient organic cofactors (such as the pyridine ring of the cofactor pyridoxal phosphate; see Figure 3.19) and cationic metal centers. Electrophilic centers also can be formed *in situ* during catalytic turnover. A common example of this is the formation of a protonated Schiff base (an imine) following nucleophilic attack of a carbonyl by an amine:

$$(R_1)(R_2)C{=}O + H_2\ddot{N}R \rightleftharpoons (R_1)(R_2)C{=}NR + H_2O \overset{H^+}{\rightleftharpoons} (R_1)(R_2)C{=}\overset{+}{N}HR$$

The protonated nitrogen can now act as a strong electron-withdrawing group to stabilize a negative charge on one of the alpha carbons and to activate the carbonyl carbon toward nucleophilic attack (Fersht, 1985). Examples of some enzymatic reactions involving electrophilic catalysis are provided in Table 4.6.

4.4 Summary

In this chapter we have explored the chemical nature of enzyme catalysis. We have seen that enzymes function by enhancing the rate of chemical reaction by lowering the energy barrier to attainment of the reaction transition state. The active site of enzymes provides the structural basis for this transition state stabilization through a number of chemical mechanisms, including general acid–base catalysis and nucleophilic and electrophilic catalysis. The structural architecture of the active site further dictates the specificity of the enzyme for substrate. A structural complementarity exists between the substrate ground and/or transition state and the enzyme active site. Several hypotheses have been presented to describe this structural complementarity and to explain its role in enzyme specificity and rate enhancement.

References and Further Reading

Bairoch, A. (1993) *Nucleic Acid. Res.* **21**, 3155.

Carter, P. and Wells, J. A. (1988) *Nature*, **332**, 564.

Fersht, A. (1985) *Enzyme Structure and Mechanism*, Freeman, New York.

Fischer, E. (1894) *Berichte*, **27**, 2985.

Florkin, M., and Stotz, E. H., Eds. (1973) *Comprehensive Biochemistry*, Vol. 13, Elsevier, New York.

Goldsmith, J. O., and Kuo, L. C. (1993) *J. Biol. Chem.* **268**, 18481.

Hammes, G. G. (1982) *Enzyme Catalysis and Regulation*, Academic Press, New York.

Jencks, W. P. (1975) *Adv. Enzymol.* **43**, 219.

Koshland, D. E. (1958) *Proc. Natl. Acad. Sci. U.S.A.* **44**, 98.

Lerner, R. A., Benkovic, S. J., and Schultz, P. G. (1991) *Science*, **252**, 659.

Liao, D., Breddam, K., Sweet, R. M., Bullock, T., and Remington, S. J. (1992) *Biochemistry*, **31**, 9796.

Menger, F. M. (1992) *Biochemistry*, **31**, 5368.

Pauling, L. (1948) *Nature*, **161**, 707.

Perona, J. J., and Craik, C. S. (1995) *Protein Sci.* **4**, 337.

Schechter, I., and Berger, A. (1967) *Biochem. Biophys. Res. Commun.* **27**, 157.

Schowen, R. L. (1978) in *Transition States of Biochemical Processes* (Gandous, R. D. and Schowen, R. L., Eds.), Chapter 2, Plenum, New York.

Segel, I. H. (1975) *Enzymes Kinetics*, Wiley, New York.

Storm, D. R., and Koshland, D. E. (1970) *Proc. Natl. Acad. Sci. U.S.A.* **66**, 445.

Walsh, C. (1979) *Enzyme Reaction Mechanisms*, Freeman, New York.

Webb, E. C. (1992) *Enzyme Nomenclature*, Academic Press, San Diego, CA.

Wilson, C., and Agard, D. A. (1991) *Curr. Opinions Struct. Biol.* **1**, 617.

Yagisawa, S. (1995) *Biochem. J.* **308**, 305.

CHAPTER

5

Steady State Kinetics of Single Substrate Enzyme Reactions

Enzyme-catalyzed reactions can be studied in a variety of ways to explore different aspects of catalysis. Enzyme–substrate and enzyme–inhibitor complexes can be rapidly frozen and studied by spectroscopic means. Many enzymes have been crystallized and their structures determined by x-ray diffraction methods. More recently, enzyme structures have been determined by multidimensional NMR methods. Kinetic analysis of enzyme-catalyzed reactions, however, is the most commonly used means of elucidating enzyme mechanisms and, especially when coupled with protein engineering, identifying catalytically relevant structural components. In this chapter we explore the use of steady state enzyme kinetics as a means of defining the catalytic efficiency and substrate affinity of simple enzymes. As we shall see, the term "steady state" refers to experimental conditions in which the ES complex can build up to an appreciable "steady state" level. These conditions are easily obtained in the laboratory and allow for convenient interpretation of the time courses of enzyme reactions. All the data analysis described in this chapter rest on the ability of the scientist to conveniently measure the initial velocity of the enzyme-catalyzed reaction under a variety of conditions. We shall assume here that some convenient method for determining the initial velocity of the reaction exists; in Chapter 6 we address specifically how initial velocities are measured and describe a variety of experimental methods for performing such measurements.

5.1 The Time Course of Enzyme-Catalyzed Reactions

If one mixes an enzyme with its substrate in solution and then (by some convenient means) measures the amount of substrate remaining and/or the amount of product produced over time, *progress curves* similar to those shown in Figure 5.1 will result. Note that the substrate depletion curve is the mirror image of the product appearance curve. Early in the progression, substrate loss and product appearance change rapidly with time; as time goes on, however, these rates diminish, reaching zero when all of the substrate has been converted to product by the enzyme. Such time courses are well modeled by a first-order kinetic equation:

$$[S] = [S_0]e^{-kt} \tag{5.1}$$

where [S] is the substrate concentration remaining at time t, $[S_0]$ is the starting substrate concentration, and k is the rate constant for the reaction. The velocity v of such a reaction is thus given by:

$$v = -\frac{d[S]}{dt} = \frac{d[P]}{dt} = k[S_0]e^{-kt} \tag{5.2}$$

Let us look more carefully at the product appearance profile for an enzyme-catalyzed reaction (Figure 5.2). If we restrict our attention to the very

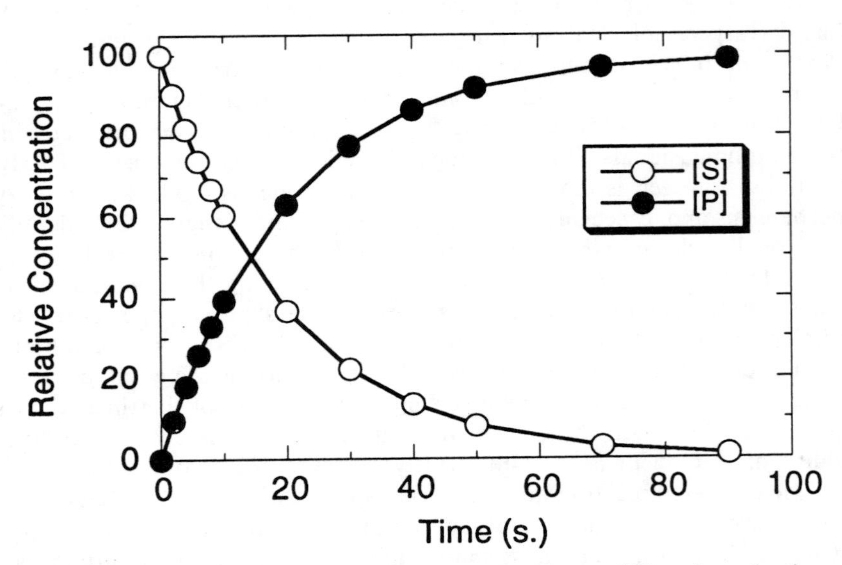

Figure 5.1 Reaction progress curves for the loss of substrate [S] and production of product [P] during an enzyme-catalyzed reaction.

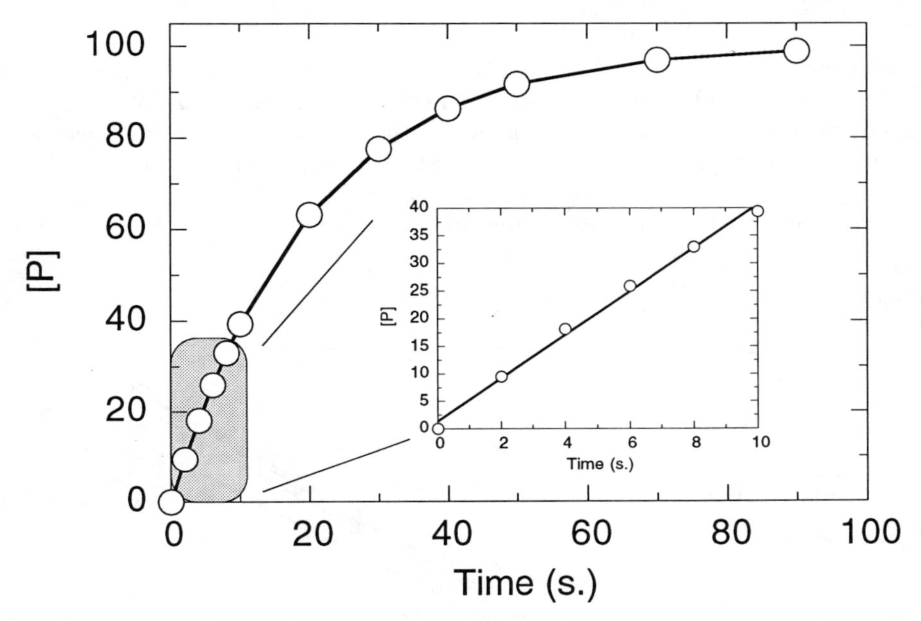

Figure 5.2 Reaction progress curve for the production of product during an enzyme-catalyzed reaction. Inset highlights the early time points, where the initial velocity can be determined from the slope of the linear plot of [P] versus time.

early portion of this plot (shaded area in Figure 5.2), we see that the increase in product formation (and substrate depletion as well) tracks approximately linear with time. For this limited time period, the *initial velocity* (v_0) can be approximated as the slope (change in y over change in x) of the linear plot of [S] or [P] as a function of time:

$$v_0 = -\frac{\Delta[\mathrm{S}]}{\Delta t} = \frac{\Delta[\mathrm{P}]}{\Delta t} \tag{5.3}$$

Experimentally one finds that the time course of product appearance and substrate depletion is well modeled by a linear function up to the time when about 10% of the initial substrate concentration has been converted to product. We shall see in Chapter 6 that the length of time over which an enzyme catalyzed reaction will display linear kinetics can be altered by varying solution conditions. For the rest of this chapter we shall assume the reaction velocity is measured during this early phase of the reaction, which means that from here $v = v_0$, the initial velocity.

5.2 Effects of Substrate Concentration on Velocity

Referring back to Equation 5.2, one would expect the velocity of a first-order reaction to depend linearly on the initial substrate concentration. When early studies were performed on enzyme-catalyzed reactions, however, scientists found instead that the reactions followed the substrate dependence plotted in Figure 5.3. Figure 5.3A illustrates the time course of the enzyme-catalyzed reaction observed at different starting concentrations of substrate; the velocities

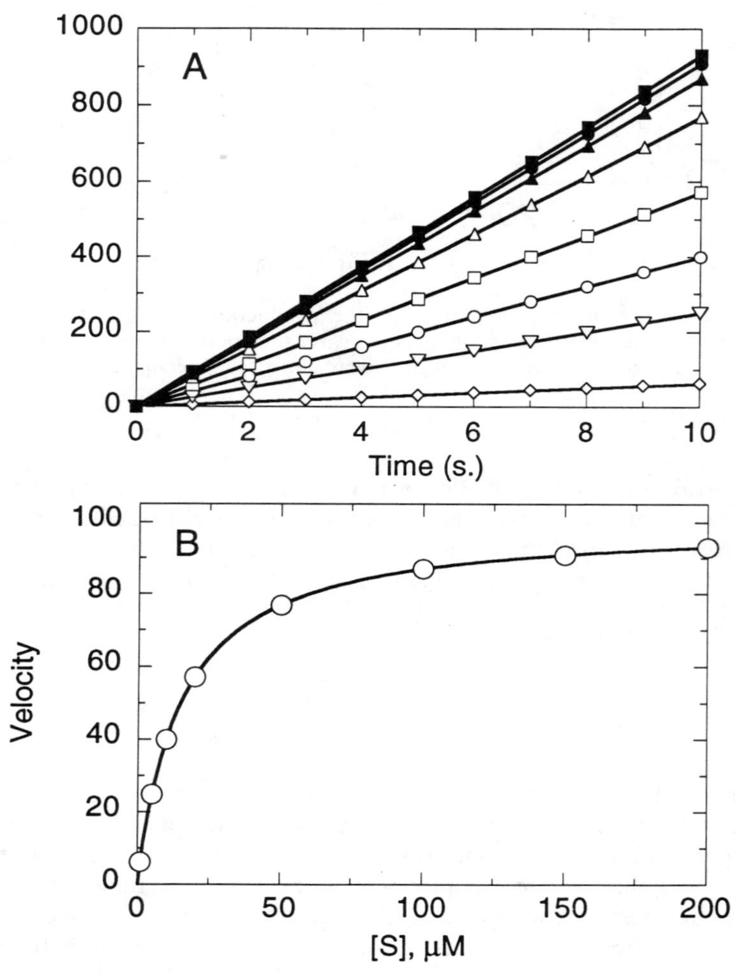

Figure 5.3 (A) Progress curves for a set of enzyme-catalyzed reactions with different starting concentrations of substrate. (B) Plot of the reaction velocities, measured as the slopes of the lines from (A), as a function of [S].

for each experiment are measured as the slopes of the [P] versus time plots. Figure 5.3B replots these data as the initial velocity v as a function of [S], the starting concentration of substrate. Rather than yielding the linear relationship expected for first-order kinetics, the second plot shows that the velocity appears to be saturable at high substrate concentrations. This result puzzled early enzymologists. Three distinct regions of the curve in question can be identified: at low substrate concentrations the velocity appears to display first-order behavior, tracking linearly with substrate concentration. At very high concentrations of substrate, the velocity switches to zero-order behavior, displaying no dependence on substrate concentration. In the intermediate region, the velocity displays a curvilinear dependence on substrate concentration. How can one rationalize these experimental observations?

A qualitative explanation for the substrate dependence of enzyme-catalyzed reaction velocities was provided by Brown (1902). At the same time that the kinetic characteristics of enzyme reactions were being explored, evidence for complex formation between enzymes and their substrates was accumulating. Brown thus argued that enzyme-catalyzed reactions could best be described by the following reaction scheme:

$$E + S \underset{k_{-1}}{\overset{k_1}{\rightleftharpoons}} ES \xrightarrow{k_2} E + P$$

This scheme (which we encountered in Chapter 4) predicts that the reaction velocity will be proportional to the concentration of the ES complex as: $v = k_2[ES]$. Suppose that we held the total enzyme concentration constant at some low level and varied the concentration of S. At low concentrations of S the concentration of ES would be directly proportional to [S]; hence the velocity would depend on [S] in an apparent first-order fashion.

At very high concentrations of S, however, practically all of the enzymes will be present in the form of the ES complex. Under such conditions, the velocity will depend on the rate of the chemical transformations that convert ES to EP and the subsequent release of product to re-form free enzyme. Adding more substrate under these conditions would not effect a change in reaction velocity; hence the slope of the plot of the velocity versus [S] would approach zero (as seen in Figure 5.3B). The complete [S] dependence of the reaction velocity predicted by the model of Brown resembles a rectangular hyperbola, as in Figure 5.3B.

Although Brown's model provided a useful qualitative picture of enzyme reactions, to be fully utilized by experimental scientists, it needed to be put into a rigorous mathematical framework. This was accomplished first by Henri (1903) and later by Michaelis and Menten (1913). Ironically, despite their acknowledgment of the prior work of Henri, Michaelis and Menten are most widely recognized for this contribution. The basic rate equation derived in Section 5.3 is commonly referred to as the Michaelis–Menten equation. Several writers have recently taken to referring to the equation as the Henri–

Michaelis–Menten equation, in an attempt to correct this oversight of Henri's contributions. The reader should be aware, however, that the majority of the scientific literature continues to use the traditional terminology.

5.3 The Steady State Kinetic Approach

The original derivations by Henri and by Michaelis and Menten depended on a rapid equilibrium approach to enzyme reactions. Most experimental enzymology, however, relies on measurements made under steady state conditions, as we shall define later. Therefore, we present here a steady state approach to enzyme kinetics that is based on the original derivation by Briggs and Haldane (1925). Our final kinetic equation is nevertheless very similar to that originally described by Henri and by Michaelis and Menten, and is thus referred to as the Henri–Michaelis–Menten equation.

To derive the rate equation for a simple enzyme-catalyzed reaction under steady state conditions, we must first make a few simplifying assumptions. These are as follows:

1. We assume that early in the reaction (conditions under which we are measuring initial velocity), all the enzyme molecules can be accounted for by either free enzyme or enzyme–substrate complexes. Thus, the total enzyme concentration $[E_t]$ is given by:

$$[E_t] = [E] + [ES] \tag{5.4}$$

2. Since enzymes are catalytic, to effect catalysis, we require very small concentrations of these proteins relative to the concentration of substrate. Typically substrate concentrations are in the micromolar-to-millimolar range, while enzyme concentrations are in the nanomolar range or lower. Thus we assume that the substrate concentration greatly exceeds the enzyme concentration: $[S] \gg [E_t]$.
3. Under steady state conditions and early in the reaction, the free enzyme and enzyme–substrate concentrations are in equilibrium, and essentially do not change over the short period of time of our measurement. We can thus make the approximation that:

$$\frac{d[E]}{dt} = \frac{d[ES]}{dt} = 0 \tag{5.5}$$

   We also know that very little product is formed during the early phase of the reaction, where we measure initial velocity. The amount of product formed relative to the concentrations of [E], [ES], and [S] is so small that we can make the approximation that $[P] = 0$ during this time interval.

With these assumptions made, we can now work out an expression for the enzyme velocity under steady state conditions. We have said that in the initial stage of the reaction the velocity can be expressed as follows:

$$v = k_2[\mathrm{ES}] \tag{5.6}$$

Now, [ES] is dependent on the rate of formation of the complex (governed by k_1) and the rate of loss of the complex (governed by k_{-1} and k_2):

$$\frac{d[\mathrm{ES}]}{dt} = k_1[\mathrm{E}][\mathrm{S}] \quad \text{and} \quad \frac{-d[\mathrm{ES}]}{dt} = (k_{-1} + k_2)[\mathrm{ES}] \tag{5.7}$$

As we have stated, under steady state conditions [ES] remains constant over the time interval of our measurements, thus the preceding two expressions must be equal:

$$k_1[\mathrm{E}][\mathrm{S}] = (k_{-1} + k_2)[\mathrm{ES}] \tag{5.8}$$

or, rearranging:

$$[\mathrm{ES}] = \frac{[\mathrm{E}][\mathrm{S}]}{(k_{-1} + k_2)/k_1} \tag{5.9}$$

At this point let us define the term K_m as an abbreviation for the kinetic constants in the denominator on the right-hand side of Equation 5.9:

$$K_m = \frac{k_{-1} + k_2}{k_1} \tag{5.10}$$

For now we will consider K_m to be merely an abbreviation to make our subsequent mathematical expressions less cumbersome. Later, however, we shall see that K_m has a more significant meaning. Substituting Equation 5.10 into Equation 5.9 we obtain:

$$[\mathrm{ES}] = \frac{[\mathrm{E}][\mathrm{S}]}{K_m} \tag{5.11}$$

Now, referring to Equation 5.4 and rearranging, we see that $[\mathrm{E}] = [\mathrm{E_t}] - [\mathrm{ES}]$. Substituting this expression into Equation 5.11, and performing a little algebra, we can arrive at the following expression:

$$[\mathrm{ES}] = [\mathrm{E_t}]\frac{[\mathrm{S}]}{[\mathrm{S}] + K_m} \tag{5.12}$$

If we now combine the expression for [ES] derived in Equation 5.12 with the velocity expression of Equation 5.6 we obtain:

$$v = k_2[\mathrm{E_t}]\frac{[\mathrm{S}]}{[\mathrm{S}] + K_m} \tag{5.13}$$

As we observed in connection with Figure 5.3B, at very high concentrations of substrate, the reaction velocity reaches a maximum value, which we shall define as V_{max}. Under these conditions the concentration of substrate is very much greater than the numerical value of K_m. In this situation, K_m does not contribute much to Equation 5.13, and thus we can make the approximation:

$$\frac{[S]}{[S] + K_m} \cong 1 \tag{5.14}$$

hence,

$$V_{max} = k_2[E_t] \tag{5.15}$$

Combining this expression with Equation 5.13, we finally arrive at an expression very similar to that first described by Henri and Michaelis and Menten:

$$v = \frac{V_{max}[S]}{K_m + [S]} \tag{5.16}$$

This equation, which defines a rectangular hyperbola, describes the curve of experimental velocity versus [S] observed by early enzymologists and exemplified in Figure 5.3B.

In our definition of K_m we combined first-order rate constants (k_{-1} and k_2, which have units of reciprocal time) with a second-order rate constant (k_1, which has units of reciprocal molarity, reciprocal time) in such a way that the resulting K_m has units of molarity, as does [S]. If we set up our system so that the concentration of substrate exactly matches K_m, Equation 5.16 will reduce to:

$$v = V_{max}\frac{K_m}{2K_m} = \frac{V_{max}}{2} \tag{5.17}$$

This provides us with a working definition of K_m: *The K_m is the substrate concentration that provides a reaction velocity that is half the maximal velocity obtainable under saturating substrate conditions.* The K_m value is often referred to in the literature as the *Michaelis constant* and is sometimes equated with the dissociation constant for the ES complex, K_S, which is defined as k_{-1}/k_1. This, however, is not always reasonable, as we shall describe later.

5.4 Experimental Measurement of V_{max} and K_m

5.4.1 Graphical Determinations from Untransformed Data

The kinetic parameters V_{max} and K_m are determined graphically with initial velocity measurements obtained at varying substrate concentrations. The graphical methods are best illustrated by working through examples with some numerical data. Suppose that we measured an enzyme-catalyzed reaction under steady state conditions and recorded the initial velocity over a very broad range of substrate concentrations. Table 5.1 provides data of this type,

Table 5.1 Initial Velocity as a Function of Substrate Concentration for Some Model Enzyme-Catalyzed Reactions

[S] (μM)	v (μM product formed s)	$1/v$ ($\mu M^{-1} \cdot s$)	1/[S] (μM^{-1})
0.2	1	1.0000	5.0000
0.4	3	0.3333	2.5000
0.6	4	0.2500	1.6667
0.8	5	0.2000	1.2500
1.0	6	0.1667	1.0000
5.0	23	0.0435	0.2000
10.0	42	0.0238	0.1000
15.0	50	0.0200	0.0667
20.0	58	0.0172	0.0500
25.0	60	0.0167	0.0400
30.0	67	0.0149	0.0333
35.0	71	0.0141	0.0286
40.0	70	0.0143	0.0250
45.0	74	0.0135	0.0222
50.0	77	0.0130	0.0200
100.0	90	0.0111	0.0100
300.0	93	0.0108	0.0033
600.0	97	0.0103	0.0017
900.0	101	0.0099	0.0011
1200	100	0.0100	0.0008

which the reader can use to construct plots as we proceed through our discussion of data analysis.

The first and most straightforward way of graphing the data is as a direct plot of velocity versus [S]; we shall refer to such a plot as a Michaelis–Menten plot. In Figure 5.4A, a Michaelis–Menten plot for the data in Table 5.1 is shown; the line drawn through the data was generated by a nonlinear least-squares fit of the data to Equation 5.16. With modern computer graphics programs, the reader has a wide choice of options for performing nonlinear curve fitting; some programs that are particularly well suited for enzyme studies are listed in Appendix II. The plots in this book, for example, were generated with Kaleidagraph (from Abelbeck Software), which contains a built-in iterative method for performing nonlinear curve fitting. Here both K_m and V_{max} were unknowns that were simultaneously solved for by the curve-fitting routine. The estimates of K_m and V_{max} determined in this way were 15 μM and 100 μM/s, respectively. Such direct fits of the untransformed data provide the most reliable estimates of V_{max} and K_m.

If sophisticated software is not available, the direct plot can still provide good estimates of these kinetic parameters, provided the experimenter obtains data over a broad enough range of substrate concentrations. Using a

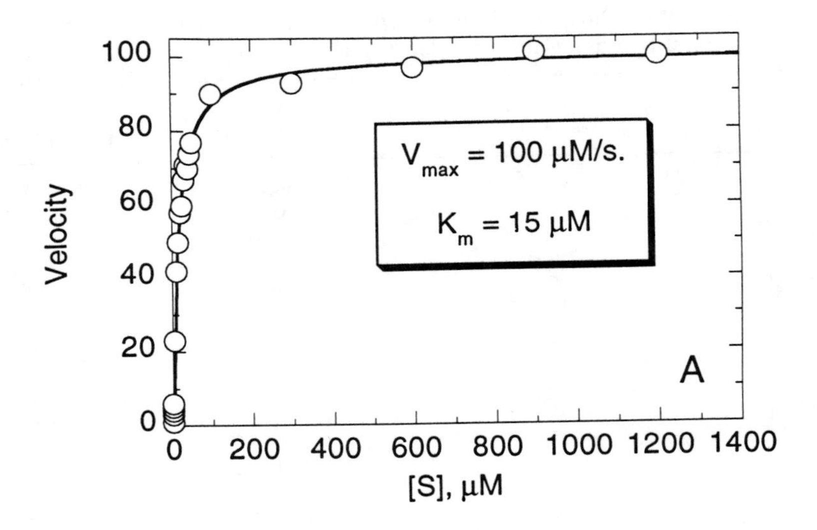

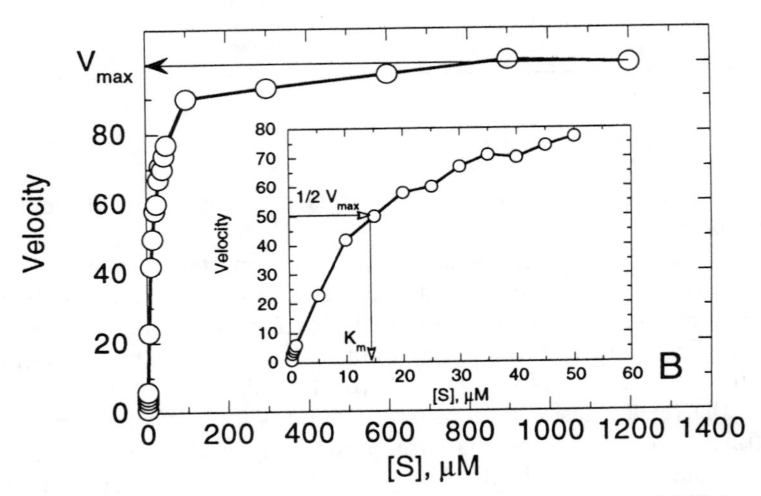

Figure 5.4 Michaelis–Menten plots of velocity as a function of [S] for the data presented in Table 5.1. (A) The line drawn through the data points represents the nonlinear least-squares best fit to Equation 5.16. (B) As in (A) but without computer-based curve fitting: the V_{max} and K_m values have been estimated by visual inspection of the graph. Inset highlights a restricted region of substrate concentration to more clearly illustrate the graphical determination of K_m.

straightedge and graph paper, one can plot the data in Table 5.1 and get a reasonable curve by simply connecting the data points by straight lines (Figure 5.4B). A horizontal line drawn at the apparent plateau value of v (defined by the limiting value of v at very high substrate concentrations) that intersects the y axis then defines V_{max}. Next two more lines are drawn: another horizontal line at the numerical value of $\frac{1}{2}V_{max}$, from the y axis to the point of intersection with the data curve, and a vertical line from this intersection point to the x axis. The point at which the vertical line intersects the x axis defines the value of K_m.

The manual method of determining V_{max} and K_m works only when there are sufficient data representing high substrate concentrations to ensure an adequate definition of V_{max}. This is sometimes experimentally infeasible, however, because of limiting substrate solubility or for other reason: certain physical characteristics of the substrate or product (see Chapter 6), for example, or because product inhibition has caused deviations from the expected Henri–Michaelis–Menten behavior. Such limitations can make it very difficult to obtain good estimates of the kinetic parameters by manual methods.

Let us assume for the moment that we have available a program for performing nonlinear curve fits. What limitations are there then to our ability to estimate V_{max} and K_m from the experimental data? The accuracy of such estimates will depend on the range of substrate concentrations over which the initial velocity has been determined. If measurements are made only at low substrate concentrations, the data will appear to be first-order (i.e., v will appear to be a linear function of [S]). This is illustrated in Figure 5.5A for the data in table 5.1 between substrate concentrations of 0.2 and 5 μM (i.e., $<0.33\,K_m$). In this concentration range, the enzyme active sites never reach

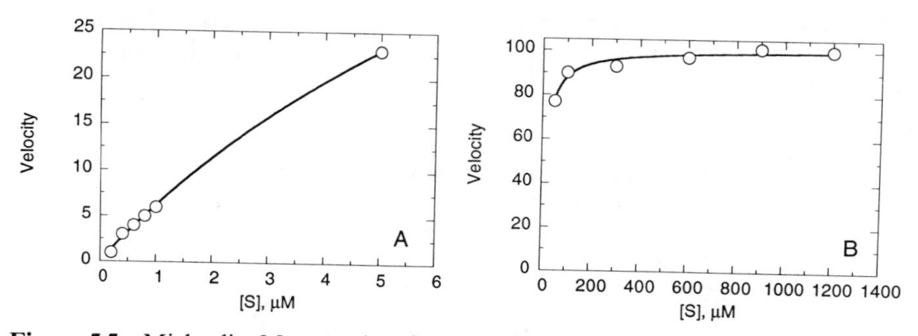

Figure 5.5 Michaelis–Menten plots for restricted data from Table 5.1. (A) The range of [S] values is inappropriately low ($\leqslant 0.33\,K_m$), hence K_m and V_{max} appear to be infinite. (B) The range of [S] values is inappropriately high, with the result that every data point represents saturating conditions. Here one can estimate V_{max} but K_m cannot be determined.

saturation, and graphically, both V_{max} and K_m appear to be infinite. On the other hand, Figure 5.5B illustrates what happens when measurements are made at very high substrate concentrations only; here the data for substrate concentrations between 50 and 1200 μM are considered (i.e., $[S] \geqslant 5K_m$). In this saturating substrate concentration range, the velocity appears to be independent of substrate concentration; while V_{max} can be estimated from these data, there is no way to determine the K_m value here.

The plots in Figure 5.5 emphasize the need for exploring a broad range of substrate concentrations to accurately determine the kinetic constants for the enzyme of interest. Again, there may be practical limits on the range of substrate concentrations over which such measurements can be performed. To best determine V_{max} and K_m, one should attempt to at least cover the range of substrate concentrations that bracket values of 0.33–2.0 K_m. Since the kinetic constants are unknowns prior to these experiments, it is common to perform initial experiments with a limited number of data points that span a broad range of substrate concentration, obtaining a rough estimate of K_m from these data. Improved estimates can then be obtained by narrowing the substrate concentration range between 0.33 and 2.0 K_m, and obtaining a larger number of data points within this range. Let's work through another example to illustrate how such experiments can be performed and analyzed.

To begin with, we might chose to make a small number of measurements over a broad range of [S]. Let's pick six points between 1 and 100 μM substrate. Table 5.2 provides the results of these measurements for our made-up example, and Figure 5.6A plots these results. From the curve fit to these data, we estimate that the K_m for this substrate is about 11 μM, and V_{max} is about 99 μM/s. With this information we might next choose to narrow the substrate concentration range to between 5 and 30 μM (roughly between 0.33 and 2.0 K_m). Within this range we increase our number of data points to eight; the results are given in Table 5.3 and plotted in Figure 5.6B. With this second data set we obtain a much better estimate of K_m, 15 μM, and our estimate of V_{max}, 101 μM/s, is comparable to that obtained from the first data set.

Table 5.2 A Preliminary Data Set for Determining K_m and V_{max}

[S] (μM)	v (μM product/s)
1	5
5	30
10	45
20	60
50	72
100	85

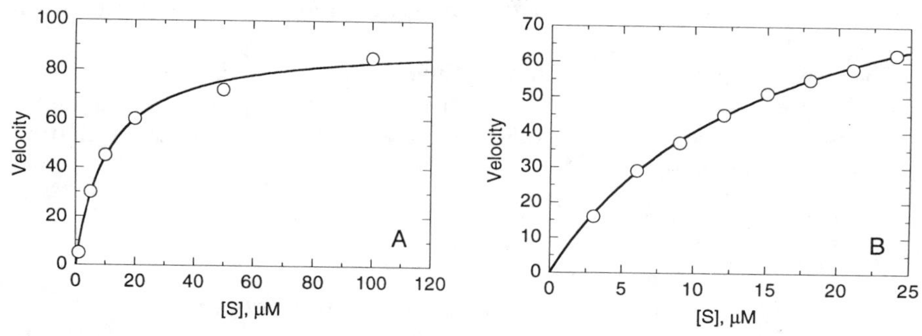

Figure 5.6 Experimental strategy for estimating K_m and V_{max}. (A) A limited data set is collected over a broad range of [S] to get rough estimates of the kinetic constants. (B) Once a rough estimate of K_m has been determined, a second set of experiments is performed with more data within the range of 0.33–2.0 K_m to obtain more precise estimates of the kinetic constants.

Table 5.3 A Second Data Set for Determining K_m and V_{max}: Sustrate Concentration Range Narrowed to 0.33–2.0 Times Initial Estimate of K_m

[S] (μM)	v (μM product/s)
3	16
6	29
9	37
12	45
15	51
18	55
21	58
24	62

5.4.2 Lineweaver–Burk Plots of Enzyme Kinetics

The widespread availability of user-friendly nonlinear curve-fitting programs is a relatively recent development. In the past, determination of the kinetic constants for an enzyme from the untransformed data was not so routine. To facilitate work in this area, scientists searched for means of transforming the data to produce linear plots from which the kinetic constants could be determined simply with graph paper and a straightedge. While today many of us have nonlinear curve-fitting programs at our disposal, there is still considerable value in these linearized plots. As we shall see in subsequent chapters, these plots are extremely useful in diagnosing the mechanistic details of multisubstrate enzymes and for determining the mode of interaction between an enzyme an inhibitor.

The most commonly used method for linearizing enzyme kinetic data is that of Lineweaver and Burk (1934). We start with the same steady state assumptions described earlier. Applying some simple algebra, we can rewrite Equation 5.16 in the following form:

$$v = \frac{V_{max}[S]}{K_m} + V_{max} \tag{5.18}$$

Now we simply take the reciprocal of Equation 5.18 to obtain:

$$\frac{1}{v} = \left(\frac{K_m}{V_{max}}\frac{1}{[S]}\right) + \frac{1}{V_{max}} \tag{5.19}$$

Comparing Equation 5.19 with the standard equation for a straight line, we have

$$y = (mx) + b \tag{5.20}$$

where m is the slope and b is the y intercept. We see that Equation 5.19 is an equation for a straight line with slope $= K_m/V_{max}$, and y intercept $= 1/V_{max}$. Thus if the reciprocal of initial velocity is plotted as a function of the reciprocal of [S], we would expect from Equation 5.19 to obtain a linear plot. For the same reasons described earlier for untransformed data, these plots work best when the substrate concentration is limited to 0.33–2.0 K_m. Within this range, good linearity is observed, as illustrated in Figure 5.7 for the data between [S] = 5 and [S] = 30 in Table 5.1. A plot like that in Figure 5.7 is known as a Lineweaver–Burk plot. The kinetic constants K_m and V_{max} can be determined from the slope and y intercepts of such a plot, as illustrated in Figure 5.7. The y intercept occurs at the point where the x-axis value is zero. Since our x axis here is 1/[S], it follows that $x = 0$ when $[S] = \infty$. It thus stands to reason that the maximal velocity will occur at infinite substrate concentrations; hence the y intercept of our plot can be equated with the reciprocal of V_{max}.

The K_m value can be determined from Lineweaver–Burk plots in two ways. First we note from Equation 5.19 that the slope is equal to K_m divided by V_{max}. If we therefore divide the slope of our line by the y-intercept value (i.e., by $1/V_{max}$), the product will be equal to K_m. Alternatively, we could extrapolate our linear fit to the point of intersecting the x axis. This x intercept is equal to $-1/K_m$; thus we could determine K_m from the absolute value of the reciprocal of the x intercept of our plot.

Note that the data points in Figure 5.7 are not evenly spaced along the x axis. This is because we are plotting the reciprocal of data that were designed to be evenly spaced along the x axis of our original untransformed plot. In using Lineweaver–Burk plots it is more convenient to have data points that are evenly spaced in units of 1/[S]. This is easily accomplished experimentally as follows. One picks a maximum value of [S] ($[S_{max}]$) to work with and

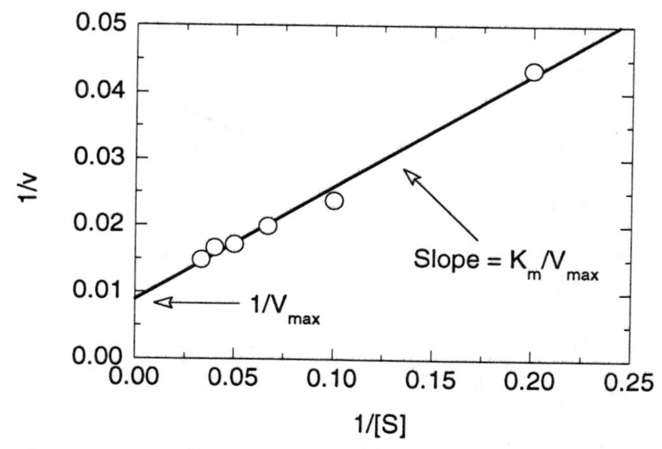

Figure 5.7 A Lineweaver–Burk double-reciprocal plot for selected data from Table 5.1.

makes a stock solution of substrate that will give this final concentration after dilution into the enzyme reaction mixture. Additional initial velocity measurements are then made by adding the same final volume to the enzyme reaction mixture from stock substrate solutions made by diluting the original stock solution by 1:2, 1:3, 1:4, 1:5, 1:6, and so on. In this way, the data points will fall along the 1/[S] axis at intervals of 1, 2, 3, 4, 5, 6,... units.

For example, let us say that we have decided to work with a maximum concentration of substrate of 30 μM in our enzyme reaction. If we prepare a 300 μM stock solution of substrate for this data point, we could dilute it 1:10 into our enzyme reaction mixture to obtain the desired final substrate concentration. If, for example, our total reaction volume were 1.0 mL, we would start our reaction by mixing 100 μL of substrate stock with 900 μL of the other components of our reaction system (enzyme, buffer, cofactors, etc.). Table 5.4 summarizes the additional stock solutions that would be needed to prepare

Table 5.4 Setup for an Experimental Determination of Enzyme Kinetics Using a Lineweaver–Burk Plot

Stock [S] (μM)	Final [S] (μM)	1/[S] (μM^{-1})	v (μM/s)	$1/v$ ($\mu M^{-1} \cdot s$)
300	30.0	0.033	66	0.015
150	15.0	0.067	49	0.020
100	10.0	0.10	41	0.024
75	7.5	0.13	33	0.030
60	6.0	0.17	29	0.034
50	5.0	0.20	24	0.042

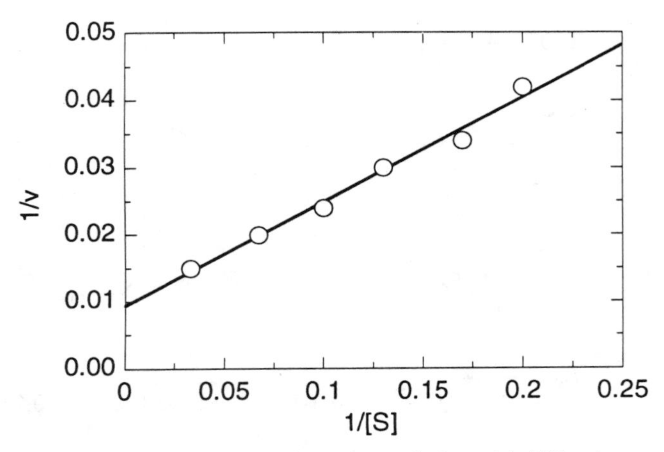

Figure 5.8 A Lineweaver–Burk double-reciprocal plot with [S] values spaced evenly along the 1/[S] axis; data from Table 5.4.

final substrate concentrations evenly spaced along a 1/[S] axis. Figure 5.8 illustrates a Lineweaver–Burk plot for these data.

Note that the form of Equation 5.19 is such that the data values in a Lineweaver–Burk plot are unevenly weighted in the analysis, with maximum weight placed on data points obtained at very high substrate concentrations. Typically the highest degree of experimental error is associated with these measurements. Hence the Lineweaver–Burk method gives greatest weight to the least precise measurements. For this reason, many scientists criticize the use of these plots for quantitative estimations of K_m and V_{max}. Nevertheless, the Lineweaver–Burk method is the most commonly used linear transformation of enzyme kinetic data. Today enzymologists use these plots to determine mechanistic details for enzyme interactions with substrates and inhibitors, as discussed in subsequent chapters. For these plots, many workers determine the values of K_m and V_{max} from nonlinear least-squares fits of the untransformed data and insert these values into Equation 5.19 to construct the linear Lineweaver–Burk plots. We shall discuss this strategy further when we describe the diagnostic uses of these plots.

5.5 Other Linear Transformations of Enzyme Kinetic Data

Despite the errors associated with the method, the Lineweaver–Burk double-reciprocal plot has become the most popular means of graphically representing enzyme kinetic data. There are, however, a variety of other linearizing transformations. We describe three other popular graphical methods for presenting enzyme kinetic data: Eadie–Hofstee, Hanes–Woolf, and Eisenthal–Cornish-Bowden direct plots.

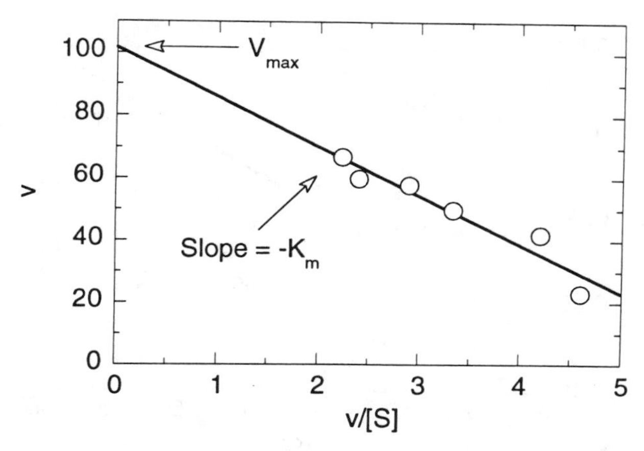

Figure 5.9 An Eadie–Hofstee plot of enzyme kinetic data; data from Table 5.1.

5.5.1 Eadie–Hofstee Plots

If we multiply both sides of Equation 5.16 by $K_m + [S]$, we obtain:

$$v(K_m + [S]) = V_{max}[S] \tag{5.21}$$

If we now divide both sides by [S] and rearrange, we obtain:

$$v = V_{max} - K_m \frac{v}{[S]} \tag{5.22}$$

Hence, if we plot v as a function of $v/[S]$, Equation 5.22 would predict a straight-line relationship with a slope of $-K_m$ and a y intercept of V_{max}. Such a plot is referred to as an Eadie–Hofstee plot. Figure 5.9 uses this format to present the data in Table 5.1 for $[S] = 5$–$30\,\mu$M.

5.5.2 Hanes–Woolf Plots

If one multiplies both sides of the Lineweaver–Burk equation (Equation 5.19) by [S], one obtains:

$$\frac{[S]}{v} = [S]\left(\frac{1}{V_{max}}\right) + \frac{K_m}{V_{max}} \tag{5.23}$$

This treatment also leads to linear plots when $[S]/v$ is plotted as a function of [S]. Figure 5.10 illustrates such a plot, known as a Hanes–Woolf plot, for selected data from Table 5.1. In this plot the slope is $1/V_{max}$, the y intercept is K_m/V_{max}, and the x intercept is $-K_m$.

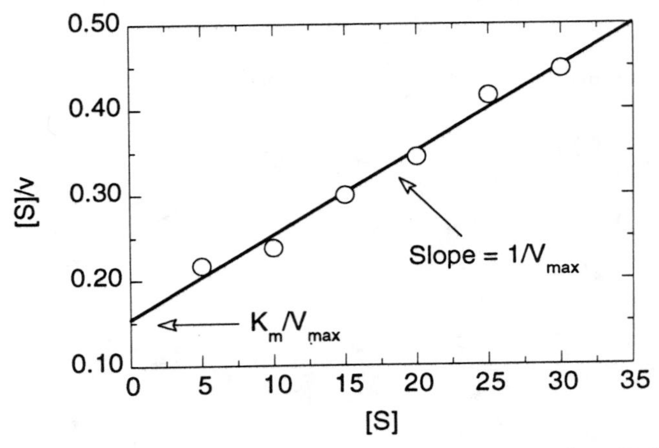

Figure 5.10 A Hanes–Woolf plot of enzyme kinetic data; data from Table 5.1.

5.5.3 Eisenthal–Cornish-Bowden Direct Plots

In our final method, pairs of v, [S] data (as in Table 5.1) are plotted as follows: values of v along the y axis and the negative values of [S] along the x axis (Eisenthal and Cornish-Bowden, 1974). For each pair, one then draws a straight line connecting the points on the two axes and extrapolates these lines

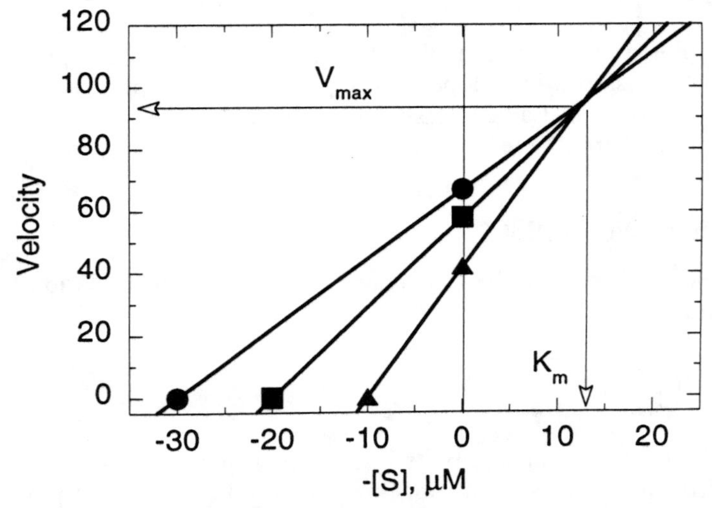

Figure 5.11 An Eisenthal–Cornish-Bowden direct plot of enzyme kinetic data; data from Table 5.1.

past there point of intersection (Figure 5.11). When a horizontal line is drawn from the point of intersection of these line to the y axis, the value at which this horizontal line crosses the y axis is equal to V_{max}. Similarly, when a vertical line is dropped from the point of intersection to the x axis, the value at which it crosses the x axis defines K_m. Plots like Figure 5.11, referred to as Eisenthal–Cornish-Bowden direct plots, are considered to give the best estimates of K_m and V_{max} of any of the linear transformation methods. Thus they are highly recommended when it is desired to determine these kinetic parameters but nonlinear curve fitting to Equation 5.16 is not feasible.

5.6 What Do K_m and V_{max} Tell Us?

We have gone to great lengths in this chapter to find ways of determining the kinetic constants K_m and V_{max} from enzyme kinetic data. What value do these constants add to our understanding of the enzyme under study?

The value of K_m varies considerable from one enzyme to another; and for any particular enzyme, K_m varies also with different substrates. We have already presented a working definition of K_m: the substrate concentration that results in half-maximal velocity for the enzyme-catalyzed reaction. An equivalent way of stating this is that the K_m represents the substrate concentration at which half the enzyme active sites in the sample are filled (saturated) by substrate molecules. We also have seen that K_m relates to the individual rate constants k_1, k_{-1}, and k_2. Recall from Equation 5.10 that K_m is equivalent to $(k_{-1} + k_2)/k_1$. Suppose that k_2, the rate constant for the ES complex going on to enzyme plus product, is much less than k_{-1}. Under these conditions, the dissociation of ES back to E + S will be much faster than product formation, and the expression for K_m will reduce to:

$$K_m = \frac{k_{-1}}{k_1} \tag{5.24}$$

The reader will immediately recognize that Equation 5.24 is an expression for the dissociation constant of the ES complex. Hence, under these very special circumstances, the K_m value is a measure of the affinity of the enzyme for the substrate. It is a common mistake among students and some professionals to assume that K_m always can be equated with the dissociation constant for ES. Yet because this assumption ignores the contribution of k_2 to K_m, it is not rigorously true in all circumstances. Note, however, that the term k_2 appears both in the expression for K_m and the expression for V_{max}. Recall from Equation 5.15 that V_{max} is equal to $k_2[E_t]$. We shall see in Chapter 6 that at constant $[E_t]$, certain solution conditions—such as pH—can have differential effects on K_m and V_{max}. In such cases the effects on K_m are typically equated with perturbation of substrate binding (i.e., an effect on k_{-1}/k_1) and the effects on V_{max} are equated with perturbation of the catalytic events (i.e., effects on k_2).

Referring again to Equation 5.15, we see that if the value of $[E_t]$ is known, the value of k_2 can be determined directly from the measured value of V_{max}. This value is sometimes also referred to as the *turnover number for* the enzyme (TN):

$$k_2 = \text{TN} = \frac{V_{max}}{[E_t]} \tag{5.25}$$

The kinetic constant k_2 is a first-order rate constant, having units of reciprocal time (e.g., min^{-1}, s^{-1}). Turnover numbers, however, are typically reported in units of molecules of product produced per unit time per molecule of enzyme present. As long as the same units are used to express the amount of product produced and the amount of enzyme present, these units will cancel and, as expected, the final units will be reciprocal time. When the V_{max} value in units of molarity per unit time, the enzyme concentration in units of molarity, and the volume of the reaction mixture are known, it is easy to convert the units back and forth. We can find the turnover number in the laboratory quite easily by measuring the velocity under saturating substrate conditions: that is, when $[S] \gg K_m$ and thus v approaches V_{max}. Under most physiological conditions in vivo, however, the ratio $[S]/K_m$ is more typically between 0.01 and 1.0. When $[S] \ll K_m$ we must change our expression for velocity to:

$$v = \frac{k_2}{K_m}[E][S] \tag{5.26}$$

Since $[S] \ll K_m$ here, the free enzyme concentration [E] will be well approximated by $[E_t]$, and Equation 5.26 can be approximated by:

$$v = \frac{k_2}{K_m}[E_t][S] \tag{5.27}$$

Recalling our definition of K_m, we note that:

$$\frac{k_2}{K_m} = \frac{k_2 k_1}{k_{-1} + k_2} \tag{5.28}$$

Under conditions of low substrate concentration, this ratio is most likely to be limited by the diffusional rate of encounter of the free enzyme with substrate, which is defined by k_1 and is typically in the range of 10^8–$10^9\ \text{M}^{-1} \cdot \text{s}^{-1}$.

We saw in Chapter 4 that multiple events can occur in the formation of product from the ES state. In a number of cases distinct intermediates are formed along the reaction pathway. Hence, it is not always rigorously correct to speak of a single rate constant for ES going to E + P. Therefore, it would be more correct to replace k_2 in all our derivations with a collection of rate constants depending on the individual reaction mechanism at hand. To simplify these analyses, it is common practice to represent the multiple kinetic steps by a single constant, referred to as k_{cat} (the subscript "cat" standing for catalysis). Thus Equation 5.25 becomes:

$$k_{cat} = \frac{V_{max}}{[E_t]} \tag{5.29}$$

and Equation 5.27 becomes:

$$v = \frac{k_{cat}}{K_m}[S][E_t] \tag{5.30}$$

The second-order rate constant defined by the ratio k_{cat}/K_m is generally considered to be a good measure of the catalytic efficiency of an enzyme.

The value of k_{cat}/K_m varies considerably from one enzyme to the next. For an enzyme in solution, the rate-determining step in catalysis will either be k_{cat} or k_1. If k_{cat} is rate limiting, the catalytic events that occur after substrate binding are slower than the rate of formation of the ES complex. If k_1 is rate limiting, the enzyme turns over essentially instantaneously once the ES complex has formed. In either case we see that the fastest rate of catalysis for an enzyme in solution is limited by the rate of diffusion of molecules in the solution. Some enzymes, such as carbonic anhydrase, display k_{cat}/K_m values of 10^8–$10^9\ M^{-1} \cdot s^{-1}$, which is at the diffusion limit. Such enzymes are said to have achieved *kinetic perfection*, because they convert substrate to product as fast as the substrate is delivered to the active site of the enzyme!

The diffusion limit would seem to set an upper limit on the value of k_{cat}/K_m that an enzyme can achieve. This is true for most enzymes in solution. However, some enzyme systems have overcome this limit by compartmentalizing themselves and their substrates within close proximity in subcellular locals where three-dimensional diffusion no longer comes into play. This can be accomplished by assembling enzymes and substrates into organized systems such as multienzyme complexes or cellular membranes. Two examples are presented here.

We consider first the respiratory electron transfer system of the inner mitochrondrial membrane. Here enzymes in a cascade are localized in close proximity to one another within the membrane bilayer. The product of one enzyme is the substrate for the next in the cascade. Because of the proximity of the enzymes in the membrane, the product leaves the active site of one enzyme and is presented to the active site of the next without the need for diffusion through solution.

The second example comes from the de novo synthetic pathway for pyrimidines. The first three steps in the synthesis of uridine monophosphate are performed by a supercomplex of three enzymes that are noncovalently associated as a multiprotein complex. This supercomplex, referred to as CAD, comprises the enzymes carbamyl phosphate synthase, aspartate transcarbamylase, and dihydroorotase. Because the active sites of the three enzymes are compartmentalized inside the supercomplex, the product of the first enzyme is immediately in proximity to the active site of the second enzyme, and so on. In this way, the supercomplex can overcome the diffusion barrier to rapid catalysis.

5.7 Measurements at Low Substrate Concentrations

In some instances the concentration range of substrates suitable for experimental measurements is severely limited, perhaps because of poor solubility or some physicochemical property of the substrate that interferes with the measurements above a critical concentration. If one is limited to measurements in which the substrate concentration is much less than the K_m, the reaction will follow first-order kinetics, and it may be difficult to find a time window over which the reaction velocity can be approximated by a linear function. In such situations one can still derive an estimate of k_{cat}/K_m by fitting the reaction progress curve to a first-order equation at a fixed substrate concentration. Suppose that we were to follow the loss of substrate as a function of time under first-order conditions (i.e., where $[S] \ll K_m$). The progress curve could be fit by the following equation:

$$[S] = [S_0]e^{-kt} \tag{5.31}$$

where [S] is the substrate concentration remaining after time t, $[S_0]$ is the starting concentration of substrate, and k, is the observed first-order rate constant. When $[S] \ll K_m$, the [S] term can be ignored in the denominator of Equation 5.16. Combining this with our definition of V_{max} from Equation 5.29 we obtain:

$$-\frac{d[S]}{dt} = \frac{k_{cat}}{K_m}[E_t][S] \tag{5.32}$$

Rearranging Equation 5.32 and integrating, we obtain:

$$[S] = [S_0]\exp\left(-\frac{k_{cat}}{K_m}[E_t]t\right) \tag{5.33}$$

Comparing Equations 5.33 and 5.31, we see that:

$$k = \frac{k_{cat}}{K_m}[E_t] \tag{5.34}$$

Thus if the concentration of enzyme used in the reaction is known, an estimate of k_{cat}/K_m can be obtained from the measured first-order rate constant of the reaction progress curve when $[S] \ll K_m$ (Chapman et al., 1993).

5.8 Deviations from Hyperbolic Kinetics

In most cases enzyme kinetic measurements fit remarkable well to the Henri–Michaelis–Menten behavior discussed in this chapter. Occasionally, however, there are deviations from the hyperbolic dependence of velocity on substrate concentration that we have come to expect. Such anomalies occur for several reasons. Some physical methods of measuring velocity, such as optical spectroscopies, can lead to experimental artifacts that have the appearance of deviations from the expected behavior, and we shall discuss these in detail in Chapter 6. Nonhyperbolic behavior also can be caused by the presence of certain types of inhibitor as well. In the most often encountered case, product inhibition, the product of the enzymatic reaction has enough affinity for the enzyme active site to be able, at high concentrations, to block the binding of substrate molecules. As the concentration of substrate is increased, the amount of product formed during the experimental measurements also increases. At very high concentrations of substrate, the observed velocity is lower than expected because of the buildup of inhibiting product.

Figure 5.12 illustrates the type of behavior one might observe for an enzyme that displays product inhibition. Usually this behavior is realized only at very high substrate concentrations. Therefore, the kinetic constants can still be determined by focusing on data points below the substrate concentration at which inhibition becomes significant. Inhibition effects at very high substrate concentrations also can be readily detected as nonlinearity in the Lineweaver–

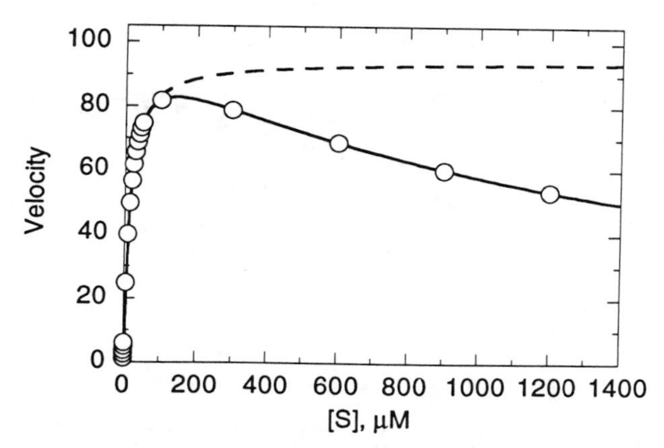

Figure 5.12 A Michaelis–Menten plot for an enzyme reaction displaying product inhibition at high concentrations of substrate: dashed line, best fit of the data at low substrate concentrations to Equations 5.16; solid line, fit of all the data to the equation $v = V_{max}/(1 + K_m/[S] + [S]/K_i)$, where K_i is the dissociation constant for the enzyme–inhibitor complex (in this case the enzyme–product complex). The constant K_i is described further in subsequent chapters.

Burk plots of the data. Here one observes a sudden and dramatic curving up of the data near the y-axis intercept.

Another cause of nonhyperbolic kinetics is the presence of more than one enzyme acting on the same substrate. Many enzyme studies are performed with only partially purified enzymes, and many clinical diagnostic tests that rely on measuring enzyme activities are performed on crude samples (of blood, tissue homogenates, etc.). When the substrate for the reaction is unique to the enzyme of interest, these crude samples can be used with good results. If, however, the sample contains more than one enzyme that can act on the substrate, deviations from the expected kinetic results occur. Suppose that our sample contains two enzymes, both capable of converting the substrate to product but with different kinetic constants. Suppose further that for one of the enzymes $V_{max} = V_1$ and $K_m = K_1$, and for the second enzyme $V_{max} = V_2$ and $K_m = K_2$. The velocity of the overall mixture then is given by:

$$v = \frac{V_1[S]}{K_1 + [S]} + \frac{V_2[S]}{K_2 + [S]} \tag{5.35}$$

This can be rearranged to give the following expression (Schultz, 1994):

$$v = \frac{(V_1K_2 + V_2K_1)[S] + (V_1 + V_2)[S]^2}{K_1K_2 + (K_1 + K_2)[S] + [S]^2} \tag{5.36}$$

Equation 5.36 is a polynomial expression, which yields behavior very different from the rectangular hyperbolic behavior we expect, as illustrated in Figure 5.13.

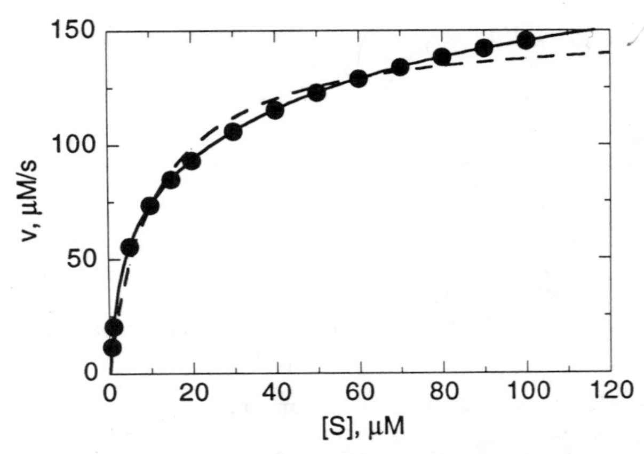

Figure 5.13 Effects of multiple enzymes acting on the same substrate: dashed line, fit of the data to Equation 5.16 for a single enzyme; solid line, fit to Equation 5.36 for two enzymes acting on the same substrate with $V_1 = 120\,\mu M/s$, $V_2 = 75\,\mu M/s$, $K_1 = 65\,\mu M$, and $K_2 = 3\,\mu M$.

One last example of deviation from hyperbolic kinetics is that of enzymes displaying cooperativity of substrate binding. In the derivation of Equation 5.16 we assumed that the active sites of the enzyme molecules behave independently of one another. As we saw in Chapter 4, sometimes proteins occur as multimeric assemblies of subunits. Some enzymes occur as homomultimers, each subunit containing a separate active site. It is possible that the binding of a substrate molecule at one of these active sites could influence the affinity of the other active sites in the multisubunit assembly (see Chapter 11 for more details). This effect is known as *cooperativity*. It is said to be *positive* when the binding of a substrate molecule to one active site increases the affinity for substrate of the other active sites. On the other hand, when the binding of substrate to one active site lowers the affinity of the other active sites for the substrate, the effect is termed *negative cooperativity*. The number of potential substrate binding sites on the enzyme and the degree of cooperativity among them can be quantified by the Hill coefficient, h. The influence of cooperativity on the measured values of velocity can be easily taken into account by modifying Equation 5.16 as follows:

$$v = \frac{V_{\max}[\mathrm{S}]^h}{K' + [\mathrm{S}]^h} \tag{5.37}$$

where K' is related to K_m but also contains terms related to the effect of substrate occupancy at one site on the substrate affinity of other sites (see Chapter 11). Figure 5.14 illustrates how positive cooperativity can affect the Michaelis–Menten and Lineweaver–Burk plots of an enzyme reaction.

The velocity data for cooperative enzymes can be presented in a linear form by use of Equation 5.38:

$$\log\left(\frac{v}{V_{\max} - v}\right) = h\log[\mathrm{S}] - \log(K') \tag{5.38}$$

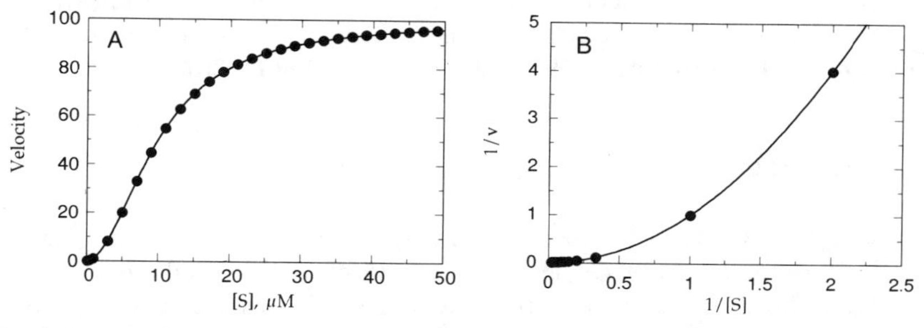

Figure 5.14 Effects of positive cooperativity on the kinetics of an enzyme-catalyzed reaction. (A) Data graphed as a Michaelis–Menten (i.e., direct) plot. (B) Data from (A) replotted as a Lineweaver–Burk double-reciprocal plot.

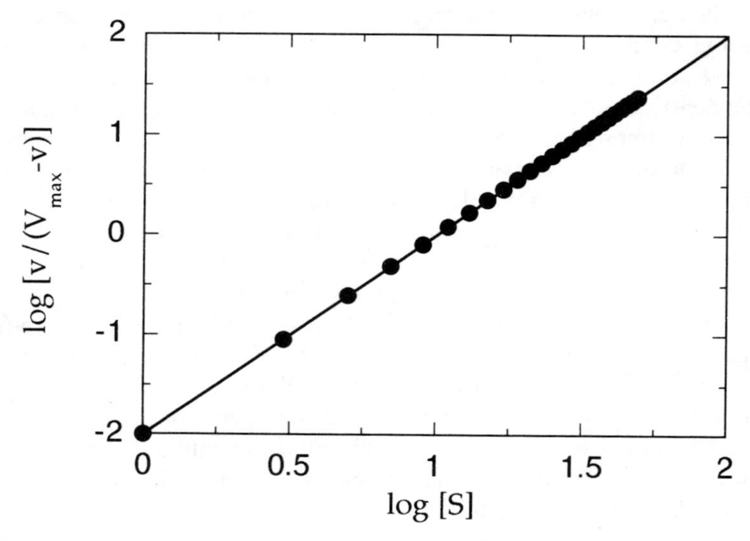

Figure 5.15 Hill plot for the data from Figure 5.14: $\log[v/(V_{max} - v)]$ versus $\log[S]$. The slope of the best fit line provides an estimate of the Hill coefficient *h*, and the *y* intercept provides an estimate of $-\log(K')$.

Thus, a plot of $\log(v/(V_{max} - v)$ as a function of $\log[S]$ should yield a straight line with a slope of *h*, and a *y* intercept of $-\log(K')$, as illustrated in Figure 5.15. The utility of plots such as that in Figure 5.15 is limited, however, by the need to know V_{max} a priori and because the linear relationship described by Equation 5.38 holds over only a limited range of substrate concentrations (in the region of $[S] = K'$). Hence, whenever possible, it is best to determine V_{max}, *h*, and K_m for cooperative enzymes from direct nonlinear curve fits to Equation 5.37.

These examples illustrate the more commonly encountered deviations from hyperbolic kinetics. A number of other causes of deviations are known, but they are less common. A more comprehensive discussion of such deviations can be found in the texts by Segal (1975) and Bell and Bell (1988).

5.9 Summary

In this chapter we have described the kinetics of enzyme-catalyzed reactions under steady state conditions, where there is a large excess of substrate relative to enzyme and the measurements are made early in the reaction progress curve (i.e., when product accumulation is limited). These measurements permit us to obtain estimates of the kinetic constants K_m and k_{cat}, which together provide a measure of the catalytic efficiency of an enzyme. Several graphical methods for determining these kinetic constants were presented here.

We have limited our discussion to steady state enzyme kinetics. Such measurements are the easiest to perform in a standard laboratory and provide a wealth of information. Rapid kinetic techniques can also be performed on enzyme-catalyzed reactions to obtain more detailed information on the individual rate constants for different steps in the reaction sequence. A discussion of these more advanced techniques is beyond the scope of an introductory text. A number of excellent texts are devoted to the subject, however. The recent review article by Johnson (1992) provides a good starting point for those interested in reading more about rapid kinetic methods.

References and Further Reading

Bell, J. E., and Bell, E. T. (1988) *Proteins and Enzymes*, Prentice-Hall, Englewood Cliffs, NJ.

Briggs, G. E., and Haldane, J. B. S. (1925) *Biochem. J.* **19**, 383.

Brown, A. J. (1902) *J. Chem. Soc.* **81**, 373–388.

Chapman, K. T., Kopka, I. E., Durette, P. I., Esser, C. K., Lanza, T. J., Izquierdo-Martin, M., Niedzwiecki, L., Chang, B., Harrison, R. K., Kuo, D. W., Lin, T.-Y., Stein, R. L., and Hagmann, W. K. (1993) *J. Med. Chem.* **36**, 4293–4301.

Cleland, W. W. (1967) *Adv. Enzymol.* **29**, 1–65.

Cornish-Bowden, A., and Wharton, C. W. (1988) *Enzyme Kinetics*, IRL Press, Oxford.

Eisenthal, R., and Cornish-Bowden, A. (1974) *Biochem. J.* **139**, 715–720.

Henri, V. (1903) *Lois Générales de l'action des diastases*, Hermann, Paris.

Johnson, K. A. (1992) *Enzymes*, **XX**, 1–61.

Lineweaver, H., and Burk, J. (1934) *J. Am. Chem. Soc.* **56**, 658–666.

Michaelis, L., and Menten, M. L. (1913) *Biochem. Z.* **49**, 333–369.

Schulz, A. R. (1994) *Enzyme Kinetics from Diastase to Multi-enzyme Systems,* Cambridge University Press, New York.

Segal, I. H. (1975) *Enzyme Kinetics*, Wiley, New York.

CHAPTER

6

Experimental Measures of Enzyme Activity

Steady state enzyme kinetics offers a wealth of information on the mechanistic details of enzyme catalysis and on the interactions of enzymes with ligands, such as substrates and inhibitors. Chapter 5 provided the basis for determining the kinetic constants K_m and k_{cat} from initial velocity measurements taken at varying substrate concentrations. The determination of these kinetic constants rests on the ability to measure accurately the initial velocity of an enzyme-catalyzed reaction under well-controlled conditions.

In this chapter we describe some of the experimental methods used to determine reaction velocities. We shall see that numerous strategies have been developed for following over time the loss of substrate or appearance of products that results from enzyme turnover. The velocity of an enzymatic reaction is sensitive to many solution conditions, such as pH, temperature, and solvent isotopic composition; these conditions must be well controlled if meaningful data are to be obtained. Controlled changes in these solution conditions and measurement of their effects on the reaction velocity can provide useful information about the mechanism of catalysis as well. Like all proteins, enzymes are sensitive to storage conditions and can be denatured easily by mishandling. Therefore we also discuss methods for the proper handling of enzymes to ensure their maximum catalytic activity and stability.

6.1 Initial Velocity Measurements

6.1.1 Direct, Indirect, and Coupled Assays

To measure the velocity of a reaction, it is necessary to follow some signal that reports product formation or substrate depletion over time. The type of signal varies from assay to assay but usually relies on some unique physicochemical property of the substrate or product and/or the analyst's ability to separate the substrate from the product. Generally, most enzyme assays rely on one (or more) of the following broad classes of detection and separation methods to follow the course of the reaction:

Spectroscopy
Polarography
Radioactive decay
Electrophoretic separation
Chromatographic separation
Immunological reactivity

These methods can be used in *direct assays*: the direct measurement of the substrate or product concentration as a function of time. For example, the enzyme cytochrome *c* oxidase catalyzes the oxidation of the heme-containing protein cytochrome *c*. In its reduced (ferrous iron) form, cytochrome *c* displays a strong absorption band at 550 nm, which is significantly diminished in intensity when the heme iron is oxidized (ferric form) by the oxidase. One can thus measure the change in light absorption at 550 nm for a solution of ferrous cytochrome *c* as a function of time after addition of cytochrome *c* oxidase; the diminution of adsorption at 550 nm that is observed is a direct measure of the loss of substrate (ferrous cytochrome *c*) concentration (Figure 6.1).

In some cases the substrate and product of an enzymatic reaction do not provide a distinct signal for convenient measurement of their concentrations. Often, however, product generation can be coupled to another, nonenzymatic, reaction that does produce a convenient signal; such a strategy is referred to as an *indirect assay*. Dihydroorotate dehydrogenase (DHODase) provides an example of the use of an indirect assays. This enzyme catalyzes the conversion of dihydroorotate to orotic acid in the presence of the exogenous cofactor ubiquinone. During enzyme turnover, electrons generated by the conversion of dihydroorotate to orotic acid are transferred by the enzyme to a ubiquinone cofactor to form ubiquinol. It is difficult to measure this reaction directly, but the reduction of ubiquinone can be coupled to other nonenzymatic redox reactions.

Several redox active dyes are known to change color upon oxidation or reduction. Among these, 2,6-dichlorophenolindophenol (DCIP) is a convenient dye with which to follow the DHODase reaction. In its oxidized form DCIP is bright blue, absorbing light strongly at 610 nm. Upon reduction, however, this absorption band is completely lost. DCIP is reduced stoichiometrically by

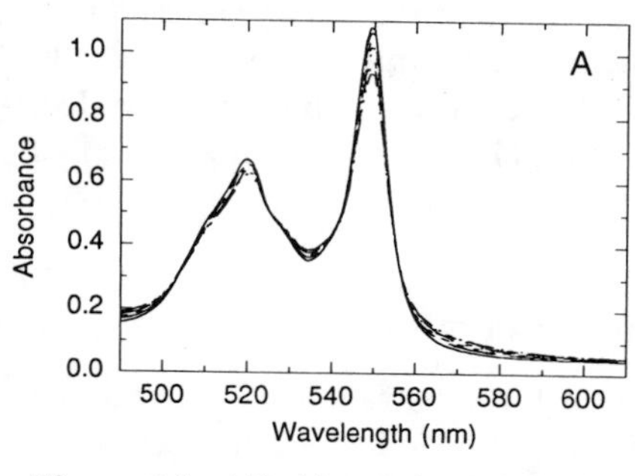

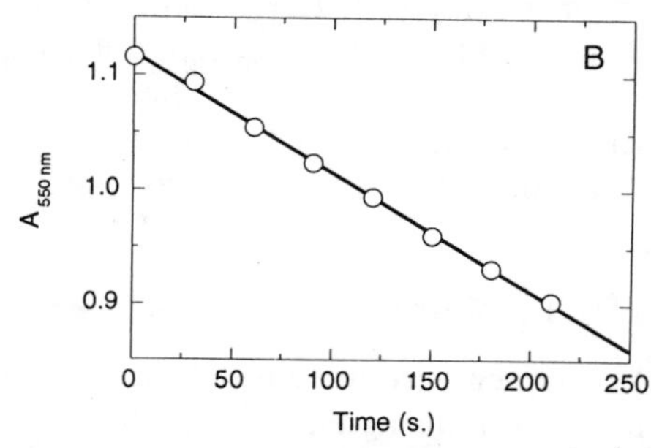

Figure 6.1 (A) Absorption of ferrocytochrome *c* as a function of time after addition of the enzyme cytochrome *c* oxidase. As the cytochrome *c* iron is oxidized by the enzyme, the absorption feature at 550 nm decreases. (B) Plot of the absorption at 550 nm for the spectra in (A) as a function of time. Note that in this early stage of the reaction ($\leqslant 10\%$ of the substrate has been converted), the plot yields a linear relationship between absorption and time. The reaction velocity can thus be determined from the slope of this linear function.

ubiquinol, which is formed during DHODase turnover. Hence, it is possible to measure enzymatic turnover by having an excess of DCIP present in a solution of substrate (dihydroorotate) and cofactor (ubiquinone), and following the loss of 610 nm absorption with time after addition of enzyme to initiate the reaction.

A third method for following the course of an enzyme-catalyzed reaction is referred to as the *coupled assay*. Here the enzymatic reaction of interest is paired with a second enzymatic reaction which can be conveniently measured. In a typical assay, the product of the enzyme reaction of interest is the substrate for the enzyme reaction to which it is coupled for convenient measurement. An example of this strategy is the measurement of activity for hexokinase, the enzyme that catalyzes the formation of glucose-6-phosphate and ADP from glucose and ATP. None of these products or substrates provides a particularly convenient means of measuring enzymatic activity. However, the product glucose-6-phosphate is the substrate for the enzyme glucose-6-phosphate dehydrogenase, which, in the presence of $NADP^+$, converts this molecule to 6-phosphogluconolactone. In the course of the second enzymatic reaction, $NADP^+$ is reduced to NADPH, and this cofactor reduction can be monitored easily by light absorption at 340 nm.

Because coupled reactions entail multiple enzymes, these assays present a number of potential problems that are not encountered with direct or indirect assays. For example, to obtain meaningful data on the enzyme of interest in coupled assays, it is imperative that the reaction of interest be the rate-limiting reaction. Otherwise, velocity changes that accompany changes in reaction conditions may not reflect accurately effects on the target enzyme. This is also true when one wishes to study the effects of inhibitors on the activity of the enzyme of interest. The presence of multiple enzymes can introduce ambiguities in interpreting the results of such experiments (e.g., which enzyme is really being inhibited?). Hence, whenever possible these assays should be avoided. Rudolph et al. (1979), Cleland (1979), and Tipton (1992) provide more detailed discussions of coupled enzyme assays.

6.1.2 Continuous Versus End Point Assays

In chapter 5 we introduced the progress curves for substrate loss and product formation during enzyme catalysis (Figures 5.1 and 5.2). In all our discussions here we shall restrict our attention to the early portion of the progress curves, where substrate loss and product formation track linearly with time. Of course, the duration of this linear phase must itself be determined experimentally. It is critical to verify that the time interval over which the reaction velocity is to be measured displays good signal linearity with time. Once this time period has been established, the researcher has two options for obtaining a velocity measurement. First, the signal could be monitored at discrete intervals over the entire linear time period, or some convenient portion thereof. This strategy, referred to as a *continuous assay*, provides the safest means of accurately determining reaction velocity from the slope of a plot of signal versus time.

It is not always convenient to assay samples continuously, however, especially when one is using separation techniques, such as high performance liquid chromatography (HPLC) or electrophoresis. In such cases a second strategy, called *end point* or *discontinuous assay*, is often employed. Having established a linear time period for an assay, one measured the signal at a single specific time point within the linear time period. The reaction velocity is then determined from the difference in signal at that time point and at the initiation of the reaction, divided by the time:

$$v = \frac{\Delta I}{t} = \frac{I_t - I_0}{t_{\text{reading}}} \tag{6.1}$$

where I_t is the intensity of the signal being measured at time t, I_0 is the intensity of the signal at time zero, and t_{reading} is the time interval between initiation of the reaction and measurement of the signal.

In many instances it is much easier to take a single reading than to make multiple measurements during a reaction. Inherent in the use of end point readings, however, is the danger of assuming that the signal will track linearly with time over the period chosen, under the conditions of the measurement. Changes in temperature, pH, substrate concentration, and enzyme concentration, as well as the presence of certain types of inhibitor, can dramatically change the linearity of the signal over a fixed time window. Hence, end point reading can be misleading. Whenever feasible, then, one should use continuous assays to monitor substrate loss or product formation. When this is impractical, end point reading can be used, but cautiously, with careful controls.

6.1.3 Initiating, Mixing, and Stopping Reactions

In a typical enzyme assay, all but one of the components of the reaction mixture are added to the reaction vessel, and the reaction is started at time zero by adding the missing component, which can either be the enzyme or the substrate. The choice of the initiating component will depend on the details of the assay format and the stability of the enzyme sample to the conditions of the assay. In either case, the other components should be mixed well and equilibrated in terms of pH, temperature, and ionic strength. The reaction should then be initiated by addition of a small volume of a concentrated stock solution of the missing component. A small volume of the initiating component is used to ensure that its addition does not significantly perturb the conditions (temperature, pH, etc.) of the overall reaction volume. Unless the reaction mixture and initiating solutions are well matched in terms of buffer content, pH, temperature, and other factors, the initiating solution should not be more than about 5–10% of the total volume of the reaction mixture.

Samples should be mixed rapidly after addition of the initiating solution, but vigorous shaking or vortex mixing is denaturing to enzymes and should be avoided. Mixing must, however, be complete; otherwise there will be artifactual deviations from linear initial velocities as mixing continues during the measurements. One way to achieve gentle, but complete, rapid mixing is to add the initiating solution to the side of the reaction vessel as a "hanging drop" above the remainder of the reaction mixture, as illustrated in Figure 6.2. With small volumes (say, $<50\,\mu L$), the surface tension will hold the drop in place above

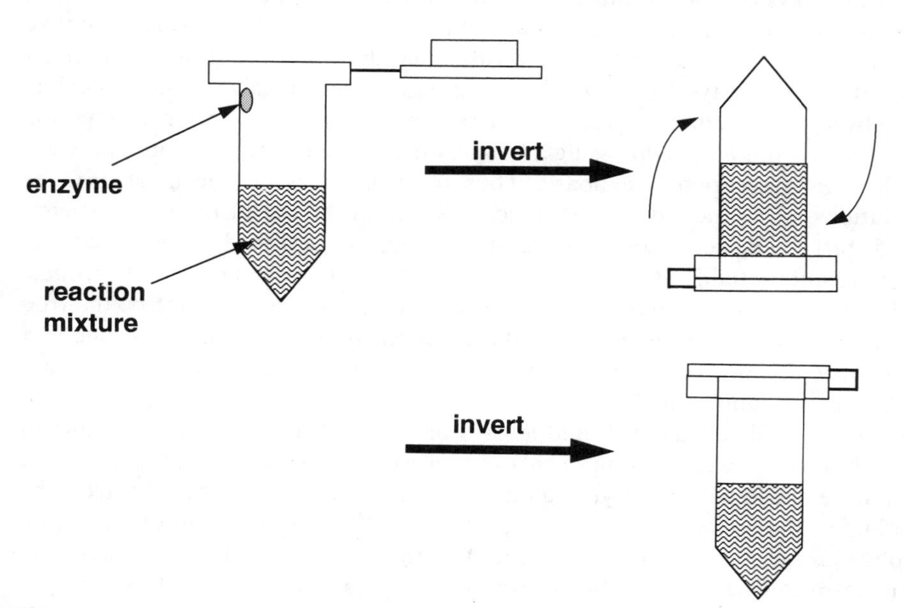

Figure 6.2 A common strategy for initiating an enzymatic reaction in a microcentrifuge tube.

the reaction mixture. At time zero the reaction is initiated by gently inverting the closed vessel two or three times to mix the solutions. Figure 6.2 illustrates this technique for a reaction taking place in a microcentrifuge tube. Using these tubes it is also convenient to place the initiating solution in the tube cap, which is then closed, permitting the solutions to be mixed by inversion as illustrated in Figure 6.2. For optical spectroscopic assays (see Section 6.2.2), the reaction can be initiated directly in the spectroscopic cuvette by the same technique, using a piece of parafilm and one's thumb to seal the top of the cuvette during the inversions.

Regardless of how the reaction and initiating solutions are mixed, the mixing must be achieved in a short period of time relative to the interval between measurements of the reaction's progress. With a little practice one can achieve this mixing in 10 seconds or less using the inversion method just described. This is usually fast enough for assays in which measurements are to be made in 1-minute or longer time intervals. There are a number of parameters, such as temperature and enzyme concentration (see Section 6.1.4), that can be adjusted to ensure that the reaction velocity is slow enough to allow mixing of the solutions and making of measurements on a convenient time scale. In some rare cases, the enzymatic velocity is so rapid that it cannot be conveniently measured in this way. Then one must resort to specialized rapid mixing and detection methods, such as stopped-flow techniques (Roughton and Chance, 1963; Kyte, 1995); these methods are also used to measure pre–steady state enzyme kinetics (Kyte, 1995).

For assays in which samples are removed from the reaction vessel at specific times for measurement, one can start the timer at the point of mixing and make measurements at known time intervals after the initiation point. For many spectroscopic assays, however, it is necessary to continuously measure a change in absorption of fluorescence with time. For most modern spectrometers, the detection is initiated by pressing a button on an instrument panel or depressing a key on a computer keyboard. Thus to start an assay one must mix the solutions, place the cuvette, or optical cell, in the holder of the spectrometer, and start the detection by pressing the appropriate button. The delay between mixing and actually starting a measurement can be as much as 20 seconds. Thus the time point recorded by the spectrometer as zero will not be the true zero point (i.e., mixing point) of the reaction. Again, with practice one can minimize this delay time, and in most cases the assay can be set up to render this error insignificant.

As we shall see, there should always be two control measurements: one in which all the reaction components except for the enzyme are present, and a separate one in which everything but substrate is present. (In these controls the volumes that would have been contributed by the enzyme or substrate solutions are made up for with buffer.) With these two control measurements one can calculate what the absorption or fluorescence should be for the reaction mixture at the true time zero. If the first spectrometer reading (i.e., the point recorded as time zero by the spectrometer) is significantly different from this calculated value, it is necessary to correct the time points recorded by the spectrometer for the time delay between the start of mixing and the initiation of the detection device. A laboratory timer or stopwatch can be used to determine the time gap.

Many nonspectroscopic assays require measurement times that are long compared to the rate of the enzymatic reaction being monitored. Suppose, for example, that we wished to measure the amount of product formed every 5 minutes over the course of a 30-minute enzymatic reaction, and assay for product by an HPLC method. The HPLC measurement itself might take 20–30 minutes to complete. If the enzymatic reaction is continuing during the measurement time, the amount of product produced during specific time intervals cannot be determined accurately. In such cases it is necessary to quench or stop the reaction at a specific time, to prevent further enzymatic production of product or utilization of substrate.

Methods for stopping enzymatic reactions usually involve denaturation of the enzyme by some means, or rapid freezing of the reaction solution. Examples of quenching methods include immersion in a dry ice–ethanol slurry to rapidly freeze the solution, and denaturation of the enzyme by addition of strong acid or base, addition of electrophoretic sample buffer, or immersion in a boiling water bath. In addition to these methods, one can add reagents that interfere in a specific way with a particular enzyme. For example, the activity of many metalloenzymes can be quenched by adding an excess of a metal chelating agent, such as ethylenediaminetetraacetic acid (EDTA).

Three points must be considered in choosing a quenching method for an enzymatic reaction. First, the technique used to quench the reaction must not interfere with the subsequent detection of product or substrate. Second, it must be established experimentally that the quenching technique chosen does indeed completely stop the reaction. Finally, the volume change that occurs upon addition of the quenching reagent to the reaction mixture must be accounted for. Similarly, measuement of product or substrate concentration must be corrected for the dilution effects of quencher addition.

6.1.4 The Importance of Running Controls

Regardless of the detection method used to follow an enzymatic reaction, it is always critical to perform control measurements in which enzyme and substrate are separately left out of the reaction mixture. These control experiments permit the analyst to correct the experimental data for any time-dependent changes in signal that might occur independent of the action of the enzyme under study, and to correct for any static signal due to components in the reaction mixture. To illustrate these points, let us follow a hypothetical enzymatic reaction, by tracking the light absorption decrease at some wavelength, as substrate is converted to product. Let us say that there is some low rate of spontaneous product formation in the absence of the enzyme and that the enzyme itself imparts a small but measurable absorption at the

Table 6.1 Volumes of Stock Solutions to Prepare Experimental and Control Samples for a Hypothetical Enzyme-Catalyzed Reaction

	Volumes Added (μL)		
Stock Solutions	Experimental	No Substrate Control	No Enzyme Control
10 × Substrate	100	0	100
10 × Buffer	100	100	100
Distilled water	790	890	800
Enzyme	10	10	0
Total volume	1.0 mL	1.0 mL	1.0 mL

analytical wavelength. We might set up an experiment in which all the reaction components are placed in a cuvette and the reaction is initiated by adding a small volume of enzyme stock solution. For this illustration, let us say that the reaction mixture is prepared by addition of the volumes of stock solutions listed in Table 6.1. The strategy for preparing the reaction mixtures in Table 6.1 is typical of what one might use in a real experimental situation.

Figure 6.3A illustrates the time courses we might see for the hypothetical solutions of Table 6.1. For our experimental run, the true absorption readings are displaced by about 0.1 unit as a result of absorption of the enzyme itself ("No Substrate" trace in Figure 6.3A). To correct for this, we substract this constant value from all our experimental data points. If we were to now determine the slope of our corrected experimental trace, however, we would be overestimating the velocity of our reaction because such a slope would reflect both the catalytic conversion of substrate to product and the spontaneous absorption change seen in our "No Enzyme" control trace. To correct for this, we subtract these control data points from the experimental points at each measurement time to yield the difference plot in Figure 6.3B. Measuring the slope of this difference plot yields the true reaction velocity. As illustrated in Figure 6.3A, the correction for the spontaneous absorption change may appear at first glance to be trivial. However, the velocity measured for the uncorrected data differs from the corrected velocity by more than 10% in this example. In some cases the background signal change is even more substantial. Hence, the types of control measurement discussed here are essential for obtaining meaningful velocity measurements for the catalyzed reaction under study.

6.2 Detection Methods

A wide variety of physicochemical methods have been used to follow the course of enzymatic reactions. Some of the more common techniques are described

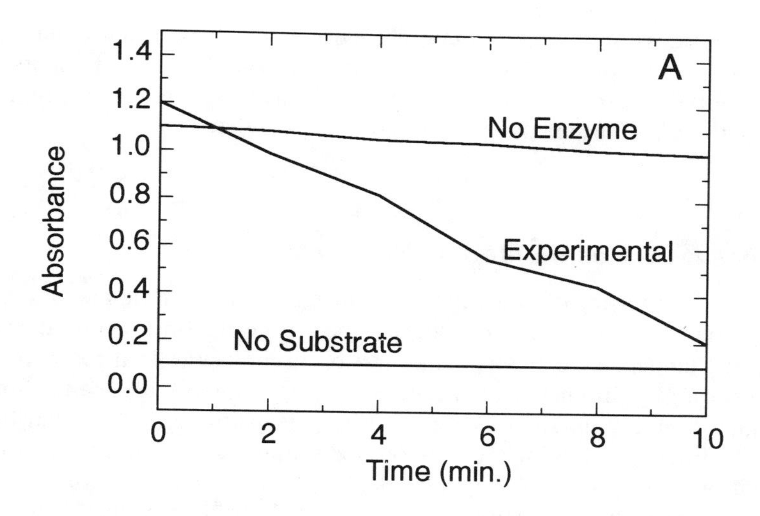

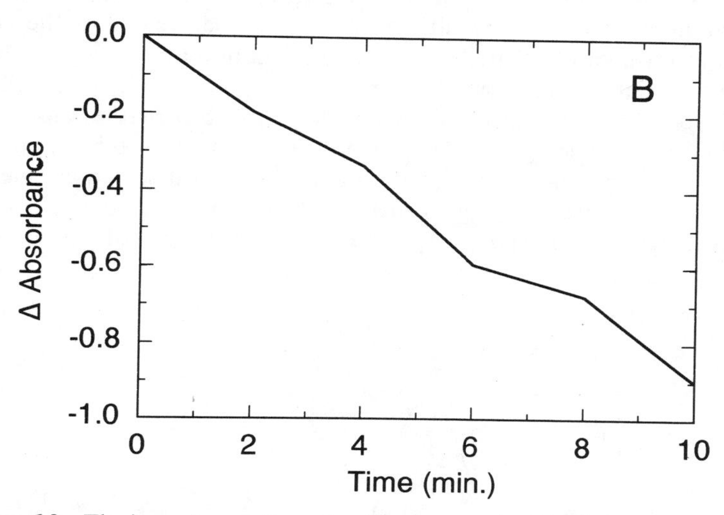

Figure 6.3 The importance of running blank controls. (A) Time courses of absorption for some hypothetical enzymatic reaction (experimental trace), along with two control samples: one with all the reaction mixture components except enzyme, the other with all the reaction mixture components except substrate. In this example, the enzyme absorbs minimally at the analytical wavelength, but enough to displace the time zero measurement by about 0.1 absorption unit. (B) The two control readings at each time point have been substracted from the experimental measurement to yield the true time course of the reaction.

here, but our discussion is far from comprehensive. Any signal that differentiates the substrates or products of the reaction from the other components of the reaction mixture can, in principle, form the basis of an enzyme assay; the only limit is the creativity of the investigator.

6.2.1 Assays Based on Optical Spectroscopy

Two very common means of following the course of an enzymatic reaction are absorption spectroscopy and fluorescence spectroscopy. Both methods are based on the changes in electronic configuration of molecules that result from their absorption of light energy of specific wavelengths. Molecules can absorb electromagnetic radiation, such as light, causing transitions between various energy levels. Energy in the infrared region, for example, can cause transitions between vibrational levels of a molecule. Microwave energy can induce transitions among rotational energy levels, while radiofrequency energy, which forms the basis of NMR spectroscopy, can cause transitions among nuclear spin levels. The energy differences between electronic levels of a molecule are so large that light energy in the ultraviolet and visible regions of the electromagnetic spectrum are required to induce transitions among these states. If a molecule is irradiated with varying wavelengths of light (of similar intensity), only light of specific wavelengths will be strongly absorbed by the sample. At these wavelengths, the energy of the light matches the energy gap between two electronic states of the molecule, and the light is absorbed to induce such a transition. Figure 6.4 presents a hypothetical absorption spectrum of a molecule in which light of wavelength λ_{max} induces an electron redistribution to

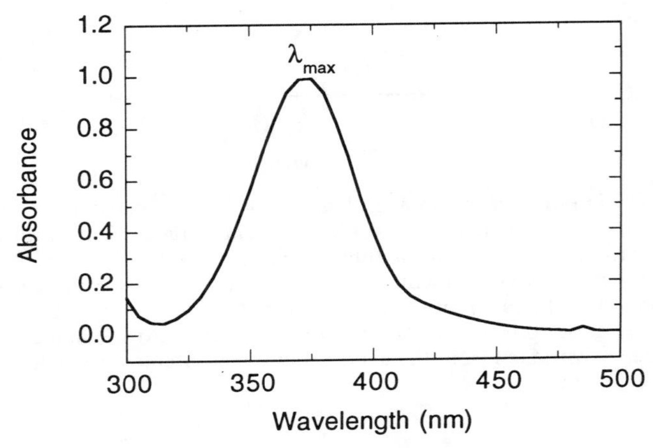

Figure 6.4 A typical absorption spectrum of a molecule with an absorption maximum at 375 nm.

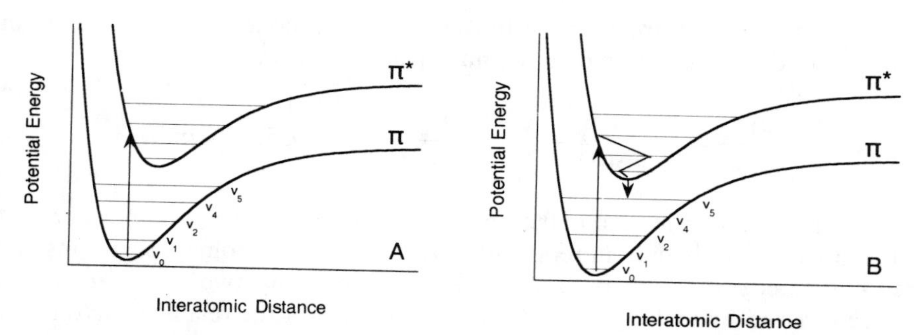

Figure 6.5 Energy level diagrams for (A) a light-induced transition from a π electronic ground state to an excited π^* state of a molecule and (B) the relaxation of the excited state molecule back to the ground state by photon emission (fluorescence).

bring the molecule from its ground π state to an excited π^* state. This process is illustrated as a potential energy diagram in Figure 6.5A.

6.2.2 Absorption Measurements

The value of absorption spectroscopy as an analytical tool is that the absorption of a molecule at a particular wavelength can be related to the concentration of that molecule in solution, as described by Beer's law:

$$A = \varepsilon c l \tag{6.2}$$

where A is the absorbance of the sample at some wavelength, c is the sample concentration in molarity units, l is the path length of sample the light beam traverses (in centimeters), and ε is an intrinsic constant of the molecule, known as the extinction coefficient, or molar absorptivity.

Since absorbance is a unitless quantity, ε must have units of reciprocal molarity times reciprocal path length (typically expressed as $M^{-1} \cdot cm^{-1}$ or $mM^{-1} \cdot cm^{-1}$)

Thus if we know the value of ε for a particular molecule, and the path length of the cuvette, we can calculate the concentration of that molecule in a solution by measuring the absorption of that solution. As we have seen in the examples of Figures 6.1 and 6.3, we can use the unique absorption features of a substrate or product of an enzymatic reaction to follow the course of such a reaction. Using Beer's law, we can convert our measured ΔA values at different time points to the concentration of the molecule being followed, and thus report the reaction velocity in terms of change in molecular concentration as a function of time.

For example, let us say that the substrate of the reaction illustrated in Figure 6.3 had an extinction coefficient of $2.5\,mM^{-1} \cdot cm^{-1}$ and that these measurements were made in a cuvette having a path length of 1 cm. We see from Figure 6.3B that, after making our corrections to the data, the absorption of our reaction mixture changes by 0.89 over the course of 10 minutes, or 0.089/min. Our reaction velocity is thus given as follows:

$$v = -\frac{d[S]}{dt} = -\frac{\Delta A}{\varepsilon l \Delta t} = -\frac{0.89}{2.5 \times 1 \times 10} = -0.0356\,\text{mM/min} \qquad (6.3)$$

Note that the units here, molarity per unit time (i.e., moles per liter per unit time) are commonly used in reporting enzyme velocities. Some workers instead report velocity in units of moles per unit time. The two units are easily interconverted by making note of the total volume of the reaction mixture for which the velocity is being measured.

6.2.3 Choosing an Analytical Wavelength

The wavelength used for following enzyme kinetics should be one that gives the greatest difference in absorption between the substrate and product molecules of the reaction. In many cases, this will correspond to the wavelength maximum of the substrate or product molecule. However, when there is significant spectral overlap between the absorption bands of the substrate and product, the most sensitive analytical wavelength may *not* be the same as the wavelength maximum. This concept is illustrated in Figure 6.6. In inspecting the spectra for the substrate and product of the hypothetical enzymatic reaction of Figure 6.6A, note that at the wavelength maximum for each molecule, the other molecule displays significant absorption. The wavelength at which the largest difference in signal is observed can be determined by calculating the difference spectrum between these two spectra, as illustrated in Figure 6.6B. Thus, in our hypothetical example, the wavelength maxima for substrate and product are 374 and 385 nm, respectively, but the most sensitive wavelengths for following loss of substrate or formation of product would be 362 and 402 nm, respectively.

6.2.4 Optical Cells

Absorption measurements are most commonly performed with a standard spectrophotometer, and the samples are contained in cuvettes. These specialized cells come in a range of path lengths and are constructed of various optical materials. Disposable plastic cuvettes that hold 1 or 3 mL samples are commercially available. Although these disposable units are very convenient and reduce the chances of sample-to-sample cross-contamination, their use should be restricted to the visible wavelength range ($\approx$ 350–800 nm). For measurements at wavelengths less than 350 nm, high quality quartz cuvettes must be used, since both plastic and glass absorb too much light in the ultraviolet. Quartz cuvettes are available from numerous manufacturers in a variety of sizes and configurations.

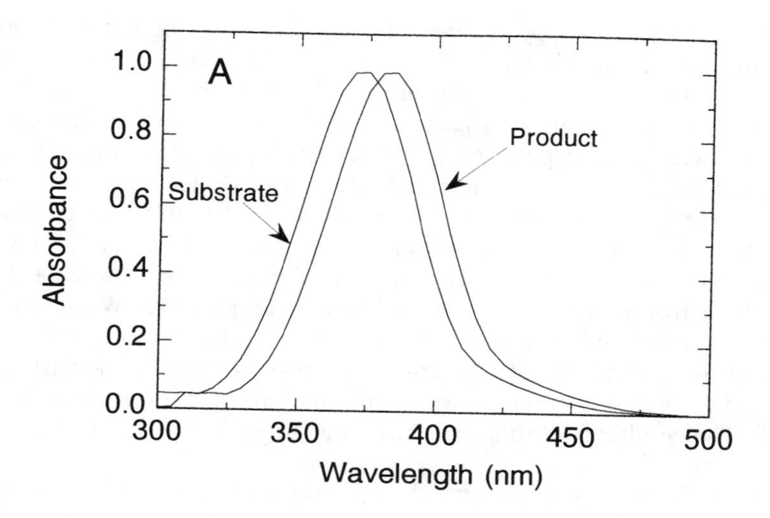

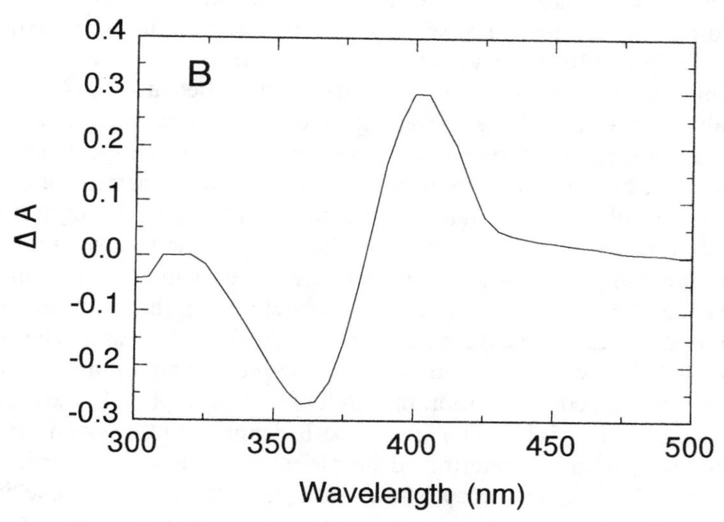

Figure 6.6 Example of the use of difference spectroscopy. (A) Absorption spectra of the substrate and product of some enzymatic reaction. Because of the high degree of spectral overlap between these two molecules, it would be difficult to quantify changes in one component of a mixture of the two species. (B) The mathematical difference spectrum (product minus substrate) for the spectra in (A). The difference spectrum highlights the differences between the two spectra, making quantitation of changes more straightforward.

Regardless of the type of cuvette selected, its path length must be known to ensure correct application of Beer's law to the measurements. Usually, the manufacturer provides this information at the time of purchase. If, for any reason, the path length of a particular cuvette is not known with certainty, it can be determined experimentally by measuring the absorption of any stable chromophoric solution in a standard 1 cm path length cuvette and then measuring the absorption of the same sample in the cuvette of unknown path length. The ratio of the two absorption readings ($A_{\text{unknown}}/A_{1\,\text{cm}}$) yields the path length of the unknown cuvette in centimeters. A good standard solution for this purpose can be prepared as follows (Haupt, 1952). Weigh out 0.0400 g of potassium chromate and 3.2000 g of potassium hydroxide and dissolve both in 700 mL of distilled water. Transfer to a one-liter volumetric flask and bring the total volume to 1000 mL with additional distilled water. Mix the solution well. The resulting solution will have an absorption of 0.991 at 375 nm in a 1 cm cuvette.

Many workers now perform absorption measurements with 96-well microtiter plate readers. These devices have remarkable sensitivity (many can measure changes in absorption of as little as 0.001) and can greatly increase productivity by allowing up to 96 samples to be assayed simultaneously. Most plate readers are based on optical filters rather than monochromators, hence providing only a finite number of analytical wavelengths for measurements. It is usually not difficult, however, to purchase an optical filter for a particular wavelength and install it in the plate reader. Because most commercially available microtiter plates are constructed of plastic, plate reader measurements are normally restricted to the visible region of the spectrum.

Recently, however, monochromator-based plate readers have come on the market that allow one to make measurements at any wavelength between 250 and 750 nm. The manufacturers of these instruments also provide special quartz-bottom plates for performing measurements in the ultraviolet. As with nonstandard cuvettes, it is necessary to determine the path length through which one is making measurements in a microtiter plate well. The path length in this situation will depend on the total volume of reaction mixture in the well. For example, a 200 μL solution in a well of a 96-well plate has an approximate path length of 0.7 cm. The exact path length under given experimental conditions should be determined empirically, as described earlier. Also, filter-based plate readers may show an apparent extinction coefficient of a molecule that differs from corresponding measurements made with a conventional spectrometer. Such discrepancies are due in part to the broader bandwidths of the optical filters used in plate readers. The absorption readings thus survey a wider range of wavelengths than the corresponding spectrometer measurements. Therefore, it is important to determine a standard curve of absorption as a function of chromophore concentration using the plate reader under the same conditions as the experimental measurements.

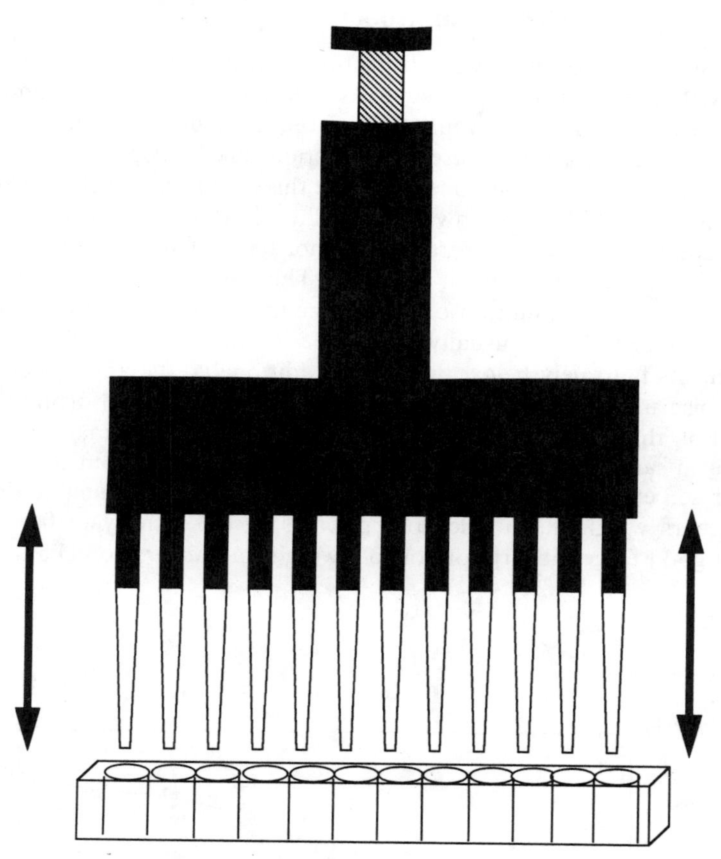

Figure 6.7 Schematic illustration of a multichannel pipettor, suitable for simultaneously mixing up to 12 samples in a 96-well microtiter plate.

Mixing of samples can be problematic in a 96-well plate, since the inversion method (Figure 6.2) cannot conveniently be applied. Instead, many commercial plate readers have automatic built-in shaking devices that allow for sample mixing. Another way to ensure good mixing is to alter the general method just described as follows. The components of the reaction mixture, except for the initiating reagent, are mixed together in sufficient volume for all the wells of the plate to be used. This solution is placed in a multichannel pipette reservoir. The initiating solution is added to the individual wells of the plate, and the remainder of the reaction solution is added to the wells using a multichannel pipette. The solutions are mixed by repeatedly pulling up and dispensing the reaction mixture with the pippettor (Figure 6.7). An entire row of 12 wells can be mixed in this way in less than 10 seconds.

6.2.5 Errors in Absorption Spectroscopy

A common error associated with absorption measurements is deviation from Beer's law. The form of Beer's law suggests that the absorption of a sample will increase linearly with the concentration of the molecule being analyzed, and indeed, this is the basis for the use of absorption spectroscopy as an analytical tool. Experimentally, however, one finds that this linear relationship holds only over a finite range of absorption values. As illustrated in Figure 6.8, absorption readings greater that 1.0 in general should not be trusted to accurately reflect the concentration of analyte in solution. Thus, experiments should be so designed that the maximum absorption to be measured is less than 1.0. With a few preliminary trials, it usually is possible to adjust conditions so that the measurements fall safely below this limit. Additionally, the amount of instrumental noise in a measurement is affected by the overall absorption of the sample. For this reason it is more difficult to measure a small absorption change for a sample of high absorption. Empirically it turns out that the best compromise between minimizing this noise and having a reasonable signal to follow occurs when the sample absorption is in the vicinity of 0.5. This is usually a good target absorption for following small absorption changes.

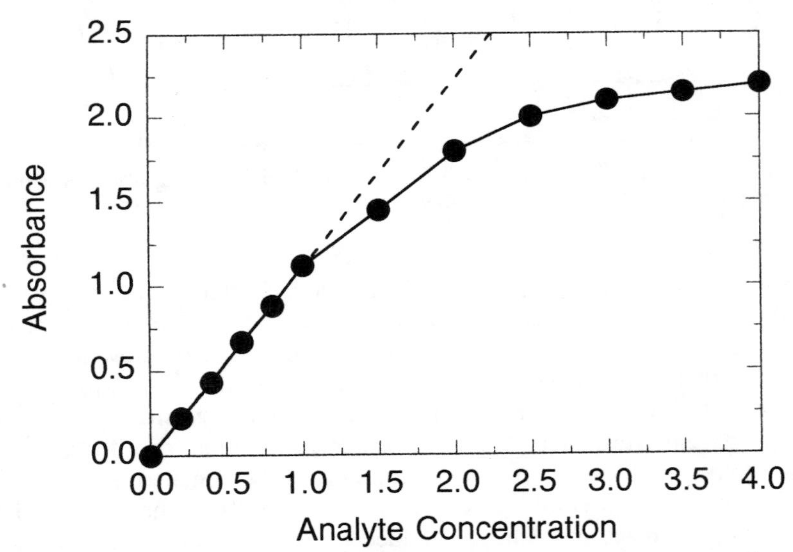

Figure 6.8 Deviation from Beer's law. Over a small concentration range, the absorption at some analytical wavelength tracks linearly with analyte concentration, as expected from Beer's law (Equation 6.2). When the analyte concentration increases to the point at which $A = 1.0$, however, significant deviations from this straight-line behavior begin to appear.

The lamps used to generate the UV and visible light for absorption spectrometers must be given ample time to warm up. The light intensity from these sources varies considerably shortly after the lamps are turned on, but stabilizes after about 30–90 minutes. Since the amount of warm-up time needed to stabilize the lamp output will vary from instrument to instrument, and from lamp to lamp within the same instrument, it is best to determine the required warm-up time for one's own instrument. This is easily done by measuring the signal from a sample of low absorption (say, $\approx 0.05–0.1$) as a function of time after turning the lamp on, and noting how long it takes for the signal to reach a stable, constant reading.

Another source of error in absorption measurements is sample turbidity. Particulate matter in a solution will scatter light that is detected as increased absorption by the sample. If settling of such particles occurs during kinetic measurements, it can lead to significant noise in the data, and in severe cases there will appear to be an additional kinetic component of the data. The best way to avoid these complications is to ensure that the sample is free of particles by filtering all the solutions through $0.2\,\mu m$ filters or by centrifugation (see Copeland, 1994, for further details).

6.2.6 Fluorescence Measurements

Light of an appropriate wavelength can be absorbed by a molecule to cause an electronic transition from the ground state to some higher lying excited state, as we have discussed. Because of its highly energetic nature, the excited state is short-lived (excited state lifetimes are typically less than 50 ns), and the molecule must find a means of releasing this excess energy to return to the ground state electronic configuration. Most of the time this excess energy is released through the dissipation of heat to the surrounding medium. Some molecules, however, can return to the ground state by emitting this excess energy in the form of light. Fluorescence, the most common and easily detected of these emissive processes, involves singlet excited and ground electronic states. The energetic processes depicted in Figure 6.5 are characteristic of molecular fluorescence. First, light of an appropriate wavelength is absorbed by a molecule, exciting it to a higher lying electronic state (Figure 6.5A). The molecule then decays through the various high energy vibrational substates of the excited electronic state by heat dissipation, finally relaxing from its lowest vibrational level to the ground electronic state with release of a photon (Figure 6.5B).

Because of the differences in equilibrium interatomic distances between the ground and excited states, and because of the loss of energy during the decay throughout the higher energy vibrational substates, the emitted photon is far less energetic than the corresponding light energy required to excite the molecule in the first place. For these reasons, the fluorescence maximum of a molecule is always at longer wavelength (less energy) than the absorption maximum; this difference in wavelength between the absorption and fluorescence maxima of a molecule is referred to as the Stokes shift. For example, the amino acid tryptophan absorbs light maximally at about 280 nm and fluoresces strongly between 325 and 350 nm (Copeland, 1994). To take advantage of this behavior, fluorescence instruments are designed to excite a sample in a cuvette with light at the wavelength of maximal absorption and detect the emitted light at a different (longer) wavelength. To best detect the emitted light with minimal interference from the excitation light beam, most commercial fluorometers are designed to collect the emitted light at an angle of 90° from the excitation beam path. Thus, unlike cells for absorption spectroscopy, fluorescence cuvettes must have at least two optical quality windows at right angles to each other; most fluorescence cuvettes have all four sides made of polished optical surfaces.

The strategies for following enzyme kinetics by fluorescence are similar to what has been described for absorption spectroscopy. Many enzyme substrates and products are naturally fluorescent and provide convenient signals with which to follow their loss or production in solution. If these molecules are not naturally fluorescent, it often is possible to covalently attach a fluorescent group without significantly perturbing the interactions with the enzyme under study. Fluorescence measurements offer two key advantages over absorption measurements for following enzyme kinetics. First, fluorescence instruments are very sensitive, permitting the detection of much lower concentration changes in substrate or product. Second, since many fluorophores have large Stokes shifts, the fluorescence signal is typically in an isolated region of the spectrum, where interferences from signals due to other reaction mixture components are minimal.

Fluorescence signals track linearly with the concentration of fluorophore in solution over a finite concentration range. In principle, fluorescence signals should vary with fluorophore concentration by a relationship similar to Beer's law, where the extinction coefficient is replaced by the molar quantum yield (ϕ). In practice, however, it is difficult to calculate sample concentrations by means of applying tabulated values of ϕ to experimental fluorescence measurements. This limitation is in part due to the nature of the instrumentation and the measurements (see Lackowicz, 1983, for more details). Thus, to convert fluorescence intensity measurements into concentration units, it is necessary to prepare a standard curve of fluorescence signal as a function of fluorophore concentration, using a set of standard solutions for which the fluorophore concentration has been determined independently. The standard curve data points must be collected at the same time as the experimental measurements, however, since day-to-day variations in lamp intensity and other instrumental factors can greatly affect fluorescence measurements.

Sometimes the fluorophore is generated only as a result of the enzymatic reaction, and it is difficult to obtain a standard sample of this molecule for construction of a standard curve. In such cases it may not be possible to report velocity in true concentration units, and units of relative fluorescence per unit time must be used, instead. It is still important to quantify this fluorescence relative to some standard fluorescent molecule, to permit comparison of relative fluorescence measurements from one day to the next and from one laboratory to another. A good standard for this purpose is quinine sulfate. A dilute solution of quinine sulfate in an aqueous sulfuric acid solution can be excited at any wavelength between 240 and 400 nm to yield a strong fluorescence signal that maximizes at 453 nm (Fletcher, 1969). Russo (1969) suggests the following protocol for preparing a quinine sulfate solution as a standard for fluorescence spectroscopy:

- Weigh out 5 mg of quinine sulfate dihydrate and dissolve in 100 mL of 0.1 NH_2SO_4.
- Measure the absorption of the sample at 366 nm, and adjust the concentration with 0.1 NH_2SO_4 so that the solution has an absorption of 0.40 at this wavelength in a 1 cm cuvette.
- Dilute a sample of this solution 1:10 with 0.1 NH_2SO_4 and use the solution to record the fluorescence spectrum.

The relative fluorescence of a sample can then be reported as the fluorescence intensity of the sample at some wavelength, divided by the fluorescence intensity of the quinine sulfate standard at 453 nm, when both sample and standard are excited at the same wavelength using the same fluorometer. Of course, both sample and standard measurements must be made under the same set of experimental conditions (monochrometer slit width, lamp voltage, dwell time, etc.), and the second set should be made soon after the first.

6.2.7 Internal Fluorescence Quenching and Energy Transfer

If a molecule absorbs light at the same wavelength at which another molecule fluoresces, the fluorescence from the second molecule may be absorbed by the first molecule, leading to a diminution or *quenching* of the observed fluorescence intensity from the second molecule. (Note that this is only one of numerous means of quenching fluorescence; see Lackowicz, 1983, for a more comprehensive treatment of fluorescence quenching.) The first molecule may then decay back to its ground state by radiationless decay (e.g., heat dissipation) or may itself fluoresce at some characteristic wavelength. We refer to the first process as "quenching," because the net effect is a loss of fluorescence intensity. The second situation is referred to as "energy transfer" because here excitation at the absorption maximum of one molecule leads to fluorescence by the other molecule (Figure 6.9).

Both quenching and energy transfer depend on several factors, including the spatial proximity of the two molecules. This property has been exploited to develop fluorescence assays for proteolytic enzymes based on synthetic peptide substrates. The basic strategy is to incorporate a fluorescent group (the donor) into a synthetic peptide on either the N- or C-terminal side of the scissile peptide bond that is recognized by the target enzyme. A fluorescence quencher or energy acceptor molecule (both referred to hereafter as the acceptor

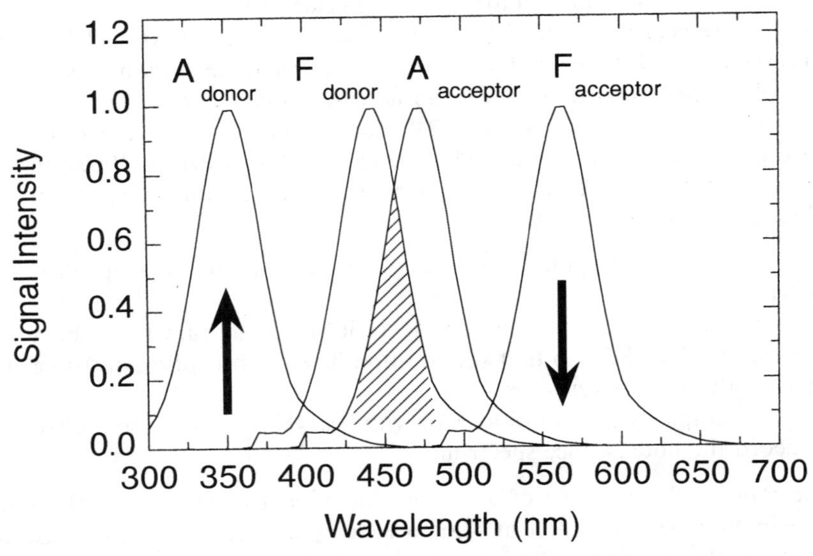

Figure 6.9 Resonance energy transfer. In an energy transfer experiment, the sample is excited with light at the wavelength of the donor absorption band (A_{donor}) to induce fluorescence of the donor molecule at the wavelengths indicated by F_{donor}. The absorption band of the acceptor molecule ($A_{acceptor}$) occurs in a wavelength range at which it overlaps with the fluorescence band of the donor; the area of overlap between these two features is shown by the hatched region. Because of this spectral overlap (and other factors) the light that would have been emitted as F_{donor} is reabsorbed by $A_{acceptor}$. This indirect excitation of the acceptor molecule can lead to fluorescence by the acceptor at the wavelengths corresponding to $F_{acceptor}$. Experimentally, excitation occurs at the wavelength indicated by the up-pointing arrow, and the fluorescence signal is measured at the wavelength indicated by the down-pointing arrow.

molecule) is also incorporated into the peptide on the other side of the scissile bond. When the peptide is intact, the donor and acceptor molecules are covalently associated and remain a relatively fixed distance apart, able to energetically interact. Once hydrolyzed by the enzyme, however, the two halves of the peptide will diffuse away from each other, thus eliminating the possibility of any interaction between the donor and acceptor. The observed effect of this hydrolysis will be an increase in the fluorescence from the donor molecule, and, in the case of energy transfer, a concomitant decrease in the fluorescence of the acceptor molecule with excitation under the absorption maximum of the donor.

These approaches have been used to follow hydrolysis of peptide substrates for a large variety of proteases (e.g., see Matayoshi et al., 1990; Knight et al., 1992). One example will suffice to illustrate the basic approach used in these assays. Knight et al. (1992) described the incorporation of the fluorescent molecule 7-methoxycoumarine-4-yl acetyl (MCA) at the N-terminus of a peptide designed to be a substrate for the matrix metalloprotease stromelysin; then, immediately after the scissile Gly-Leu peptide bond that is hydrolyzed by the enzyme, the quencher N-3-(2,4-dinitrophenyl)-L-2,3-diaminopropionyl (DPA) was incorporated, as well. The complete peptide sequence is:

MCA-Pro-Leu-Gly-Leu-DPA-Ala-Arg-NH_2

MCA absorbs maximally at 328 nm and fluoresces maximally at 393 nm. The DPA group has a strong absorption band at 363 nm with a prominent shoulder at 410 nm. This shoulder overlaps with the fluorescence band of MCA and leads to significant fluorescence quenching; a 1 μM solution of MCA-Pro-Leu-Gly (the product of enzymatic hydrolysis) was found to be 130 times more fluorescent than a comparable solution of the MCA-Pro-Leu-Gly-Leu-DPA-Ala-Arg-NH_2, with excitation and emission at 328 and 393 nm, respectively (Knight et al., 1992). Enzymatic hydrolysis of this peptide results in separation of the MCA and DPA groups, hence a large increase in MCA fluorescence. This fluorescence increase could be followed over time as a measure of the reaction velocity, allowing the investigators to establish the values of k_{cat}/K_m of this substrate for several members of the matrix metalloprotease family. This assay also has been used recently to determine the potency of potential inhibitors of stromelysin by measuring the effects of the inhibitors on the initial velocity of the enzyme reaction (Copeland et al., 1995).

6.2.8 Errors in Fluorescence Measurements

Most of the caveats described for absorption spectroscopy hold for fluorescence measurements as well. Samples must be free of particulate matters, since light scattering is a severe problem in fluorescence. Many of the commonly used fluorophores emit light in the visible region but must be excited at wavelengths in the near-ultraviolet, necessitating the use of quartz cuvettes for these measurements. Also, any fluorescence due to buffer components and so on must be measured and corrected for to ensure that meaningful data are obtained.

In addition to these more common considerations are several sources of error unique to fluorescence measurements. First, many fluorescent molecules are prone to photodecomposition after long exposure to light. Hence, fluorescent substrates and reagents should be stored in amber glass or plastic, and the containers should be wrapped in aluminum foil to minimize exposure to environmental light. Second, the quantum yield of fluorescence for any molecule is highly dependent on sample temperature. We shall see shortly that temperature effects enzyme kinetics directly, but this effect is distinct from the general influence of temperature on fluorescence intensity. In general the fluorescence signal increases with decreasing temperature, as competing nonradiative decay mechanisms for return to the ground state become less efficient. Hence, good temperature control of the sample must be maintained. Most commercial fluorometers provide temperature control by means of jacketed sample holders that attach to circulating water baths.

Finally, a major source of error in fluorescence measurements is light absorption by the sample at high concentrations. Individual molecules in a sample may be excited by the excitation light beam and caused to fluoresce. To be detected, these emitted photons must traverse the rest of the sample and escape the cuvette to impinge on the surface of the detection device (typically a photomultiplier tube or diode array). Any such photon will be lost from detection, however, if before escaping the sample it encounters another molecule that is capable of absorbing light at that wavelength. As sample concentration increases, the likelihood of such encounters and instances of light reabsorption increases exponentially. This phenomenon, referred to as the *inner filter effect* (Figure 6.10), can dramatically reduce the fluorescence signal observed from a sample. Consider Figure 6.11, which plots the apparent fluorescent product yield after a fixed amount of reaction time as a function of substrate concentration for the fluorogenic MCA/DPA peptide described in Section 6.2.7 in an assay for stromelysin activity. Instead of the rectangular hyperbolic fit expected from the Henri–Michaelis–Menten equation (Chapter 5), we observe an initial increase in fluorescence yield with increasing substrate, followed by a rapid diminution of signal as the substrate concentration is further increased. At first glance, this behavior might appear to be the result of substrate or product inhibition, as described in Chapter 5, In this case, however, the loss of fluorescence at higher substrate concentrations is an

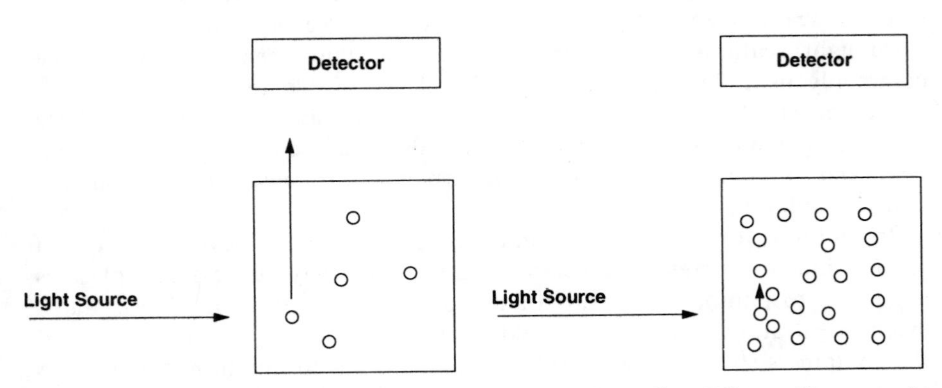

Figure 6.10 Schematic diagram illustrating the inner filter effect. When a dilute sample (left) of a fluorescent molecule is excited at an appropriate wavelength, a detector stationed at 90° relative to the excitation source will detect the emitted light that emerges from the sample container. If, however, the sample is very concentrated, emitted light from one molecule in a sample can encounter and be reabsorbed by another molecule before emerging from the sample compartment (right). The reabsorbed photons, of course, will not be detected. The likelihood of this self-absorption, or inner filter effect, increases with increasing sample concentration.

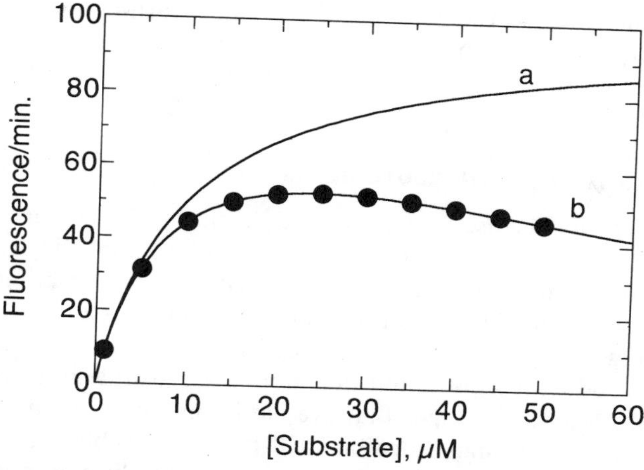

Figure 6.11 Errors in enzyme kinetic measurements due to fluorescence inner filter effects. Here the rate of fluorescence from a fluorescent peptide substrate of stromelysin is plotted as a function of substrate concentration. Instead of observing the rectangular hyperbolic behavior expected from the Henri–Michaelis–Menten equation (curve a), we see a diminution of the expected signal at high substrate concentrations (curve b). One might interpret this artifact as substrate or product inhibition, but in this case the deviation is due to the inner filter effects that become significant at high substrate concentrations. The correct interpretation can be reached by measuring the fluorescence of the higher substrate sample at several dilutions, as discussed in the text.

artifact of the inner filter effect. This can be verified by remeasuring the fluorescence of the higher substrate samples after a large dilution with buffer. If, for example, the sample were diluted 20-fold with buffer, the observed fluorescence would not be 20-fold less than that of the undiluted sample; rather, it would show much higher fluorescence than expected on the basis of the dilution factor.

The inner filter effect can be corrected for if the absorption of the sample is known at the excitation and emission wavelengths used in the fluorescence measurement. The true, or corrected, fluorescence F_{corr} can be calculated from the observed fluorescence F_{obs} as follows (Lackowicz, 1983):

$$F_{corr} = F_{obs} \times 10^{\left(\frac{A_{ex} + A_{em}}{2}\right)} \tag{6.4}$$

where A_{ex} and A_{em} are the sample absorptions at the excitation and emission wavelengths, respectively. This correction works only over a limited sample absorption range. If the sample absorption is greater than about 0.1, the correction will not be adequate. Hence, a good rule of thumb is to begin with samples that have absorption values of about 0.05 at the excitation wavelength.

The sample concentration can be adjusted from this starting point to optimize the signal-to-noise ratio, with care taken to not introduce a significant inner filter effect.

6.2.9 Radioisotopic Measurements

The basic strategy for the use of radioisotopes in enzyme kinetic measurements is to incorporate into the structure of the substrate a radioactive species that will be retained in the product molecule after catalysis. Using an appropriate technique for separating the substrate from the product (see Section 6.3 on separation methods), one can measure the amount of radioactivity in the substrate and product fractions, and thus quantify substrate loss and product production. Most of the isotopes that are used commonly in enzyme kinetic measurements decay through emission of β particles (Table 6.2). The decay process follows first-order kinetics, and the loss (or disintegration) of the starting material is thus associated with a characteristic half-life for the parent isotope. The standard unit of radioactivity is the curie (Ci), which was originally defined as the rate at which one gram of radium-226 decays completely. Relating this to other isotopes, a more useful working definition of the curie is the quantity of any substance that decays at a rate of 2.22×10^{12} disintegrations per minute (dpm).

Solutions of *p*-terphenyl or stilbene, in xylene or toluene, will emit light when in contact with a radioactive solute. This light emission, known as *scintillation*, is most commonly measured with a scintillation counter, an instrument designed around a photomultiplier tube or other light detector. Radioactivity on flat surfaces, such as thin layer chromatography (TLC) plates and gels can be measured by scintillation counting after scraping or cutting out the portion of the surface containing the sample and immersing it in scintillation fluid. Another common means of detecting radioactivity on such surfaces entails placing the surface in contact with a sheet of photographic film. The radioactivity darkens the film, making a permanent record of the location of the radioactive species on the surface. This technique, known as *autoradiogra-*

Table 6.2 Properties of Radioisotopes Commonly Used in Enzyme Kinetic Assays

Isotope	Decay Process	Half-life
Carbon-14	${}_{6}^{14}\mathrm{C} \rightarrow {}_{-1}\beta^0 + {}_{7}^{14}\mathrm{N}$	5700 years
Phosphorus-32	${}_{15}^{32}\mathrm{P} \rightarrow {}_{-1}\beta^0 + {}_{16}^{32}\mathrm{S}$	14.3 days
Sulfur-35	${}_{16}^{35}\mathrm{S} \rightarrow {}_{-1}\beta^0 + {}_{17}^{35}\mathrm{Cl}$	87.1 days
Tritium	${}_{1}^{3}\mathrm{H} \rightarrow {}_{-1}\beta^0 + {}_{2}^{3}\mathrm{He}$	12.3 years

phy, is one of the oldest methods known for detecting radioactivity. Today computer-interfaced phosphor imaging devices also are commonly used for locating and quantifying radioactivity on two-dimensional surfaces (dried gels, TLC plates, etc.

Radioactivity in a sample is quantified by measuring the dpm's of a sample using one of the methods just described. Regardless of how the detection is accomplished, no detector is 100% efficient. Therefore any instrumental readings obtained experimentally will differ from the true dpm's of the sample. The experimental units of radioactivity are referred to as counts per minute (cpm's: events detected or counted by the instrument per minute). For example, a 1 μCi sample would display 2.22×10^6 dpm. If the detector used to measure this sample had an efficiency of 50%, the experimental value obtained would be 1.11×10^6 cpm. To convert this experimental reading into true dpm's, it would be necessary to measure a standard sample of the isotope of interest, of known dpm's. This would permit the efficiency of the instrument to be calibrated and the cpm values of samples to be converted readily into dpm units.

When radiolabeled substrates are used in enzyme kinetic studies, the labeled substrate is usually mixed with "cold" (i.e., unlabeled) substrate to achieve a particular total substrate concentration without having to use high quantities of radioactivity. It is important, however, to quantify the proportion of radiolabeled molecules in the substrate sample. Quantification is commonly expressed in terms of the specific radioactivity of the sample. (*Note*: This "specific activity" refers to the radioactivity of the sample and should not be confused with the specific activity of an enzyme sample, which is defined later.) Specific radioactivity is given in units of radioactivity per mass or molarity of the sample. Common units of specific radioactivity are: dpm/μmol and μCi/mg. With the specific radioactivity of a substrate sample defined, one can easily convert into velocity units radioactivity measurements taken during an enzymatic reaction.

Table 6.3 Results of a Hypothetical Reaction of Dihydroorotate Dehydrogenase with [^{14}C] Dihydroorotate Substrate

Incubation Time (min)	Radioactivity (cpm)
10	980
20	2010
30	3050
40	3900
50	5103
60	5952

A worked example will illustrate the foregoing concept. Suppose that we wish to study the conversion of dihydroorotate to orotic acid by the enzyme dihydroorotate dehydrogenase. Let us say that we have obtained a ^{14}C-labeled version of the substrate dihydroorotate and have mixed it with cold dihydroorotate to prepare a stock substrate solution with a specific activity of 1000 cpm/nmol. The final concentration is 1 mM substrate in a reaction mixture with a total volume of 100 μL. Let us say that we initiate the reaction with enzyme and allow the reaction mixture to incubate at 37°C. Every 10 minutes we remove 10 μL of the reaction mixture and add this to 10 μL of 6 N HCl to denature the enzyme, thus stopping the reaction. The total 20 μL is then spotted onto a TLC plate and the product is separated from the substrate. Let us say that we scraped the product spot from the TLC plate and measured its radioactivity by scintillation counting (in this hypothetical assay, a control sample of 1 nmol of substrate in the reaction mixture buffer is spotted onto the TLC plate, the substrate spot is scraped off and counted, and the reading is 1000 cpm.) Tables 6.3 and Figure 6.12 show the results of this sequence of steps.

From Figure 6.12 we see that the slope of our plot of cpm versus time is 100 cpm/min. Since our substrate had a specific radioactivity (SRA) of 1000 cpm/nmol, this slope value can be converted directly into a velocity value of 0.1 nmol of product/min. Since the volume of reaction mixture spotted onto the TLC plate per measurement was 10 μL (i.e., 1×10^{-5} L), the velocity in molarity units is obtained by dividing the velocity in nanomoles of product per minute by 1×10^{-5} L to yield a velocity of 10 μM/min. These calculations are

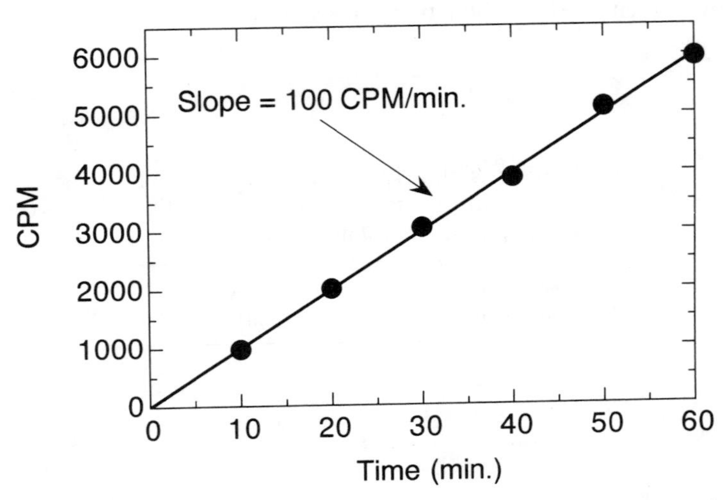

Figure 6.12 Radioassay for dihydroorotate dehydrogenase, measuring the incorporation of ^{14}C into the product, orotic acid.

summarized as follows:

$$\frac{\text{Slope}}{\text{SRA}} = \frac{\text{cpm/min}}{\text{cpm/nmol}} = \frac{\text{product mass}}{\text{unit time}} = \text{nmol/min}$$

$$\frac{\text{nmol/min}}{\text{L}} = \frac{\text{product mass/unit time}}{\text{reaction volume}} = \frac{\text{molarity}}{\text{unit time}} = \mu\text{M/min}$$

As our example illustrates, good bookkeeping is essential in these assays. The amount of total substrate used will be dictated by the purpose of the experiment and its K_m for the enzyme. The specific radioactivity, on the other hand, should be adjusted to ensure that the amount of radioactivity used is the minimum that will provide good signal-over-background readings. Guidlines for sample preparations using different radioisotopes can be found in the reviews by Oldham (1968, 1992). The other point illustrated in our example is that good postreaction separation of labeled substrate and product molecules is critical to the use of radiolabels for following enzyme kinetics.

Radiolabeled substrates are commonly used in conjuction with chromatographic and electrophoretic separation methods. When the substrate and/or product is a protein, as in some assays for kinases and proteases, bulk, precipitation or capture on nitrocellulose membranes can be used to separate the macromolecule from the other solution components. Some of these methods are discussed separately in Section 6.3.1.

6.2.10 Errors in Radioactivity Measurements

Aside from errors associated with bookkeeping, the most commonly encountered cause of inaccurate radioactivity measurements is self-absorption. When the separation method used in conjunction with the assay involves a solid separation medium, such as paper or thin layer plate chromatography, gel electrophoresis, or capture on activated charcoal, the solid material in the sample may absorb some of the emitted radiation, preventing the signals from reaching the detection device. This self-absorption is best corrected for by measuring all samples and standards at a constant density in terms of milligrams of material per milliliters. Segal (1976) suggests using an inert material such a gelatin to adjust the density of all samples for this purpose. Because scintillation counting measures light emission, the same interferences discussed for fluorescence measurements can occur. In particular, if the sample is highly colored, quenching of the signal due to the equivalent of an inner filter effect may be observed. When possible, this should be corrected for by adjusting the optical density of the samples and standards with a similarly colored inert material (Segal, 1976).

6.2.11 Other Detection Methods

Absorption, fluorescence and radioactivity are by far the most common means of following enzyme kinetics, but a wide variety of other techniques have been utilized as well. Immunologic detection, for example, has been applied to follow proteolytic cleavage of a protein substrate by Western blotting, using antibodies raised against that protein substrate. Recently, antibodies have been developed that react exclusively with the phosphorylated forms of peptides and proteins; these reagents have been widely used to follow the enzymatic activity of the kinases and phosphatases by means of Western and dot blotting as well as ELISA-type assays. Reviews of immunologic detection methods can be found in Copeland (1994) and in Harlow and Lane (1988).

Polarographic methods have also been used extensively to follow enzyme reactions. Many oxidases utilize molecular oxygen during their turnover, and the accompanying depletion of dissolved O_2 from the solutions in which catalysis occurs can be monitored with an O_2-specific electrode. Very sensitive pH electrodes can be used to follow proton abstraction or release into solution during enzyme turnover. Enzymes that perform redox chemistry as part of their catalytic cycle can also be monitored by electrochemical means. Reviews of these methods can be found in the recent text by Eisenthal and Danson (1992).

The variety of detection methods that have been applied to enzyme activity measurements is too broad to be covered comprehensively in any one volume. Our discussion here should provide the reader with a good overview of the more common techniques employed in this field. The references given can provide more in-depth accounts. Another very good source for new and interesting enzyme assay methods, the journal *Analytical Biochemistry* (Academic Press), has historically been a repository for papers dealing with the development and improvement of enzyme assays. Finally, the series *Methods in Enzymology* (Academic Press) comprises volumes dedicated to in-depth reviews of varying topics in enzymology. This series details assay methods for many of the enzymes one is likely to work with and very frequently will indicate at least an assay for a related enzyme that can serve as a starting point for development of an individual assay method.

6.3 Separation Methods in Enzyme Assays

For many enzyme assays the detection methods described thus far are applicable only after the substrate or product has been separated from the rest of the reaction mixture components, as occurs when the method of detection does not, by itself, discriminate between the analyte and other species. For example, if the optical properties of the substrate and products of an enzymatic reaction are similar, measuring the spectrum of the reaction mixture alone will not provide a useful means of monitoring changes in the concentration of the individual components. This section briefly describes some of the common separation techniques that are applied to enzyme kinetics. These techniques are usually combined with one of the detection methods already covered, to develop a useful assay for the enzyme of interest.

6.3.1 Separation of Proteins from Low Molecular Weight Solutes

A number of assay strategies involve measuring the incorporation of a radioactive or optical label into a protein substrate, or the release of a labeled peptide fragment from the protein. For these assays it is convenient to separate the protein from the bulk solution prior to detection. This can be accomplished in several ways. Proteins can be precipitated out of solution by addition of strong acids or organic denaturants, followed by centrifugation. A 10% solution of trichloroacetic acid (TCA) is commonly used for this purpose (see Copeland, 1994, for details). For dilute protein solutions ($< 5\,\mu g$ total), the TCA is often supplemented with the detergent deoxycholate (DOC) to effect more efficient precipitation. Proteins can also be precipitated by high concentrations of ammonium sulfate; most proteins precipitate from solution when the ammonium sulfate concentration is at 80% of saturation. Organic solvents such as acetone, acetonitrile, methanol, or combinations of these solvents also are used to denature and precipitate proteins. For example, mixing $100\,\mu L$ of an aqueous protein solution with $900\,\mu L$ of a 1:1 acetone/acetonitrile mixture will precipitate most proteins with good efficiency. In addition to these general precipitation methods, specific proteins can be separated from solution with an immobilized antibody (e.g., an antibody linked to an agarose bead) that has been raised against the target protein (Harlow and Lane, 1988).

Proteins also can be separated from low molecular weight solutes by selective binding to nitrocellulose membranes. Nitrocellulose and certain other membrane materials bind proteins strongly, while passing the other components of the solution. Hence, one can capture the protein molecules in a solution by filtration or centrifugation through a nitrocellulose membrane. For example, many kinase assays are based on the enzymatic incorporation of ^{32}P into a protein substrate. After incubation, the protein substrate is captured on a disk filter of nitrocellulose. After the filter has been washed to remove adventitiously bound ^{32}P, the radioactivity that is retained on the filter can be measured by scintillation counting. Of course the kinase, being a protein itself, is also captured on the nitrocellulose by this method. Since, however, the mass of enzyme in a typical assay is very small relative to that of the protein substrate, the background due to any radiolabel on the enzyme is insignificant and can be subtracted out by performing the appropriate control measurements. A similar strategy can be used with nominal molecular weight cutoff filters (Figure 6.13A). These filters, which come in a variety of formats, are constructed of a porous material whose pore size distribution permits passage

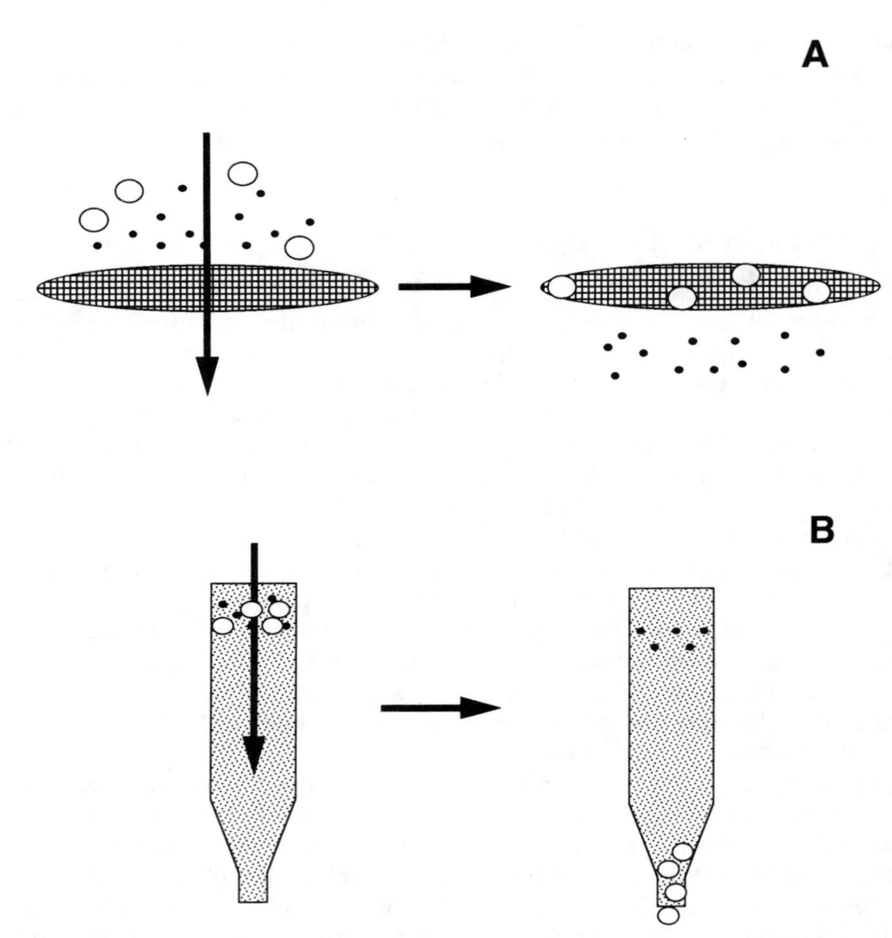

Figure 6.13 Strategies for separating proteins from small molecular weight components in solution. (A) filter binding method and (B) size exclusion chromatography.

only of molecules with molecular weight below some critical value; larger molecular weight species, such as proteins, are retained. One word of caution is in order with regard to these filters: the molecular weight cutoffs quoted by the manufacturers represent the median value of a normal distribution of filtrate molecular weights; thus to avoid significant losses, it is best to use a filter that has a much lower molecular weight cutoff than the molecular weight of the protein being studied. Manufacturers' descriptions of individual filters should be carefully read before use.

Another method for separating proteins from low molecular weight molecules is size exclusion chromatography (Figure 6.13B). Small disposable size exclusion columns, commonly referred to as *desalting columns*, are commercially available for removing low molecular weight solutes from protein solutions.

The resins used in these columns are chosen to ensure that macromolecules, such as proteins, will elute at the void volume of the column; low molecular weight solutes, such as salts, elute much later. Desalting columns typically are run by gravity, since the large molecular weight differences between the proteins and small solutes allows for separation without the need for high chromatographic resolution.

These and other methods for separating macromolecules from low molecular weight solutes have been described in greater detail in Copeland (1994) and references therein.

6.3.2 Chromatographic Separation Methods

The three most commonly used chromatographic separation methods in modern enzymology laboratories are paper chromatography, TLC, and HPLC. Before HPLC instrumentation became widely available, the paper and thin layer modes of chromatography were very commonly used for the separation of low molecular weight substrates and products of enzymatic reactions. Today these methods have largely been replaced by HPLC. An exception consists of separations involving radiolabeled low molecular weight substrates and products. Since paper and TLC separating media are disposable, and the separation can be performed in a restricted area of the laboratory, these methods are still preferable for work involving radioisotopes.

The theory and practice of paper chromatography and TLC probably are familiar to most readers of this text from courses in general and organic chemistry. Basically, separation is accomplished through the differential interactions of molecules in the sample with ion exchange or silica-based resins that are coated onto paper sheets or plastic or glass plates. A capillary tube is used to spot samples onto the medium at a marked location near one end of the sheet, which is placed in a developing tank with some solvent system (typically a mixture of aqueous and organic solvents) in contact with the end of the sheet closest to the spotted samples (Figure 6.14, steps 1 and 2). The tank is sealed, and the solvent moves up the sheet through capillary action, bringing different solutes in the sample along at different rates depending on their degree of interaction with the stationary phase media components. After a fixed time the sheet is removed from the tank and dried. The locations of solutes that have migrated during the chromatography are observed by autoradiography, by illuminating the sheet with ultraviolet light, or by spraying the sheet with a chemical (e.g., ninhydrin) that will react with specific solutes to form a colored spot (Figure 6.14, step 3). The spot locations are then marked on the sheet, and the spots can be cut out or scraped off for counting in a scintillation counter. Alternatively, the radioactivity of the entire sheet can be quantified by two-dimensional radioactivity scanners, as described earlier.

In our discussion of radioactivity assays, we used the example of a TLC-based assay for following the conversion of [^{14}C]dihydroorotate to [^{14}C]orotic acid by the enzyme dihydroorotate dehydrogenase. Figure 6.15

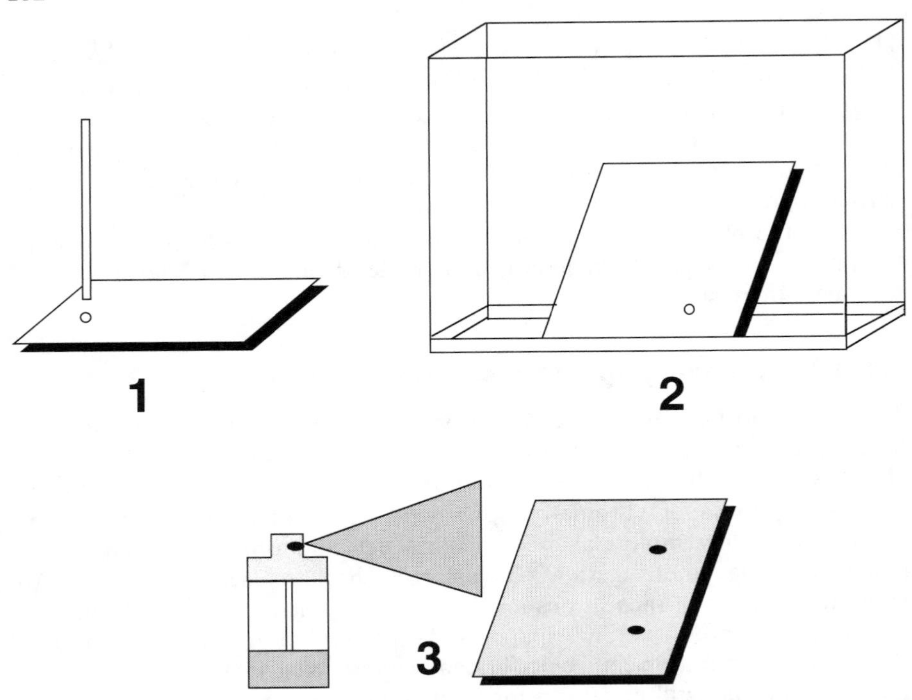

Figure 6.14 Schematic diagram of a TLC-based enzyme assay. In step 1, a sample of reaction mixture is spotted onto the TLC plate. Next the plate is dried and placed in a development tank (step 2) containing an appropriated mobile phase. After the chromatography, the plate is removed from the tank and dried again. The locations of substrate and/or product spots are then detected by, for example, spraying the plate with an appropriate visualizing stain (step 3), such as ninhydrin.

shows the separation of these molecules on TLC and their detection by autoradiography. This figure, along with the example given in Section 6.2.9, well illustrates the use of TLC-based assays. More complete descriptions of the uses of paper chromatography and TLC in enzyme assays can be found in the reviews by Oldham (1968, 1977).

HPLC has been used extensively to separate low molecular weight substrates and products, as well as the peptide-based substrates and products of proteolytic enzymes. The introduction of low compressibility resins, typically based on silica, has made it possible to run liquid chromatography at greatly elevated pressures. At these high pressures (as much as 5000 psi), resolution is greatly enhanced, thus much faster flow rates can be used, and the time required for a chromatographic run is shortened. With modern instrumentation, a typical HPLC separation can be performed in less than an hour. The three most commonly used separation mechanisms used in enzyme assays are reversed phase, ion exchange, and size exclusion HPLC.

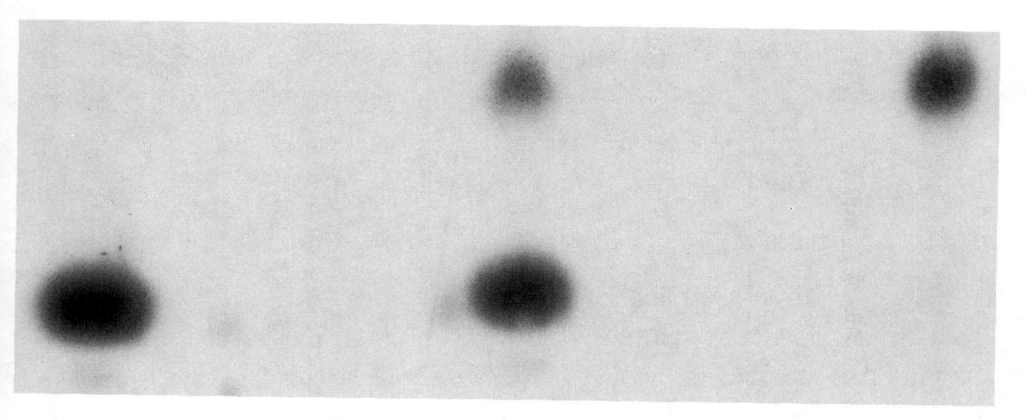

Figure 6.15 Autoradiograph of a TLC plate demonstrating separation of ^{14}C-labeled dihydroorotate and orotic acid, the substrate and product of the enzyme dihydroorotate dehydrogenase: left lane [^{14}C]dihydroorotate alone; right lane, [^{14}C]orotic acid alone; middle lane, a mixture of the two radiolabeled samples (demonstrating the ability to separate the two components in a reaction mixture).

In reversed phase HPLC, separation is based on the differential interactions of molecules with the hydrophobic surface of an alkyl silane based stationary phase. Samples are typically applied to the column in a polar solvent to maximize hydrophobic interactions with the column stationary phase. The less polar a particular solute is, the more it is retained on the stationary phase. Retention is also influenced by the carbon content per unit volume of the stationary phase. Hence a C_{18} column will typically retain nonpolar molecules more than a C_8 column, and so on. The stationary phase must therefore be selected carefully, based on the nature of the molecules to be separated. Molecules that have adhered to the stationary phase are eluted from the column in solvents of lower polarity, which can effectively compete with the analyte molecules for the hydrophobic surface of the stationary phase. Typically methanol, acetonitrile, acetone, and mixtures of these organic solvents with water are used for elution. Isocratic and gradient elutions are both commonly used, depending on the details of the separation being attempted.

A typical reversed phase separation might involve application of the sample to the column in 0.1% aqueous trifluoroacetic acid (TFA) and elution with a gradient from 100% of this solvent to 100% of a solvent composed of 70% acetonitrile, 0.085% TFA, and water. As the percentage of the organic solvent increases, the more tightly bound, hydrophobic molecules will begin to elute. As the various molecules in the sample elute from the column, they can be detected with an in-line absorption or fluorescence detector (other detection methods are used, but these two are the most common). The detector response to the elution of a molecule will produce on the strip chart a Gaussian–Lorentzian band of signal as a function of time. The length of time between

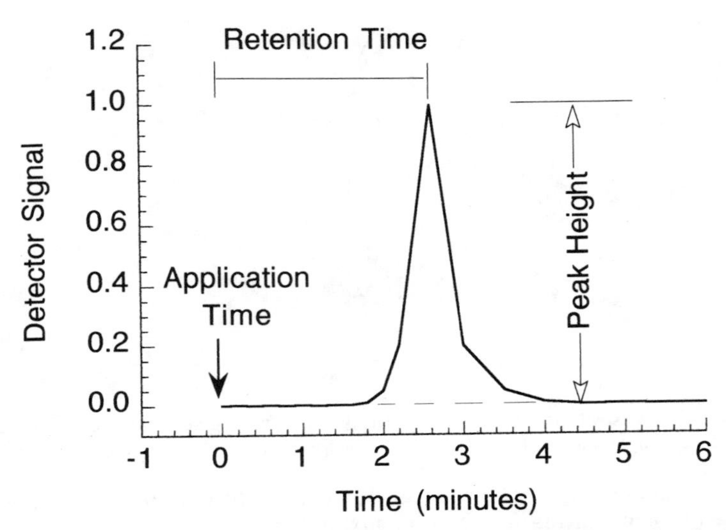

Figure 6.16 Typical signal from an HPLC chromatograph of a molecule. The sample is applied to the column at time zero and elutes, depending on the column and mobile phase, after a characteristic retention time. The concentration of the molecule in the sample can be quantified by the integrated area under the peak, or from the peak height above baseline, as defined in this figure.

application of the sample to the column and appearance of the signal maximum, the *retention time*, is characteristic of a particular molecule on a particular column under specified conditions (Figure 6.16).

To quantify substrate loss or product formation by HPLC, one typically measures the integrated area under a peak in the chromatograph and compares it to a calibration curve of the area under the peak as a function of mass for a standard sample of the analyte of interest. Let us again use the reaction of dihydroorotate dehydrogenase as an example. The substrate, dihydroorotate, and the product, orotic acid, can be purchased commercially in high purity. Ittarat et al. (1992) developed a reversed phase HPLC assay for following dihydroorotate dehydrogenase activity based on separation of dihydroorotate and orotic acid on a C_{18} column using isocratic elution with a mixed mobile phase (water/buffer/methanol) and detection by absorption at 230 nm. When a pure sample of dihydroorotate (DHO) was injected onto this column and eluted as described earlier, the resulting chromotograph displayed a single peak that eluted 4.9 minutes after injection. A pure sample of orotic acid (OA), on the other hand, displayed a single peak that eluted after 6.8 minutes under the same conditions. Using the pure samples, these workers next measured the area under the peaks for injections of varying concentrations of DHO and OA and, from these resulting data, constructed calibration curves for each of the analytes.

Note that the area under a peak will correlate directly with the mass of the analyte injected onto the column. Hence calibration curves are usually constructed with the *y* axis representing integrated peak area in some units of area [mm^2, absorption units (AU), etc.] and the *x* axis representing the injected mass of analyte in nanograms, micrograms, nanomoles, and so on. Since the volume of sample injected is known, it is easy enough to convert these mass units into standard concentration units. In this way, Ittarat et al. (1992) determined that the area under the peaks tracked linearly with concentration for both DHO and OA over a concentration range of 0–200 μM. With these results in hand, it was possible to measure the concentrations of substrate (DHO) and product (OA) in samples of a reaction mixture containing dihydroorotate dehydrogenase and a known starting concentration of substrate, as a function of time after initiating the reaction. From a plot of DHO or OA concentration as a function of reaction time, the initial velocity of the reaction could thus be determined.

With modern HPLC instrumentation, integration of peak area is performed by built-in computer programs for data analysis. If a computer-interfaced instrument is lacking, two commonly used alternative methods are available to quantify peaks from strip-chart recordings. The first is to measure the peak height rather than integrated area as a measure of analyte mass. This is done by drawing with a straightedge a line that connects the portions of the baseline on either side of the peak of interest. Next one draws a straight line, perpendicular to the *x* axis of the recording, from the peak maximum to the line drawn between the baseline points. The length of this perpendicular line can be measured with a ruler and recorded as the peak height (Figure 6.16). This procedure is repeated with each standard sample to construct the calibration curve.

The second method involves estimating the integrated area of the peak by again drawing a line between the baseline points. The two sides of the peak and the drawn baseline define an approximately triangular area, which is carefully cut from the strip-chart paper with scissors. The excised piece of paper is weighted on an analytical balance, and its mass is taken as a reasonable estimate of the relative peak area.

Obviously, the two manual methods just described are prone to greater error than the modern computational methods. Nevertheless, these traditional methods were used long before the introduction of laboratory computers and can still be used successfully when a computer is not readily available.

While reversed phase is probably the chromatographic mode most commonly employed in enzyme assays, ion exchange and size exclusion HPLC are also widely used. In ion exchange chromatography the analyte binds to a charged stationary phase through electrostatic interactions. These interactions can be disrupted by increasing the ionic strength (i.e., salt concentration) of the mobile phase; the stronger the electrostatic interactions between the analyte and the stationary phase, the greater the salt concentration of the mobile phase required to elute the analyte. Hence, multiple analytes can be separated and quantified by their differential elution from an ion exchange column.

The most common strategy for elution is to load the sample onto the column in a low ionic strength aqueous buffer and elute with a gradient from low to high salt concentration (typically NaCl or KCl) in the same buffer system. In size exclusion chromatography (also known as gel filtration), analyte molecules are separated on the basis of their molecular weights. This form of chromatography is not commonly used in conjunction with enzyme assays, except for the analysis of proteolytic enzymes when the substrate and products are peptides or proteins. For most enzymes that catalyze the reactions of small molecules the molecular weight differences between substrates and products tend to be too small to be measured by this method.

Size exclusion stationary phases are available in a wide variety of molecular weight fractionation ranges. In choosing a column for size exclusion, the ideal is to select a column for which the molecular weights of the largest and smallest analytes (i.e., substrate and product) span much of the fractionation range of the stationary phase. At the same time, the higher molecular weight analyte must lie well within the fractionation range and must not be eluted in the void volume of the column. By following these guidelines, one will obtain good separation between the analytes on the column and be able to quantify all the analyte peaks. For example, a column with a fractionation range of 8000 to 500 would be ideal to study the hydrolysis by a protease of a 5000-dalton peptide into two fragments of 2000 and 3000 Da, since all three analytes would be well resolved and within the fractionation range of the column. On the other hand, a column with a fractionation range of 5000 to 500 would not be a good choice, since the substrate molecular weight is near the limit of the fractionation range; thus the substrate peak would most likely elute with the void volume of the column, potentially making quantitation difficult. Size exclusion column packing is available in a wide variety of fractionation ranges from a number of vendors (e.g., BioRad, Pharmacia). Detailed information to guide the user in choosing an appropriate column packing and in handling and using the material correctly is provided by the manufacturers.

The analysis of peaks from ion exchange and size exclusion columns is identical to that described for reversed phase HPLC. More detailed descriptions of the theory and practice of these HPLC methods can be found in a number of texts devoted to this subject (Hancock, 1984; Oliver, 1989).

6.3.3 Electrophoretic Methods in Enzyme Assays

Electrophoresis is most often used today for the separation of macromolecules in hydrated gels of acrylamide or agarose. The most common electrophoretic technique used in enzyme assays is sodium dodecyl sulfate/polyacrylamide gel electrophoresis (SDS-PAGE), which serves to separate proteins and peptides on the basis of their molecular weights. In SDS-PAGE, samples of proteins or peptides are coated with the anionic detergent SDS to give them similar anionic charge densities. When such samples are applied to a gel, and an electric field is applied across the gel, the negatively charged proteins will migrate toward the positively charged electrode. Under these conditions, the migration of molecules toward the positive pole will be retarded by the polymer matrix of the gel, and the degree of retardation will depend on the molecular weight of the species undergoing electrophoresis. Hence, large molecular weight species will be most retarded, showing minimal migration over a fixed period of time, while smaller molecular weight species will be less retarded by the gel matrix and will migrate further during the same time period. This is the basis of resolving protein and peptide bands by SDS-PAGE. Examples of the use of SDS-PAGE can be found for assays of proteolytic enzymes, kinases, DNA-cleaving nucleases, and similar materials.

The purpose of the electrophoresis in a protease assay is to separate the protein or peptide substrate of the enzymatic reaction from the products. The fractionation range of SDS-PAGE varies with the percentage of acrylamide in the gel matrix (see Copeland, 1994, for details). In general, acrylamide percentages between 5 and 15% are used to fractionate globular proteins of molecular weights between 10,000 and 100,000 Da. Higher percentage acrylamide gels are used for separation of lower molecular weight peptides (typically 20–25% gels). In a typical experiment, the substrate protein or peptide is incubated with the protease in a small reaction vial, such as a microcentrifuge tube. After a given reaction time, a volume of the reaction mixture is removed and mixed with an equal volume of 2X SDS-containing sample buffer to denature the proteins and coat them with anionic detergent (Copeland, 1994). This buffer contains SDS to unfold and coat the proteins, a disulfide bond reducing agent (typically mercaptoethanol), glycerol to give density to the solution, and a low molecular weight, inert dye to track the progress of the electrophoresis in the gel (typically bromophenol blue). The sample mixture is incubated at boiling water temperature for 1–5 minutes and loaded onto a gel of appropriate percentage acrylamide to effect separation. Current is applied to the gel from a power source, and the electrophoresis is allowed to continue for some fixed period of time until the bromophenol blue dye front reaches the bottom of the gel (for a 10% gel, a typical electrophoretic run would be performed at 120 V constant voltage for 1.5–3 h, depending on the size of the gel).

After electrophoresis, protein or peptide bands are visualized with a peptide-specific stain, such as Coomassie Brilliant Blue or silver staining (Copeland, 1994; Hames and Rickwood, 1990). A control lane is always run containing the substrate protein or peptide alone, loaded at the same concentration as the starting concentration of substrate in the enzymatic reaction. When available, a second control lane should be run containing samples of the expected product(s) of the enzymatic reaction. A third control lane, containing commercial molecular weight markers (a collection of proteins of known molecular weights), is commonly run on the same gel also. The amount of substrate remaining and product formed for a particular reaction can be quantified by densitometry from the stained bands on the gel. A large number of commercial

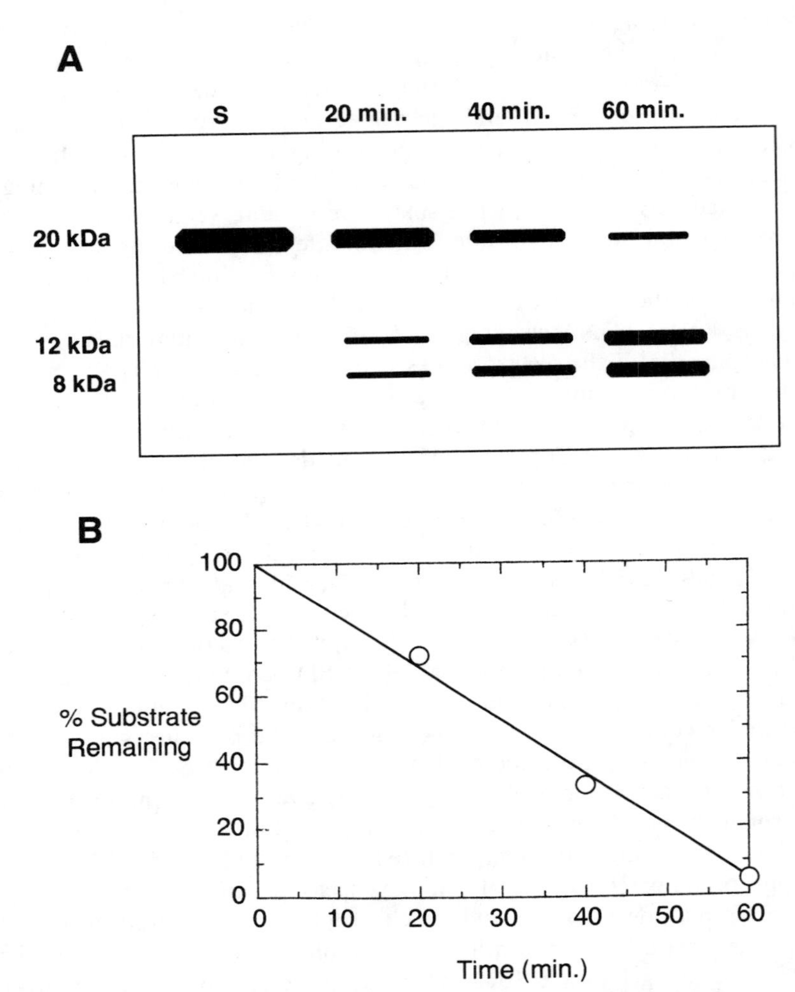

Figure 6.17 Schematic diagram of a protease assay based on SDS-PAGE separation of the protein substrate (20 kDa) and products (12 and 8 kDa) of the enzyme. (A) A typical SDS-PAGE result of such an experiment: the loss of substrate could be quantified by dye staining or other visualization methods, combined with such techniques as densitometry or radioactivity counting. (B) The time course of substrate depletion based on staining of the substrate band in the gel and quantitation by densitometry.

densitometers are available for this purpose (from BioRad, Pharmacia, Molecular Devices, and other manufacturers).

Figure 6.17 illustrates a hypothetical protease assay using SDS-PAGE. In this example, the protease cleaves a protein substrate of 20 kDa to two unequal fragments (12 and 8 kDa). As the reaction time increases, the amount of substrate remaining diminishes, and the amount of products formed increases. By scanning the gel with a densitometer, the relative amounts of both substrate and products can be quantified by ascertaining the degree of staining of these bands. As illustrated in Figure 6.17, it is fairly easy to perform this type of relative quantitation. To convert these densitometry units into concentration units of substrate or product is, however, less straightforward. For substrate loss, one can run a similar gel with varying loads of the substrate (at known concentrations) and establish a calibration curve of staining density as a function of substrate concentration. One can do the same for the product of the enzymatic reaction when a genuine sample of that product is available. For synthetic peptide, this is easily accomplished. A standard sample for protein products can sometimes be obtained by producing the product protein recombinantly in a bacterial host. This is not always a convenient option, however, and in such cases one's report may be limited to relative concentrations based on the intensity of staining.

The foregoing assay would work well for a purified protease sample, where the only major protein bands on the gel were from substrate and product. When samples are crude enzymes—for example, early in the purification of a target enzyme—contaminating protein bands may obscure the analysis of the substrate and product bands on the gel. A common strategy in these cases is to perform Western blotting analysis using an antibody that recognizes specifically the substrate or product of the enzymatic reaction under study. Detailed protocols for Western blotting have been described (Copeland, 1994; Harlow and Lane, 1988; see also technical bulletins from manufacturers of electrophoretic equipments such as BioRad, Pharmacia, and Novex).

Briefly, in Western blotting an SDS-PAGE gel is run under normal electrophoretic conditions. Afterward, the gel is soaked in a buffer designed to optimize electrophoretic migration of proteins out of the gel matrix. The gel is then placed next to a sheet of nitrocellulose (or some other protein binding surface), and protein bands are transferred electrophoretically from the gel to the nitrocellulose. After transfer, the remaining protein binding sites on the nitrocellulose are blocked using a large quantity of some nonspecific protein (nonfat dried milk, gelatin, or bovine serum albumin typically is used for this purpose). After blocking the nicrocellulose is immersed in a solution of an antibody that specifically recognizes the protein or peptide of interest (i.e., in our case, the substrate or product of the enzymatic reaction). This antibody, referred to as the *primary antibody*, is obtained by immunizing an animal (typically a mouse or a rabbit) with a purified sample of the protein of peptide or interest (see Harlow and Lane, 1988, for details).

After treatment with the primary antibody, and further blocking with nonspecific protein, the nitrocellulose is treated with a *secondary antibody* that recognizes primary antibodies from a specific animal species. For example, if the primary antibody is obtained from immunized rabbits, the secondary antibody will be an anti-rabbit antibody. The secondary antibody carries a label that provides a simple and sensitive method of detecting the presence of the antibody. Secondary antibodies bearing a variety of labels can be purchased. A popular strategy is to use a secondary antibody that has been covalently labeled with biotin. Biotin binds tightly and specifically to streptavidin, which is commercially available as a conjugate with enzymes such as horse radish peroxidase or alkaline phosphatase. The biotinylated secondary antibody adheres to the nitrocellulose at the binding sites of the primary antibody. The location of the secondary antibody on the nitrocellulose is then detected by treating the nitrocellulose with a solution containing a streptavidin-conjugated enzyme. After the streptavidin–enzyme conjugate has been bound to the blot, the blot is treated with a solution containing chromophoric substrates for the enzyme linked to the streptavidin. The products of the enzymatic reaction form a highly colored precipitate on the nitrocellulose blot wherever the enzyme–streptavidin conjugate is present. In this round-about fashion, the presence of a protein band of interest can be specifically detected from a gel that is congested with contaminating proteins.

Another use of SDS-PAGE in enzyme assays is for following the incorporation of phosphate into a particular protein or peptide that results from the action of a specific kinase. There are two common strategies for following kinase activity by gel electrophoresis. In the first, the reaction mixture includes a ^{32}P-labeled phosphate source (e.g., ATP as a cosubstrate of the kinase) that incorporates the radiolabel into the products of the enzymatic reaction. After the reaction has been stopped, the reaction mixture is fractionated by SDS-PAGE. The resulting gel is dried, and the ^{32}P-containing bands are located on the gel by autoradiography or by digital radioimaging of the dried gel. The second strategy uses commercially available antibodies that specifically recognize proteins or peptides that have phosphate modifications at specific types of amino acid residue. Antibodies can be purchased that recognize phosphotyrosine or phosphoserine/phosphothreonine, for example. These antibodies can be used as the primary antibody for Western blot analysis as described earlier. Since the antibodies recognize only the phosphate-containing proteins or peptides, they provide a very specific measure of kinase activity.

Aside from their use in quantitative kinetic assays, electrophoretic methods have been used in enzymology to identify protein bands associated with specific enzymatic activities after fractionation on gels. This technique, which relies on specific staining of enzyme bands in the gel, based on the enzymatic conversion of substrates to products, can be a very powerful tool for the initial identification of a new enzyme or for locating an enzyme during purification attempts. For these methods to work, one must have a staining method that is specific to the enzymatic activity of interest, and the enzyme in the gel must be in its native (i.e., active) conformation.

Since SDS-PAGE is normally denaturing to proteins, measures must be taken to ensure that the enzyme will be active in the gel after electrophoresis: either the electrophoretic method must be altered so that it is not denaturing or a way must be found to renature the unfolded enzyme in situ after electrophoresis.

Native gel electrophoresis is commonly used for these applications. In this method, SDS and disulfide reducing agents are excluded from the sample and the running buffers, and the protein samples are not subjected to denaturing heat before application to the gel. Under these conditions most proteins will retain their native conformation within the gel matrix after electophoresis. The migration rate during electrophoresis, however, is no longer dependent solely on the molecular weight of the proteins under native conditions. In the absence of SDS, the proteins will not have uniform charge densities; hence, their migration on the electric field will depend on a combination of their molecular weights, total charge, and general shape. It is thus not appropriate to compare the electrophoretic mobility of proteins under the denaturing and native gel forms of electrophoresis.

Sometimes enzymes can be electrophoresed under denaturing conditions and subsequently refolded or renatured within the gel matrix. In such cases the gel is usually run under nonreducing conditions (i.e., without mercaptoethanol or other disulfide reducing agents in the sample buffer), since proper reformation of disulfide bonds is often difficult inside the gel. A number of methods for renaturing various enzymes after electrophoresis have been reported, and these have been reviewed by Mozhaev et al. (1987). The following protocol, provided by NOVEX, Inc., has been found to work well for many enzymes in the author's laboratory. One must realize, however, that not all enzymes will be successfully renatured after the harsh treatments of electrophoretic separation. Hence, the appropriateness of these methods must be determined empirically for each enzyme individually.

General Protocol for Renaturation of Enzymes After SDS-PAGE

1. After electrophoresis, soak gel for 30 minutes at room temperature, with gentle agitation, in 100 mL of 2.5% (v:v) Triton X-100 in distilled water.
2. Decant the solution and replace with 100 mL of an aqueous buffer containing 1.21 g/L Tris base, 6.30 g/L Tris HCl, 11.7 g/L NaCl, 0.74 g/L $CaCl_2$, and 0.02% (w:v) Brig 35 detergent. Equilibrate the gel in this solution for 30 minutes at room temperature, with gentle agitation. Replace the solution with another 100 mL of the same buffer and incubate at 37°C for 4–16 hours.

The recent electrophoresis text by Hames and Rickwood (1990) provides an extensive list of enzymes ($\approx$200) that can be detected by activity staining after native gel electrophoresis and gives references to detailed protocols for each of the listed enzymes. Figure 6.18 illustrates activity staining after native gel electrophoresis for human dihydroorotate dehydrogenase (DHODase). As we have discussed before, this enzyme catalyzes the conversion of dihydroorotate to orotic acid utilizing the redox cofactor ubiquinone. As is true of many other dehydrogenases, the activity of DHODase can be coupled to the formation of

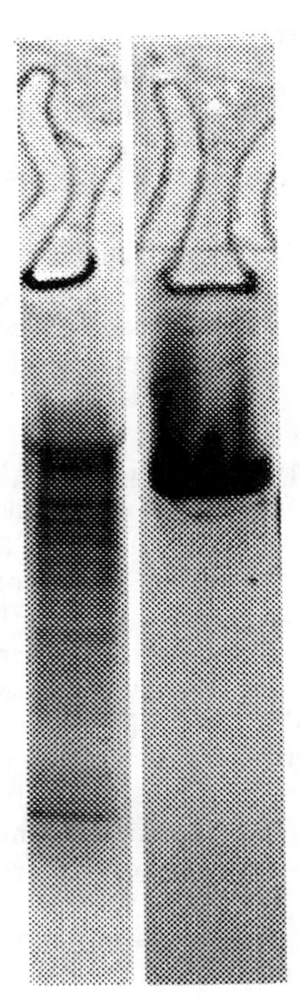

Figure 6.18 Example of activity staining of an enzyme after gel electrophoresis. *Left*: Coomassie Brilliant Blue stained native gel of a detergent extraction of human liver mitochodrial membranes; note the large number of proteins of varied electrophoretic mobility in the sample. *Right*: a native gel of the same sample (run under identical conditions as in the left lane) stained with nitroblue tetrazolium (NBT) in the presence of the substrates of the enzyme dihydroorotate dehydrogenase; the signal protein band that is stained intensely represents the active dihydroorotate dehydrogenase.

an intensely colored formazan product by reduction of the reagent nitroblue tetrazolium (NBT) or methyl thiazolyl tetrazolium (MTT); the formazan product precipitates on the gel at the sites of enzymatic activity. The left-hand panel of Figure 6.18 shows a native gel of a detergent extract of human liver mitochodrial membranes stained with Coosmassie Brilliant Blue. As one would expect, there are a large number of proteins present in this sample, displaying a congested pattern of protein bands on the gel. The right-hand panel of Figure 6.18 displays another native gel of the same sample that was soaked in a solution of 100 μM dihydroorotate, 100 μM ubiquinone, and 1 mM NBT (in a

50 mM Tris buffer, pH 7.5) after electrophoresis: There is a single dark band due to the NBT staining of the enzymatically active protein in the sample. Thus it is seen that the enzymatic activity in a complex sample can be associated with a specific protein or set of proteins. The active band(s) can subsequently be excised from the gel for further analysis, such as N-terminal sequencing or to serve as part of a purification protocol for a particular enzyme; alternatively, they can be used for the production of antibodies agaist the enzyme of interest.

In the case of proteolytic enzymes an alternative to activity staining is a technique known as gel zymography. In this method the acrylamide resolving gel is cast in the presence of a high concentration of a protein-based substrate of the enzyme of interest (casein, gelatin, collagen, etc.). The polymerized gel is thus impregnated with the protein throughout. Samples containing the proteolytic enzyme are then electrophoresed on the gel. If denaturing conditions are used, the enzymes are renatured by the protocol described earlier, and the gel is then stained with Coomassie Brilliant Blue. Because there is a high concentration of protein (i.e., substrate) throughout the gel, the entire field will

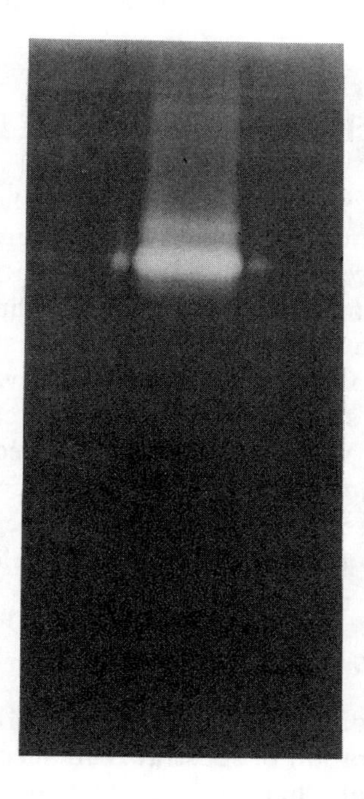

Figure 6.19 Gelatin zymography of a whole-cell lysate from Sf9 insect cells that had been infected with a baculovirus construct containing the gene for human 92 kDa gelatinase (MMP9). The location of the active enzyme is easily observed from the loss of Coomassie staining of the gelatin substrate in the gel. (Figure kindly provided by Henry George, DuPont Merck Research Laboratories.)

be stained bright blue. Where there has been significant proteolysis of the protein substrate, however, the intensity of blue staining will be greatly diminished. Hence the location of proteolytic enzymes in the gel can be determined by the *reverse staining*, (i.e., the absence of Coomassie staining), as illustrated in Figure 6.19 for the metalloprotease gelatinase (MMP9).

A related, less direct method of protease detection has also been reported. In this "sandwich gel" technique, an agar solution is saturated with the protein substrate and allowed to solidify in a petri dish or other convenient container. A standard acrylamide gel is used to electrophoresis the protease-containing sample. After electrophoresis (and renaturation in the case of denaturing gels) the substrate-containing agar is overlaid with the protease-containing acrylamide gel, and the materials are left in contact with each other for 30–90 minutes at 37°C. The sites of proteolytic activity can then be determined by treating the agar with an ammonium sulfate solution, trichloroacetic acid, or some other protein-precipitating agent. After this treatment, the bulk of the agar will turn opaque as a result of protein precipitation. The proteolysis sites, however, will appear as clear zones against the opaque field of the agar. Methods for the detection of enzymatic activity after gel electrophoresis have been recently reviewed by Hames and Rickwood (1990) and by Gabriel and Gersten (1992).

6.4 Factors Affecting the Velocity of Enzymatic Reactions

The velocity of an enzymatic reaction can display remarkable sensitivity to a number of solution conditions, (e.g., temperature, pH, ionic strength, specific cation and anion concentration). Failure to control these parameters can lead to significant errors and lack of reproducibility in velocity measurements. Hence it is important to keep these parameters constant from one measurement to the next. In some cases, the changes in velocity that are observed with controlled changes in some of these conditions can yield valuable information on aspects of the enzyme mechanism. In this section we discuss four of these parameters: enzyme concentration, temperature, pH, and solvent isotope makeup. Each of these can affect enzyme velocities in well-understood ways and can be controlled by the investigator to yield important information.

6.4.1 Enzyme Concentration

In Chapter 5, in our discussion of the Henri–Michaelis–Menten equation, we saw how the concentration of substrate can affect the velocity of an enzymatic reaction. At the end of Chapter 5 we recast this equation, replacing the term V_{max} by the product of k_{cat} and $[E_t]$, the total concentration of enzyme in the sample (Equation 5.30). From this equation we see that the velocity of an enzyme-catalyzed reaction should be linearly proportional to the concentration of enzyme present at constant substrate concentration.

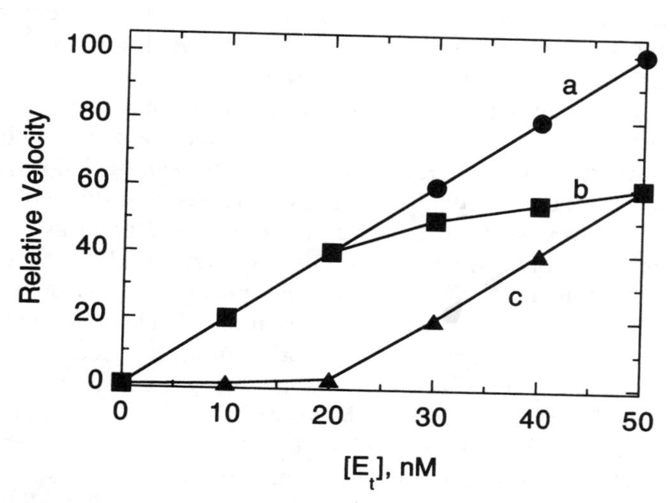

Figure 6.20 The relative velocity of an enzymatic reaction, under controlled conditions, as a function of total enzyme concentration [E_t]. The straight-line relationship of curve a is the expected behavior. Curve b illustrates the type of behavior observed when substrate depletion becomes significant at the higher enzyme concentrations. Curve c illustrates the behavior that would be observed for an enzyme sample containing a reversible inhibitor. See text for further details.

Over a finite range, a plot of velocity as a function of [E_t] should yield a straight line, as illustrated in Figure 6.20, curve a. The range over which this linear relationship will hold depends on our ability to measure the true initial velocity of the reaction at varying enzyme concentrations. Recall from Chapter 5 that initial velocity measurements are valid only in the range of substrate depletion between 0 and 10% of the total initial substrate concentration. As we add more and more enzyme, the velocity can increase to the point at which significant amounts of the total substrate concentration are being depleted during the time window of our assay. When substrate depletion becomes significant, further increases in enzyme concentration will no longer demonstrate as steep a change in reaction velocity as a function of [E_t]. As a result, we may observe a plot of velocity as a function of [E_t] that is linear at low [E_t] but curves over and may even show saturation effects at higher values of [E_t], as in curve b of Figure 6.20.

In general, as stated in Chapter 5, one should work at enzyme concentrations very much lower than the substrate concentration. This range will vary from system to system, but in a typical assay substrate is present in micromolar to millimolar concentrations, and enzyme present in picomolar to nanomolar concentrations. Within this range of $[E_t] \ll [S]$, one must make initial velocity measurements over a number of enzyme concentrations to determine the range of enzyme concentration over which substrate depletion is not significant.

Substrate depletion is not the only cause of a downward-curving velocity–$[E_t]$ plot, such as that represented by curve b of Figure 6.20. The same type of behavior also results from saturation of the detection system at the higher velocity values seen at high $[E_t]$. We have discussed some of these problems in this chapter. For example, suppose that we measure the velocity of an enzymatic reaction as an end point absorption reading, following product formation. As we increase $[E_t]$, the velocity increases, and thus the amount of product formed over the fixed time window of our end point assay increases. If the concentration of product increases until the sample absorption is beyond the Beer's law limit (see the discussion of optical methods of detection in Section 6.2.2), we observe an apparent saturation of velocity at high value of $[E_t]$. As with substrate depletion, detector saturation effects lead to down-curving velocity–$[E_t]$ plots, not as a result of any intrinsic property of our enzyme system, but rather because of failure to measure the true initial velocity of the reaction under conditions of high $[E_t]$.

Plots of velocity as a function of enzyme concentration also can display upward curvature, as illustrated in curve c of Figure 6.20. Potential causes of this type of behavior include inadequate temperature equilibration, as discussed shortly, and the presence of an inhibitor or enzyme activator in the reaction mixture. If, for example, a small amount of an irreversible inhibitor (see Chapter 9) is present in one of the components of the reaction mixture, additions of low concentrations of enzyme will result in complete inhibition of the enzyme, and no activity will be observed. The enzymatic activity will be realized in such a system only after enzyme has been added to a concentration that exceeds that of the irreversible inhibitor. Hence, at low values of $[E_t]$ one observes zero or minimal velocity, while above some critical concentration the velocity–$[E_t]$ curve is steeper. Another potential cause of upward curvature is the presence in the enzyme stock solution of an enzyme activator or cofactor that is missing from the remainder of the reaction mixture components. Suppose that the enzyme under study requires a dissociable cofactor for full activity (as we saw in Chapter 3, many enzymes fall into this category). The concentrations of the free enzyme E and free cofactor C will be in equilibrium with that of the active enzyme–cofactor complex EC, and the concentration of EC present under any set of solution conditions will be defined by the equilibrium constant K_c:

$$[EC] = \frac{[E][C]}{K_c} \tag{6.5}$$

In the enzyme stock solution, the concentrations of enzyme and cofactor will be in some specific proportion. When we dilute a sample of this stock solution into our reaction mixture, the total amount of enzyme added will be the sum of the free enzyme and enzyme-cofactor complex; that is, $[E_t] = [E] + [EC]$. Hence, the concentration of cofactor added to the reaction mixture from the enzyme stock solution will be proportional to the amount of total enzyme added; that is, $[C_t] = \alpha[E_t]$. It can be shown (Tipton, 1992) that the amount of active EC complex in the final reaction mixture will depend on the total enzyme added and the enzyme–cofactor equilibrium constant as follows:

$$[EC] = \frac{[E_t]^2}{[E_t] + (K_c/\alpha)} \tag{6.6}$$

We can see from Equation 6.6 that the amount of activated enzyme (i.e., [EC]) will not track linearly with the amount of total enzyme added at low values of $[E_t]$, and thus an upward-curving plot, as in Figure 6.20c, will result.

If one is aware of a cofactor requirement for the enzyme under study, these effects can often be avoided by supplementing the reaction mixture with an excess of the required cofactor. For example, the enzyme prostaglandin synthase is a heme-requiring oxidoreductase that binds the heme cofactor in a noncovalent, dissociable fashion. The apoenzyme (without heme) is inactive, but it can be reconstituted with excess heme to form the active holoenzyme. The activity of the enzyme can be followed by diluting a stock solution of the holoenzyme into a reaction mixture containing a redox active dye and measuring the changes in dye absorption following initiation of the reaction with arachidonic acid, the substrate of the enzyme. To observe full enzymatic activity, it is necessary to supplement the reaction mixture with heme so that the final heme concentration is in excess of the total enzyme concentration. As long as this precaution is taken, well-behaved plots of linear velocity versus $[E_t]$ are observed for prostaglandin synthase over a fairly broad range of enzyme concentrations (Copeland et al., 1994).

In summary, when one is measuring the true initial velocity of the reaction, the velocity of an enzyme-catalyzed reaction will increase linearly with enzyme concentration. Deviations from this linear behavior can be seen when the analyst's ability to measure the true initial velocity is compromised by instrumental or solution limitations. Deviations from linearity are observed also when certain inhibitors or enzyme activators are present in the reaction mixture. A more comprehensive discussion of cases of deviation from the expected linear response can be found in the text by Dixon and Webb (1979).

6.4.2 pH Effects

The pH of an enzyme solution can affect overall catalytic activity in a number of ways. Like all proteins, enzymes have a native tertiary structure that is sensitive to pH, and in general denaturation of enzymes occurs at extreme low and high pH values. There are a number of physical methods for following protein denaturation. Loss of secondary structure can be followed by circular dichroic spectroscopy, and changes in tertiary structure can often be observed

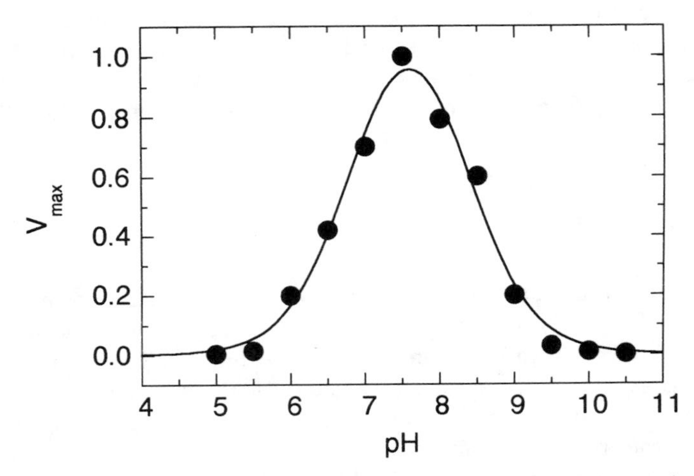

Figure 6.21 The effects of pH on the velocity of a typical enzymatic reaction.

by absorption and fluorescence spectroscopy (Copeland, 1994). Many proteins aggregate or precipitate upon pH-induced denaturation, and this behavior can be observed by visual inspection. The pH range over which the native state of an enzyme will be stable varies from one such protein to the next. While most enzymes are most stable near physiological pH ($\approx$7.4), some display maximal activity at much lower or higher pH values. The appropriate range for a specific enzyme must be determined empirically.

Typically, one finds that protein conformation can be maintained over a relatively broad pH range, say 4–5 pH units. Within this range, however, the velocity of the enzymatic reaction varies with pH. Figure 6.21 shows a typical profile of the velocity of an enzymatic reaction as a function of pH, within the pH range over which protein denaturation is not a major factor. What is most obvious from this figure is the narrow range of pH values over which enzyme catalytic efficiency is typically maximized. For most general assays of enzyme activity then, one will wish to maintain the solution pH at the optimum for catalysis. To keep within this range, the reaction mixture must be buffered by a component with a pK_a at or near the desired solution pH value.

A buffer is a species whose presence in solution resists changes in the pH of that solution due to additions of acid or base. For enzymatic studies, a number of useful buffers are available commercially; some of these are listed in Table 6.4. The buffering capacity of these and other buffers declines as one moves away from the pK_a value of the substance. In general these buffers provide good buffering capacity from one pH unit below to one pH unit above their pK_a values. Thus, for example, HEPES buffer can be used to stablilize the pH of a solution between pH values of 6.55 and 8.55, but would not be an appropriate buffer below pH 6.5 or above pH 8.6.

Table 6.4 Some Buffers That Are Useful in Enzyme Studies

Common Name	Molecular Weight	pK_a at 25°C[a]	$\Delta pK_a/\Delta°C$
MES	195.2	6.15	−0.011
PIPES	324.3	6.80	−0.0085
Imidazole	68.1	7.00	−0.020
MOPS	231.2	7.20	−0.013
TES	251.2	7.50	−0.020
HEPES	260.3	7.55	−0.014
HEPPS	252.3	8.00	−0.015
Tricine	179.2	8.15	−0.021
Tris	121.1	8.30	−0.031
CHES	207.3	9.50	−0.029
CAPS	221.3	10.40	−0.032

[a]Values listed for the buffers at infinite dilution.

The buffers listed in Table 6.4 span a broad range of pK_a values, providing a selection of single-component buffers for maintaining specific solution pH values. Note, however, that the pK_a values in Table 6.4 are for the buffers at 25°C and at infinite dilution. The temperature, buffer concentration, and overall ionic strength can perturb these pK_a values, hence altering the pH of the final solution. In most enzymatic studies, the buffers will be present at final concentrations of 0.05–0.1 M, and solution ionic strength is typically between 0.1 and 0.2 M (near physiological conditions). Typically one will have a high concentration stock solution of the pH-adjusted buffer in the laboratory that will be diluted to prepare the final reaction mixture. It is important to measure the final solution pH to determine the extent of pH change that accompanies dilution. These effects are usually relatively small, and minor adjustments can be made if necessary.

Another potential problem is a change in pK_a due to changes in solution temperature. In some cases the pH of a buffered solution can change dramatically between temperatures of 4 and 37°C. Table 6.4 lists the change in pK_a per change in degrees Celsius of the tabulated buffers. In principle one could calculate the change in solution pH that will accompany a temperature change, but this is a tedious task and often impractical. Instead, if the assays are to be run at elevated temperatures (e.g., 37°C), the pH meter should be calibrated at the assay temperature and all pH measurements performed at 37°C as well. This will ensure that the pH values measured reflect accurately the true pH values under the assay conditions. In some cases one may wish to measure enzyme activity over a range of temperatures while maintaining the pH at a fixed value (see later). For such studies it is best to use a buffer with a low $\Delta pK_a/\Delta°C$ value, to keep the change in pH over the temperature range of interest minimal. From Table 6.4, PIPES ($pK_a = 6.8$; $\Delta pK_a/\Delta°C = -0.0085$) and MOPS ($pK_a = 7.20$; $\Delta pK_a/\Delta°C = -0.013$) would be good choices for this application.

The pH dependence of the activity of an enzyme is of practical importance in optimizing assay conditions, but the dependency is largely phenomenological. On the other hand, useful mechanistic information regarding the role of acid–base groups involved in enzyme turnover can be gleaned from properly performed pH studies. By measuring the velocity as a function of substrate concentration at varying pH, one can simultaneously determine the effects of pH on the k_{cat}, K_m, and k_{cat}/K_m values for an enzyme-catalyzed reaction. If titration of ionizable groups on the substrate molecule does not occur over the pH range being studied, these pH profiles will make possible some general conclusions about the roles of acid–base groups in the enzyme molecule. In general the pH dependence of K_m reflects the involvement of acid–base groups that are essential to the substrate binding event(s) that precede catalysis. Effects of pH on k_{cat} mainly reflect acid–base group involvement in the catalytic step of substrate-to-product conversion; that is, these ionization steps occur in the enzyme–substrate complex.

Finally, a plot of k_{cat}/K_m as a function of pH is said to reflect the essential ionizing groups of the free enzyme that play a role in both substrate binding and catalytic processing (Palmer, 1985). As an example, let us consider the pH profile of the serine protease chymotrypsin. As described in Chapter 4, the active site of chymotrypsin contains a catalytic triad (Figure 6.22A) composed of Asp 102, His 57, and Ser 195. Acylation of Ser 195 to form the transition state analogue and hydrolysis of the peptidic substrate depend on hydrogen-bonding and proton transfer steps among the residues within this active site triad.

From idealized pH profiles for the k_{cat}, K_m, and k_{cat}/K_m for chymotrypsin (Figure 6.22B–D), we see that both K_m and k_{cat} display pH profiles that are well fit by the Henderson–Hasselbalch equation introduced in Chapter 2 (Equation 2.12), but with strikingly different profiles. In the case of the profile for K_m versus pH, substrate binding affinity decreases with increasing pH with an apparent pK_a value of 9.0. This pK_a has been shown to reflect ionization of an N-terminal isoleucine residue, which must be protonated for the enzyme to adopt a conformation capable of binding substrate. The value of k_{cat} for this enzyme increases with increasing pH and displays an apparent pK_a of 6.8. This pK_a value has been alternatively ascribed to Asp 102 and His 57 of the active site triad. It is now thought that this pK_a is more correctly associated with the catalytic triad as a whole, rather than an individual amino acid residue. In Figure 6.22D, the idealized pH profile of k_{cat}/K_m for chymotrypsin, we do not observe the expected "S-shaped" curve associated with the Henderson–Hasselbalch equation; instead, there is a bell-shaped curve. This plot represents the cumulative effects of two titratable groups that influence the catalytic efficiency of the enzyme in opposite ways (i.e., one group facilitates catalysis in its conjugate base form, while the other facilitates catalyses in its Brønsted–Lowry

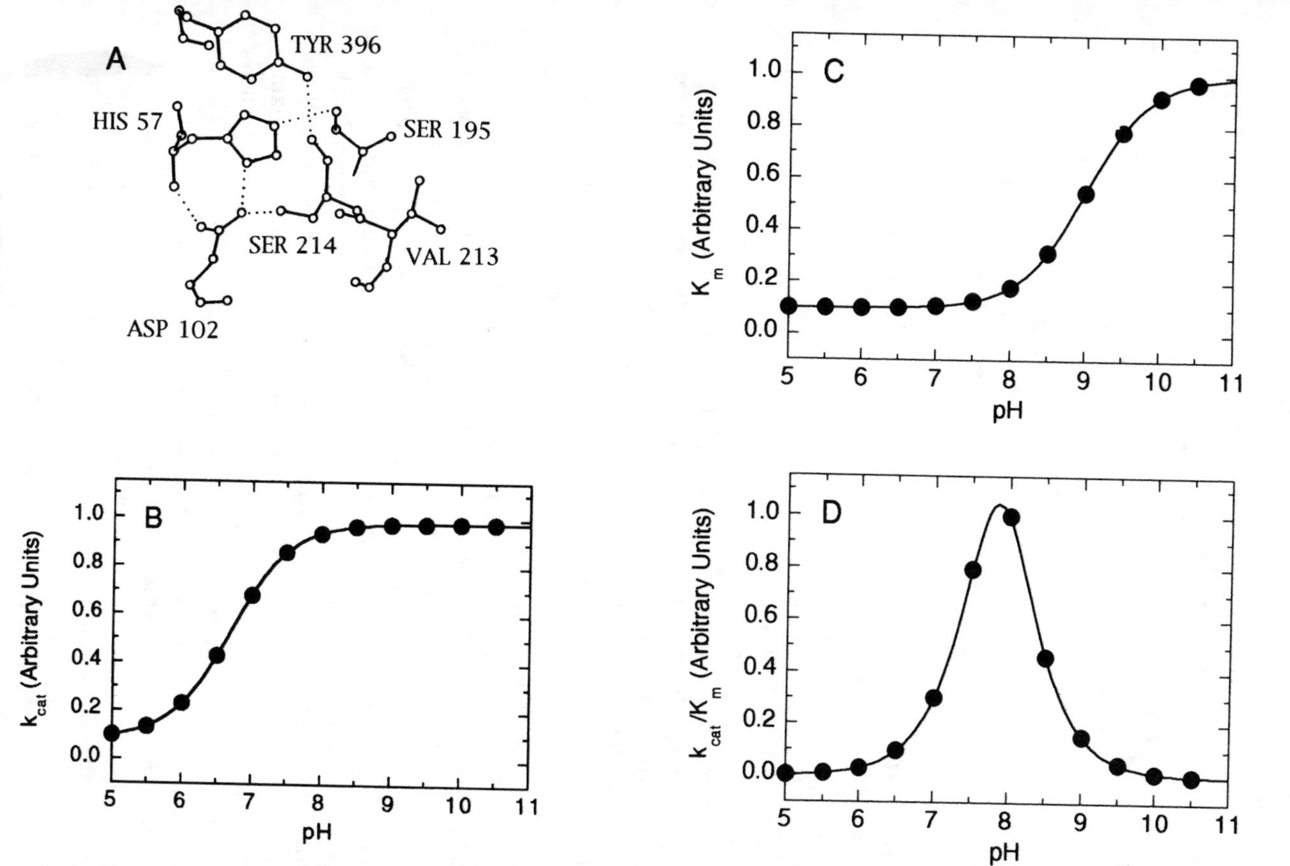

Figure 6.22 (A) Cartoon of the active site structure of α-chymotrypsin, based on the crystal structure reported by Tsukada and Blow (1985), showing the active site triad of amino acids. (B) Idealized pH profile of k_{cat} for α-chymotrypsin, showing an apparent pK_a of 6.8. (C) Idealized pH profile of K_m for α-chymotrypsin, showing an apparent pK_a of 9.0. (D) Idealized pH profile of k_{cat}/K_m for α-chymotrypsin, showing a bell-shaped curve that can be fit to Equation 6.7, with pK_a values of 6.8 and 9.0.

Table 6.5 Examples of Amino Acid Residues with Perturbed Side Chain pK_a Values

	pK_a in:		
Side Chain	Free Amino Acid	Enzyme	Enzyme
Glu	3.9	6.5	Lysozyme
His	6.8	3.4	Papain
His	6.8	5.2	Ribonuclease
Cys	8.3	4.0	Papain
Lys	10.8	5.9	Acetoacetate dehydrogenase

acid form). A pH profile such as that seen in Figure 6.22D can be fit by the following equation:

$$y = \frac{y_{\max}}{(10^{-\mathrm{pH}}/10^{-\mathrm{p}K_{a1}}) + (10^{-\mathrm{p}K_{a2}}/10^{-\mathrm{pH}}) + 1} \tag{6.7}$$

where y is the experimental measure that is plotted on the y axis (in this case $k_{\text{cat}}/K_{\text{m}}$), $y_{\max}$ is the observed maximum value of that experimental measure, and pK_{a1} and pK_{a2} refer to the pK_a values for the two relevant acid–base groups being titrated. A fit of the curve in Figure 6.33D to Equation 6.7 yields values of pK_{a1} and pK_{a2} of 6.8 and 9.0, respectively. Thus both the pK_a values that were found to influence k_{cat} and K_{m}, respectively, are reflected in the pH profile of $k_{\text{cat}}/K_{\text{m}}$ for this enzyme.

The type of data presented in Figure 6.22 is often used to predict the identities of key amino acid residues participating in acid–base chemistry during catalysis. Some caution must be exercised, however, in making such predictive statements. As we have seen for chymotrypsin, in some cases the pK_a value that is measured cannot be correctly ascribed to a particular amino acid but, rather, reflects a specific set of residue interactions in an enzyme molecule that create *in situ* a unique acid–base center. Also, the hydrophobic interior of enzyme molecules can greatly perturb the pK_a values of amino acid side chains relative to their typical pK_a values in aqueous solution. Some examples of such perturbations of amino acid side chain pK_a values are listed in Table 6.5. These examples should make it clear that one cannot rely simply on a comparison of the measured pK_a of an enzymatic kinetic parameter and the pK_a values of amino acid side chains in solution. Thus, for example, a k_{cat} pH profile for a particular enzyme that displays an apparent pK_a value of 6.8 may reflect the ionization of an essential histidine residue, but it may equally well represent the ionization of a perturbed glutamic acid residue, or yet some other residue in the specialized environment of the protein interior.

The effects of pH on the kinetic parameters k_{cat} and K_{m} also have been analyzed by plotting the value of the kinetic constant on a logarithmic scale as

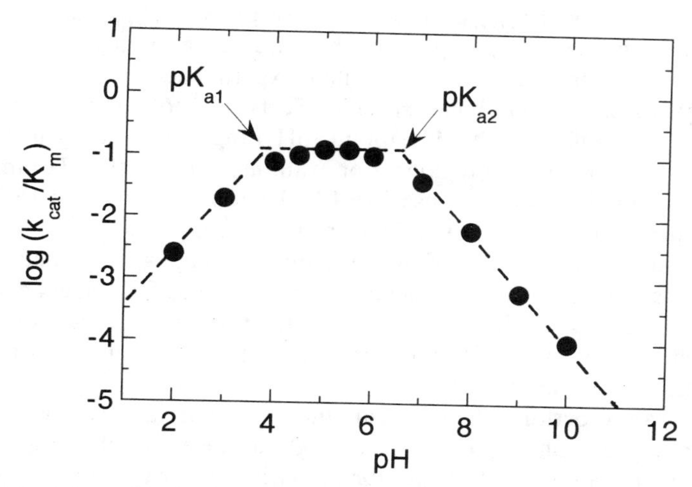

Figure 6.23. Plot of $\log(k_{\text{cat}}/K_{\text{m}})$ as a function of pH for a typical enzymatic reaction. The two pK_a values are determined from the intersection of the straight lines drawn through the data in different regions of the pH range. See text for further details.

a function of pH (Dixon and Webb, 1979). For a single titration event, such plots appear as the superposition of two linear functions, one with a slope of zero and the other with a unit slope value. Similarly, a kinetic parameter that undergoes two titration events over a specific pH range will yield a plot that appears as the superposition of three straight lines: one with a positive unit slope value, one with a slope of zero, and one with a negative unit slope value. Figure 6.23 is an example of such a plot. An advantage of these plots is that the pK_a value can be determined graphically without resorting to nonlinear curve-fitting routines; rather, the pK_a is defined by the point of intersection of the two straight lines drawn through the data points in the regions of minimal curvature of the plot. A second advantage is that the number of acid–base groups participating in the ionization event can be estimated: the slope of the line in the transition region of the plot reflects the number of ionizable groups that are titrated over this pH range. Thus, a slope of 1 indicates that a single group is being titrated, a slope of 2 indicates the involvement of two ionizable species, and so on. An expanded discussion of these plots, and examples of their application to specific enzymes, can be found in the text by Dixon and Webb (1979).

In designing experiments to measure the effects of pH on the steady state kinetics of an enzymatic reaction, it is critical for the researcher to ensure that the changes in solution pH are not made in a way that causes simultaneous changes in other solution conditions, thus confounding the analysis of the experiments. For example, a change in the species used to buffer the solution could, in principle, effect a change in the kinetics by itself. Since these studies are typically conducted over a broad range of pH values, no one buffer will have sufficient buffering capacity over the entire range of study. Hence, more than one buffering species is needed in these experiments.

One way to check that buffer-specific effects are not influencing the pH profile is to use buffers with overlapping pH ranges and perform duplicate measurements in the overlap regions. For example, to cover the pH range from 5.5 to 8.5 one might choose to use MES buffer (useful range ≈5.15–7.15) at the lower pH values and HEPES buffer (useful range ≈6.55–8.55) for the higher pH values. In the range of overlapping buffering capacity, between 6.55 and 7.15, one should make measuements with both buffers independently. If there are no significant differences between the measured values for the two buffer systems at the same pH values, it is fairly safe to assume that no major buffer-specific effects are occurring.

A better way to perform these measurements is to use a mixed buffer system that will have good buffering capacity throughout the entire pH range of study. Gomori (1992) has provided recipes for preparing buffer systems composed of two or more buffering species that are appropriate for enzyme studies and span several different ranges of pH values. For example, Gomori recommends the use of a mixed Tris–maleate buffer for work in the pH range between 5.2 and 8.6. This buffer system requires two stock solutions. The first is composed of 24.2 g of tris(hyroxymethyl)aminomethane and 23.2 g of maleic acid dissolved in 1 L of distilled water (0.2 M Tris–maleate). The second solution is 0.2 M NaOH in distilled water. To prepare a 0.05 M buffer at a given pH, 50 mL of the Tris–maleate stock is mixed with X mL of the 0.2 M NaOH stock, according to Table 6.6, and diluted with distilled water to a final volume of 200 mL. Recipes for other mixed buffer systems for use in lower and higher pH

Table 6.6 Volume of 0.2 M NaOH to be Added to 50 mL of 0.2 M Tris–Maleate Stock to Produce a 0.05 M Tris–Maleate Buffer at the Indicated pH After Dilution to 200 mL with Distilled Water

X (mL)	pH	X (mL)	pH
7.0	5.2	48.0	7.0
10.8	5.4	51.0	7.2
15.5	5.6	56.0	7.4
20.5	5.8	58.0	7.6
26.0	6.0	63.5	7.8
31.5	6.2	69.0	8.0
37.0	6.4	75.0	8.2
42.5	6.6	81.0	8.4
45.0	6.8	86.5	8.6

Source: Data from Gomori (1992).

ranges can be found in the compilation of Gomori (1992) and references therein.

6.4.3 Temperature Effects

It is often stated that the rate of a chemical reaction generally doubles with every 10°C increase in reaction temperature. Most chemical catalysts display such an increase in activity with increasing temperature, and enzymes are no exception. Enzymes, however, are also proteins, and like all proteins they undergo thermal denaturation at elevated temperatures. Hence the enhancement of catalytic efficiency with increasing temperature is compromised by the competing effects of general protein denaturation at high temperatures. For this reason the activity of a typical enzyme will increase with temperature over a finite temperature range, then diminish significantly above some critical temperature that is characteristic of the denaturation of the protein (Figure 6.24). As with the pH profile, the information gleamed from plots such as Figure 6.24 is of practical value in designing enzyme assays. One will wish to measure the activity at a temperature that supports high enzymatic activity, does not lead to significant protein denaturation, and is experimentally convenient. Balancing these factors, one finds that the majority of enzymes assays reported in the literature are conducted at either 25 or 37°C (i.e., physiological temperature).

If one restricts attention to the temperature range over which protein denaturation is not significant, an analysis of the changes in enzyme activity

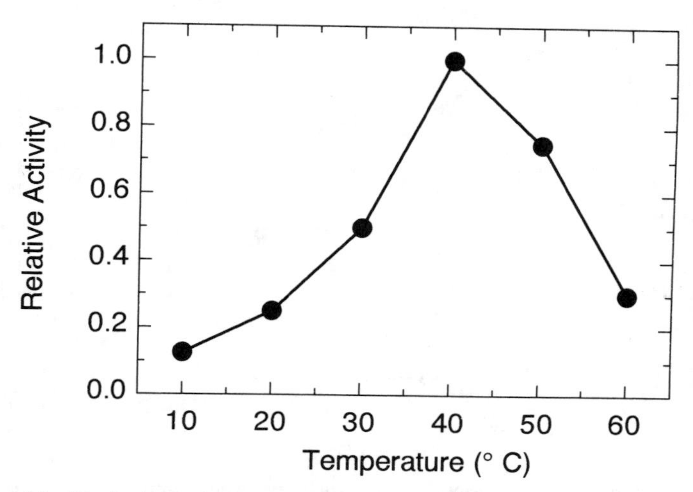

Figure 6.24 Typical profile of the relative activity of an enzyme as a function of temperature.

that accompany changes in temperature can be mechanistically informative. Recall from Chapter 2 that the rate of a reaction can be related to the activation energy for attaining the reaction transition state E_a by the Arrhenius equation (Equation 2.7). This relationship holds true for enzyme catalysis as well, as long as protein denaturation is not a complicating factor in the temperature range being studied. We can relate the kinetic constant k_{cat} to the activation energy as follows:

$$k_{cat} = A \exp\left(\frac{-E_a}{RT}\right) \tag{6.8}$$

where R is the ideal gas constant (1.98×10^{-3} kcal/mol·degree) and T is the temperature in degrees kelvin; for our purpose, we can treat the preexponential term A as a constant of proportionality (see Chapter 2 for a more explicit definition of A). Taking the $\log_{10}$ of both sides of Equation 6.8, we obtain:

$$\log(k_{cat}) = -E_a \frac{1}{2.3RT} + \log(A) \tag{6.9}$$

Thus if we plot the log (k_{cat}) of an enzymatic reaction as a function of $1/2.3RT$, Equation 6.9 predicts that we will obtain a straight-line relationship with a slope equal to $-E_a$, the activation energy, in units of kcal/mol. Note that the relationships described by Equations 6.8 and 6.9 hold for V_{max} at constant

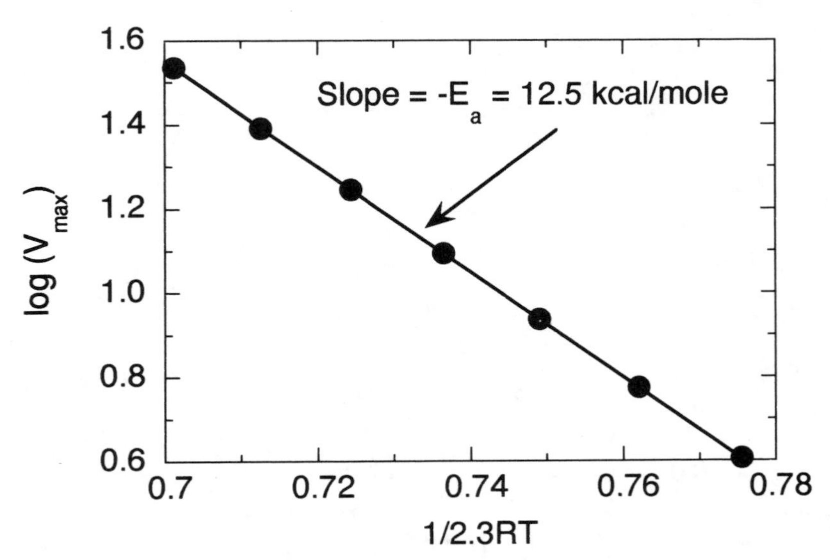

Figure 6.25 Arrhenius plot of log(V_{max}) of an enzymatic reaction as a function of $1/(2.3RT)$. The slope of the line in such a plot gives an estimate of the negative value of E_a, the activation energy of the reaction.

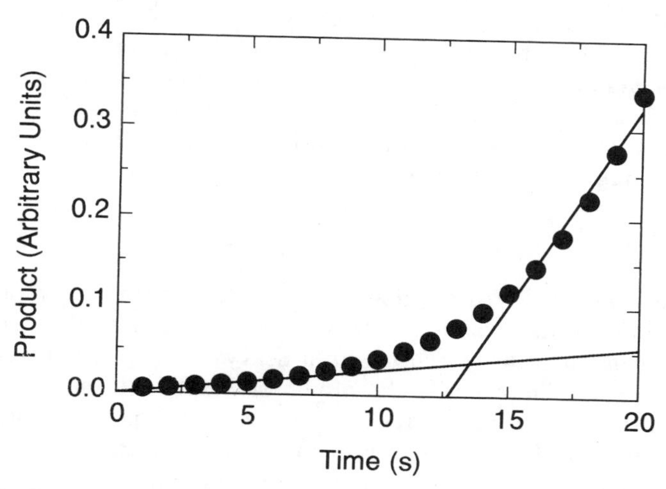

Figure 6.26 Example of a reaction progress curve showing a long lag phase before reaching the true steady state rate of reaction. Such a lag phase can be caused by several factors, including insufficient temperature equilibrium of the enzyme and reaction mixture solutions. See text for further details.

enzyme concentration as well as for k_{cat}. When an accurate estimate of the concentration of enzyme in a stock solution is lacking, one can hold the amount of that stock solution used in the assays constant and determine the activation energy of the reaction by using the measured values of V_{max} in place of k_{cat} in Equation 6.9. Figure 6.25 is such an Arrhenius plot for a hypothetical enzyme with an activation energy of 12.5 kcal/mol, measured over a temperature range of 10–40°C.

Since the temperature of the reaction mixture can have such a dramatic effect on the kinetic parameters of an enzyme-catalyzed reaction, it is critical to carefully control temperature during measurements of initial velocity. For a reaction initiated by addition of substrate, the reaction mixture and substrate solution should be equilibrated to the same temperature before mixing. For a reaction initiated by addition of enzyme, it is sometimes desirable to maximize protein stability by maintaining the enzyme stock solution at 4°C (ice temperature) prior to the assay. In such situations it is best to use a stock of enzyme so concentrated that only a small volume needs to be added to the overall reaction mixture, which should already be equilibrated at the assay temperature. In a typical assay of this type, one might add 10–50 μL of enzyme stock (at 4°C) to a 1.0 mL reaction mixture that is equilibrated at 25 or 37°C. This small volume will not significantly perturb the temperature of the overall reaction mixture, and the enzyme will come to the assay temperature during mixing. If too large a volume of enzyme stock is used for these assays, full temperature equilibrium may not be achieved during the mixing time. This will be reflected as a lag phase in the initial velocity measurements, as illustrated in Figure 6.26. Lag phases of this type can significantly compromise the accuracy of end point type assays, where the occurrence of the lag phase might be missed. Control measurement at several time points should be performed in these cases, to ensure that such effects of insufficient temperature equilibrium are not affecting the measurements.

6.4.4 Isotope Effects in Enzyme Kinetics

When an enzyme-catalyzed reaction is rate-limited by a group transfer step, a slowing down of the reaction rate will be observed if the group being transferred is isotopically enriched with a heavy isotope. Such kinetic isotope effects can be used to identify the atoms of a substrate molecule that are undergoing transfer during catalysis by an enzyme. To perform such analysis, the investigator must synthesize a version of the substrate that is isotopically labeled at a specific atom. Since protons are perhaps the most commonly transferred atoms in enzymatic reactions, we shall focus on the use of heavy isotopes of hydrogen in such studies.

Why is it that a heavier isotope leads to a dimunition of the reaction rate for proton transfer reactions? To answer this question, let us consider a reaction in which a hydrogen is transferred from a carbon atom to some general base:

$$\text{C—H :B} \rightleftharpoons [\text{C}^{\delta-} \cdots \text{H} \cdots \text{B}^{\delta+}]^{\ddagger} \rightarrow \text{C}^{-} \quad \text{HB}$$

As we saw in Chapter 2, the electronic state of the reactant can be represented as a potential energy well that has built upon it a manifold of vibrational substates (see Figure 2.9). Among these vibrational substates will be potential energies associated with the stretching of the C—H bond. The transition state of the reaction is reached by elongating this C—H bond prior to bond rupture. Thus, in going from the reactant state to the transition state, the potential energy of the C—H stretching vibration is converted to transitional energy that contributes to the overall energy of activation for the reaction.

Now the potential energy minimum (i.e., the very bottom of the well) of the reactant state is characteristic of the electronic configuration of the reactant molecule when all the atoms in the molecule are at their equilibrium distances (i.e., when the vibrations of the bonds are "frozen out"). If we were to replace the proton on the carbon with a deuteron, the electronic configuration of the molecule would not be changed, and thus the bottom of the potential well for the reactant state would be unchanged. The vibrational substates involving the C—H stretching mode would, however, be affected by the isotopic change. The potential energy of a vibrational mode is directly proportional to ν, the frequency at which the bond vibrates. In the case of a vibration that stretches a bond between two atoms, as in our C—H bond, the vibrational frequency can be expressed in terms of the force constant for that vibration (a measure similar to the tension or resistance to compression of a macroscopic spring) and the masses of the two atoms of the bond, by

$$\nu = \frac{k}{\sqrt{m_r}} \tag{6.10}$$

where k is the force constant and m_r is the reduced mass of the diatomic system involved in the vibration. The reduced mass is related to the masses of the two atoms in the system (m_1 and m_2) as follows:

$$\frac{1}{m_r} = \frac{1}{m_1} + \frac{1}{m_2} \tag{6.11}$$

The activation energy associated with the transition between the reactant state and the transition state is most correctly measured as the energetic distance between the vibrational ground state of the reactant potential well and the transition state. The energy difference between the vibrational ground state and the potential well minimum is referred to as the *zero-point energy* and is given the symbol Δe. The value of Δe is directly proportional to ν, which in turn is inversely proportional to the masses of the atoms involved in the vibration, according to Equation 6.11.

The frequency of a C—H bond-stretching vibration can be measured by infrared or Raman spectroscopy and has a typical value of about 2900 cm^{-1}. If we replace the proton with a deuteron (C—D, or C—^{2}H), this vibrational frequency shifts to roughly 2200 cm^{-1}. These frequencies correspond to Δe values of 4.16 and 3.01 kcal/mol for the C—H and C—D bonds, respectively. Therefore, the zero-point energy for a C—D bond will be 1.15 kcal/mol lower than that for a C—H bond (Figure 6.27). If all the vibrational potential energy of the reactant ground state is converted to transitional energy in achieving the transition state, this difference in zero-point energy will correspond to a 1.15 kcal/mol increase in overall activation energy for the C—D bond compared to that for the C—H bond. As we saw in Chapter 2, an increase in activation energy corresponds to a decrease in reaction rate, and thus the lowering of zero-point energy for a heavier isotope explains the reduction in reaction rate that is observed in kinetic experiments.

The effects of deuterium isotope substitution on the rate of reactions is typically expressed as the ratio V_{max}^{H}/V_{max}^{D} or k_{cat}^{H}/k_{cat}^{D}. Based on the difference in zero-point energy for a C—H and a C—D bond, we would expect the difference in activation energy for these two group transfers to be 1.15 kcal/mol if all of the vibrational potential energy is converted to transitional energy. From Equation 2.7, we would thus expect the kinetic isotope effect here to be

$$\frac{k_{cat}^{C-H}}{k_{cat}^{C-D}} = \exp\left(-\frac{\Delta\Delta G^{\ddagger}}{RT}\right) = 7 \tag{6.12}$$

Note that the isotope effect will be realized in the measured kinetics only if the hydrogen transfer step is rate limiting in the overall reaction. Also, the magnitude of the isotope effect will vary from enzyme to enzyme, depending on

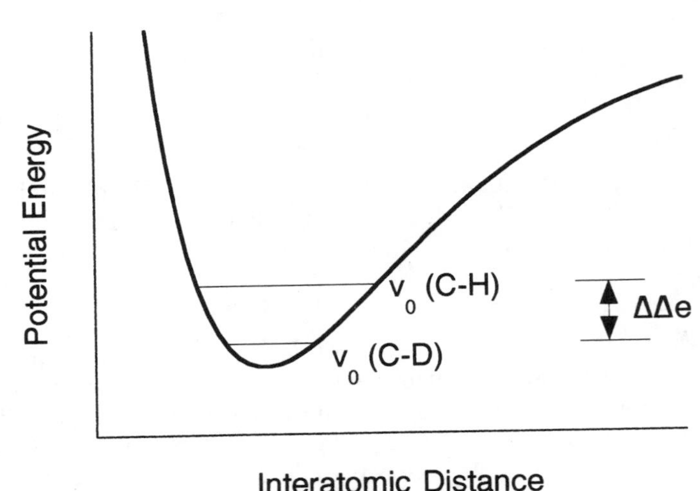

Figure 6.27 Potential energy diagram for an electronic state of a molecule illustrating the difference in zero-point energy $\Delta\Delta e$ for a C—H and a C—D bond.

the degree to which the transition state converts the vibrational potential energy of the ground reactant state to transitional energy. In proton transfer reactions one also finds that the magnitude of the kinetic isotope effect is influenced by the pK_a of the general base group that participates in the transfer step. As a rule, the largest kinetic isotope effects occur when the pK_a of the general base is well matched to that of the carbon acid of the proton donor; the magnitude of the kinetic isotope effects diminishes as the difference in these pK_a values increases.

Kinetic isotope effects can be very useful in identifying the specific atoms that participate in rate-limiting group transfer steps during catalysis. A common strategy is to synthesize substrate molecules in which a specific atom is isotopically labeled, and then to compare the rate of reaction for this substrate with that for the unlabeled molecule. When a group that participates in a rate-limiting transfer step is labeled, a kinetic isotope effect is observed. This information not only can be used to determine what groups are involved in particular transfer reactions, it also can help to identify the rate-determining steps in the catalytic mechanism of an enzyme. Comprehensive treatments of the use of kinetic isotope effects in the elucidation of enzyme mechanisms can be found in the reviews by Cleland and coworkers (Cleland et al., 1977) and by Northrop (1975).

Isotopic substitution of the solvent water hydrogens can affect the kinetics of enzyme reactions if the solvent itself serves as a proton donor during catalysis, or if the proton donor groups on the enzyme or substrate can rapidly exchange with the solvent; these effects are referred to as *solvent isotope effects.* In simple enzyme systems, the solvent isotope effects can be used to determine the number of protons transferred during the rate-determining step of catalysis.

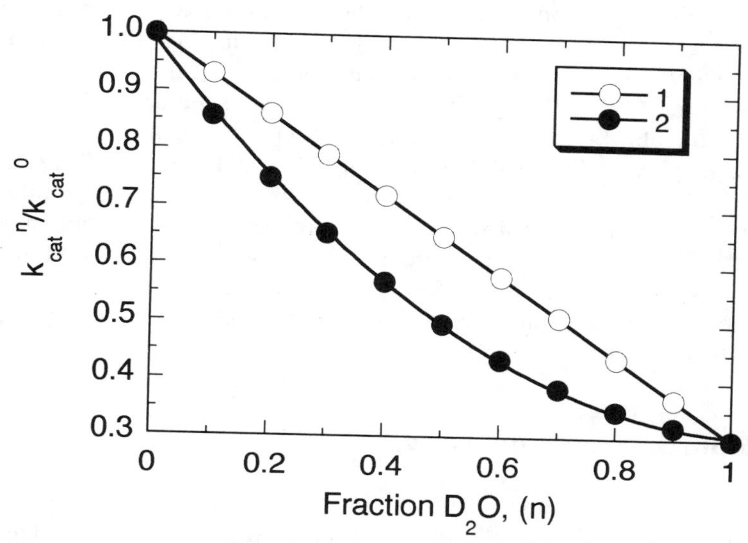

Figure 6.28 Proton inventory plot for reactions involving transfer of one (open circles) and two (solid circles) protons during the rate-limiting step in catalysis. The y axis is the ratio of k_{cat} in some mixture of D_2O and H_2O (k^n_{cat}) and the k_{cat} value in 100% H_2O (k^0_{cat}). Data for one-proton reaction fit by a linear function; data for two-proton reaction fit by a quadratic function. Reactions involving three-proton transfers in the rate-limiting step would be fit by a cubic function. Reactions involving more than three protons usually are fit by an exponential function.

This is done by measuring the velocity of the reaction as a function of the atom fraction of deuterium (n), or the percentage of D_2O in a mixed H_2O/D_2O solvent system. A plot of V_{max} or k_{cat} as a function of n, or % D_2O, will show a diminution in these kinetic parameters as the amount of D_2O in the solvent system increases. If a single proton transfer event is responsible for the solvent isotope effect, the data in such a plot will be well fit by a linear function. If two protons are transferred, the data will be best fit by a quadratic equation. For three protons, a cubic equation will be required to fit the data, and so on (Figure 6.28).

Generally, the involvement of more than three protons yields a plot that is best fit by an exponential function, which would suggest an "infinite number" of proton transfers during the rate-limiting step. This *proton inventory* method does not provide any insight into the structures or locations of the proton transferring groups, but it does allow one to quantify the number of groups participating in the rate-limiting step. Some caution must be exercised in interpreting these data. The interpretations based on curve fitting assume that a single step, the rate-limiting step, is responsible for all of the observed solvent isotope effect. In most simple enzyme systems this generally holds true. In more complex, multi–transition state systems, however, the assumption may not be valid. The validity of the data also depends on having all other solution conditions held constant as the percentage of D_2O in the solvent system is varied. One must remember, for example, that there is a difference between the true pD value of a D_2O solution and the value measured with a conventional pH meter (for a pure D_2O solution, the true pD = pH meter reading + 0.41 at 25°C); these effects must be accounted for in the preparation of solutions for the measurement of proton inventories. Two recent reviews describe the use of the proton inventory method in great detail (Schowen and Schowen, 1982; Venkatasubban and Schowen, 1984). The reader interested in applying these techniques would be well advised to refer to these more comprehensive treatments of the subject.

6.5 Reporting Enzyme Activity Data

As we saw in the preceding section of this chapter, many solution conditions can affect the overall activity of an enzyme-catalyzed reaction. Thus, for investigators in different laboratories to reproduce one another's results, it is critical that the data be reported in meaningful units and be accompanied by sufficient information on the details of the assay used. In reporting activity measurements one should always specify the buffer system used in the reaction mixture, the pH and temperature at which the assay was recorded, the time interval over which initial velocity measurements were made, and the detection method. Initial velocities and V_{max} values should always be reported in units of molarity (of substrate or product) change per unit time, while K_m and k_{cat} values be reported in molarity units and reciprocal time (min^{-1}, or s^{-1}), respectively. Turnover numbers are typically reported in terms of molarity change per unit time per molarity of enzyme, moles of substrate lost or product produced per unit time per mole of enzyme, or, equivalently, molecules of substrate lost or product produced per unit time per molecule of enzyme.

Many time it will be necessary to measure the enzymatic activity of samples that contain proteins other than the enzyme of interest. During the initial purification of an enzyme, for example, it is often helpful to follow the activity of the enzyme at various stages of the purification process, where multiple contaminating proteins will be present in the sample also. To standardize the reporting of activities in such cases, the International Union of Biochemistry (IUB) has adapted the *international unit* (IU) as the standard measure of enzyme activity: one international unit is the amount of enzyme (or crude enzyme sample) required to catalyze the transformation of one micromole of substrate per minute or, where more than one bond of each substrate molecule is attacked, one microequivalent of the group concerned, under a specific set of defined solution conditions. The definition allows the individual researcher to specify the solution conditions, but the IUB recommended that units be reported for measurements made at 25°C. The specific solution conditions have no intrinsic significance, but they must be reported to ensure reproducibility.

In crude enzyme samples the total mass of protein in the sample can be determined by a number of analytical methods (see Copeland, 1994, for details), but it is often difficult to measure specifically the mass or concentration of the enzyme of interest in such samples. To quantify the amount of enzyme present researchers often report the *specific activity* of the sample: that is, the number of international units of enzymatic activity per milligram of total protein under a specific set of solution conditions. Most typically, specific activity values are reported under conditions of saturating substrate (i.e., where $v \approx V_{max}$) and optimal solution conditions (i.e., pH, temperature, etc.) for maximal activity. As the purification of an enzyme proceeds, the specific activity of the sample will continuously increase as more and more of the total protein mass of the sample is made up by the enzyme of interest.

6.6 Enzyme Stability

One of the most common practical problems facing the experimental biochemist is the loss of enzymatic activity in a sample due to enzyme instability. Enzymes, like most proteins, are prone to denaturation under many laboratory conditions, and specific steps must be taken to stabilize these macromolecules as much as possible. Recommendations for the general handling of proteins for maximum stability have been described in detail in several texts devoted to proteins (see, e.g., Copeland, 1994). The general recommendations for the storage and handling of enzymes that follow can help to maintain the catalytic activities of these proteins.

Like all proteins, enzymes in their native states are optimally stabilized by specific solution conditions of pH, ionic strength, anion/cation composition, and so on. No generalities can be stated with respect to these conditions; the best conditions for each enzyme individually must be determined empirically. Note, however, that the solution conditions that are optimal for protein stability are not necessarily the same as those for optimal enzymatic activity. When this caveat applies, enzyme stocks should be stored under the conditions that maximally promote stability, while enzyme assays should be conducted under the conditions of optimal activity.

For long-term storage, enzymes should be kept at cryogenic temperatures in a −70°C freezer or under liquid nitrogen. Conventional freezers operate at a nominal temperature of −20°C, but most of these cycle through higher temperatures to keep them "frost free." This can lead to unintentional freeze–thaw cycling of the enzyme sample, which can be extremely denaturing. If enzymes must be stored in such a freezer, protein stability can be greatly enhanced by adding an equal volume of glycerol to the sample and mixing well. This 50% glycerol solution will maintain the enzyme sample in the liquid phase at −20°C, and thus will prevent repeated freezing and thawing.

Before freezing, the samples should be sterile-filtered through a 0.22 μm filter composed of a low protein binding material and placed in sterilized cryogenic tubes to avoid bacterial contamination. It is critical that the samples be frozen quickly and thawed quickly to avoid protein denaturation. Rapid freezing is best accomplished by immersing the sample container in a slurry of dry ice and ethanol. Rapid thawing is best done by placing the sample in a 37°C water bath until most, but not all, of the sample is in the liquid state. When there is just a small bit of frozen material remaining, the sample should be removed from the bath and allowed to continue thawing on ice (i.e., 4°C).

Repeated freeze–thaw cycles are extremely denaturing to proteins and must be avoided. Thus, a frozen enzyme sample should be thawed once and used promptly. Sample remaining at the end of the experiment should not be refrozen. An enzyme that can be maintained in stable condition for several days at 4°C, however, may be used in an experiment run soon after the first. If a particular enzyme is not stable under these conditions, any sample remaining at the end of an experiment must be discarded. To avoid wasting enzyme sample material, samples should be stored in small volume, high concentration aliquots. This way the volume of sample that is needed for each day's experiments can be thawed, while the bulk of the sample aliquots remain frozen. Once thawed, the enzyme should be kept at ice temperature (4°C) for as long as possible before equilibration to the assay temperature. Again, if the enzyme is stored at high concentration, only a small volume of the enzyme stock will be needed for dilution into the final reaction mixture. For example, a typical enzyme assay might require a final concentration of enzyme in the reaction mixture of 10 nM. Suppose that an enzyme is in long-term storage at −70°C as a 100 μM stock in 50 μL aliquots. On the day that assays are to be performed with the enzyme, a single aliquot might be thawed and diluted 1:100 with an appropriate buffer to make a 5 mL working stock of 1.0 μM enzyme. This stock would be stored on ice for the day (or potentially longer). The final reaction mixture would be prepared as a 1:100 dilution of the working stock to yield the desired final enzyme concentration of 10 nM.

Certain additives will enhance the stability of many enzymes for long-term storage at cryogenic temperatures and sometimes also for short-term storage in solution. Glycerol, sucrose, and cyclodextrans are often added to stabilize enzyme samples; the exact concentrations of these excipients that best stabilize a particular enzyme must be determined empirically. Some enzymes are greatly stabilized by the presence of cofactors, substrates, and even inhibitors that bind to their active sites. Again, the best storage conditions must be established for each enzyme individually.

Another common problem in handling enzyme solutions is the loss of enzymatic activity due to protein adsorption onto the surface of containers and pipette tips. Proteins bind avidly to glass, quartz, and polystyrene surfaces. Hence, containers made of these materials should not be used for enzyme samples. Containers and transfer devices constructed of low protein binding materials, such as polypropylene or polyethylene, should be used whenever possible; a wide variety of containers and pipette tips made of these materials are available commercially.

Even with low protein binding materials, one will still experience losses of protein due to adsorption. To minimize these effects, it is often possible to add a carrier protein to enzyme samples, as long as it has been established that the carrier protein will not interfere with the enzyme assay in any way. A carrier protein is an inert protein that is added to the enzyme solution at a concentration much higher than that of the enzyme. In this way potential protein binding surfaces become saturated with the carrier protein, hence are not available for adsorption of the enzyme of interest. It is a very common practice among enzymologists to add carrier proteins to the enzyme stock solutions and to the final reaction mixtures. Bovine serum albumin (BSA) and gelatin are the two most commonly used proteins for this purpose. Our laboratory has found that gelatin, at a concentration of 1 mg/mL, is a particularly good carrier protein for many enzymes. The lack of aromatic amino acids in the gelatin protein makes this a useful carrier protein for enzyme studies utilizing ultraviolet absorption or fluorescence spectroscopy. Both gelatin and BSA are available commercially in highly purified forms from a number of suppliers.

6.7 Summary

In this chapter we have presented an overview of some of the common methodologies for obtaining initial velocity measurements of enzymatic reactions. The most common detection methods and techniques for separating substrate and product molecules after reaction were discussed. We saw that changes in reaction conditions, such as pH and temperature, can have dramatic effects on enzymatic reaction rate. We saw further that controlled changes in these conditions can be used to obtain mechanistic information about the enzyme of interest. Finally, some advice was provided regarding storage and handling practices designed to optimally maintain the catalytic activity of enzymes in the laboratory.

References and Further Reading

Cleland, W. W. (1979) *Anal. Biochem.* **99**, 142.

Cleland, W. W. O'Leary, M. H., and Northrop, D. B. (1977) *Isotope Effects on Enzyme-Catalyzed Reactions*, University Park Press, Baltimore.

Copeland, R. A. (1994) *Methods for Protein Analysis: A Practical Guide to Laboratory Protocols*, Chapman & Hall, New York.

Copeland, R. A., Williams, J. M., Giannaras, J., Nurnberg, S., Covington, M., Pinto, D., Pick, S., and Trzaskos, J. M. (1994) *Proc. Natl. Acad. Sci. U.S.A.* **91**, 11202.

Copeland, R. A., Lombardo, D., Giannaras, J., and Decicco, C. P. (1995) *Bioorg. Med. Chem. Lett.* **5**, 1947.

Dixon, M., and Webb, E. C. (1979) *Enzymes*, 3rd ed., Academic Press, New York.

Eisenthal, R., and Danson, M. J., Eds. (1992) *Enzyme Assays: A Practical Approach*, IRL Press, Oxford.

Fletcher, A. N. (1969) *Photochem. Photobiol.* **9**, 439.

Gabriel, O., and Gersten, D. M. (1992) In *Enzyme Assays: A Practical Approach*, R. Eisenthal and M. J. Danson, Eds., IRL Press, Oxford, pp. 217–253.

Gomori, G. (1992) In *CRC Practical Handbook of Biochemistry and Molecular Biology*, G. D. Fasman, Ed., CRC Press, Boca Raton, FL, pp. 553–560.

Hames, B. D., and Rickwood, D. (1990) *Gel Electrophoresis of Proteins: A Practical Appoach*, 2nd Ed., IRL Press, London.

Hancock, W. S. (1984) *Handbook of HPLC Separation of Amino Acids, Peptides, and Proteins*, CRC Press, Boca Raton, FL.

Harlow, E., and Lane D. (1988) *Antibodies: A Laboratory Manual*, Cold Spring Harbor Laboratory, Cold Spring Harbor, NY.

Haupt, G. W. (1952) *J. Res. Natl. Bureau Stand.* **48**, 414.

Ittarat, I., Webster, H. K., and Yuthavong, Y. (1992) *J. Chromatography* **582**, 57.

Knight, C. G., Willenbrock, F., and Murphy, G. (1992) *FEBS Lett.* **296**, 263.

Kyte, J. (1995) *Mechanism in Protein Chemistry*, Garland, New York.

Lackowicz, J. R. (1983) *Principle of Fluorescence Spectroscopy*, Plenum Press, New York.

Matayashi, E. D., Wang, G. T., Krafft, G. A., and Erickson, J. (1990) *Science*, **247**, 954.

Mozhaev, V. V., Berezin, I. V., and Martinek, K. (1987) *Methods Enzymol.* **135**, 586.

Northrop, D. B. (1975) *Biochemistry*, **14**, 2644.

Oldham, K. G. (1968) *Radiochemical Methods of Enzyme Analysis*, Amersham International, Amersham, Bucks., U.K.

Oldham, K. G. (1977) In *Radiotracer Techniques and Applications*, Vol. 2, E. A. Evans and M. Muramatsu, Eds., Dekker, New York, pp. 823–891.

Oldham, K. G. (1992) In *Enzyme Assays: A Practical Approach*, R. Eisenthal and M. J. Danson, Eds., IRL Press, Oxford, pp. 93–122.

Oliver, R. W. (1989) *HPLC of Macromolecules: A Practical Approach*, IRL Press, Oxford.

Palmer, T. (1985) *Understanding Enzymes*, Wiley, New York.

Roughton, F. J. W., and Chance, B. (1963) In *Techniques of Organic Chemistry, Vol. VIII, Part II, Investigation of Rates and Mechanisms of Reactions*, S. L. Friess, E. S. Lewis, and A. Weissberger, Eds., Wiley, New York, pp. 703–792.

Rudolph, F. B., Baugher, B. W., and Beissmer, R. S. (1979) *Methods Enzymol.* **63**, 22.

Russo, S. F. (1969) *J. Chem. Educ.* **46**, 374.

Schowen, K. B., and Schowen, R. L. (1982) *Methods Enzymol.* **87**, 551.

Segel, I. H. (1976) *Biochemical Calculations*, 2nd ed., Wiley, New York.

Tipton, K. F. (1992) In *Enzyme Assays: A Practical Approach*, R. Eisenthal and M. J. Danson, Eds., IRL Press, Oxford, pp. 1–58.

Tsukada, H., and Blow, D. M. (1985) *J. Mol. Biol.* **184**, 703.

Venkatasubban, K. S., and Schowen, R. L. (1984) *CRC Critical Rev. Biochem.* **17**, 1.

CHAPTER

7

Reversible Inhibitors

The activity of an enzyme can be blocked in a number of ways. For example, inhibitory molecules can bind to sites on the enzyme that interfere with proper turnover. We have already encountered the concept of product inhibition in Chapter 5. In this case the product, having some structural resemblance to the substrate molecule, can bind to the active site of the enzyme, thus blocking the binding of further substrate molecules. This form of inhibition, in which the substrate and inhibitor compete for a common binding site on the enzyme, is known as *competitive inhibition.* While perhaps less intuitively obvious, it is also possible for inhibitory molecules to bind at sites distinct from the substrate binding site, and still block enzyme turnover in processes known as *noncompetitive, mixed,* and *uncompetitive inhibition.* In this chapter, we discuss these varied modes of inhibiting enzymes and examine kinetic methods for distinguishing among them.

There are several motivations for studying enzyme inhibition. At the basic research level, inhibitors can be useful tools for distinguishing among different potential mechanisms of enzyme turnover, particularly in the case of multisubstrate enzymes (see Chapter 10). By studying the relative binding affinity of competitive inhibitors of varying structure, one can glean information about the active site structure of an enzyme in the absence of a high resolution three-dimensional structure from x-ray crystallography or NMR spectroscopy. Inhibitors occur throughout nature, and they provide important control mechanisms in biology. Associated with many of the proteolytic enzymes involved in tissue remodeling, for example, are protein-based inhibitors of catalytic action that are found in the same tissue sources as the enzymes themselves. By balancing the relative concentrations of the proteases and their inhibitors, an organism can achieve the correct level of homeostasis. Enzyme inhibitors have a number of commercial applications as well. For example, enzyme inhibitors form the basis of a number of agricultural products, such as insecticides and weed killers of certain types. Inhibitors are extensively used to control parasites and other pest organisms by selectively inhibiting an enzyme of the pest, while sparing the enzymes of the host organism. Many of the drugs that are prescribed by physicians to combat diseases function by inhibiting specific enzymes associated with the disease process. Thus, enzyme inhibition is a major research focus throughout the pharmaceutical industry.

Inhibitors can act by irreversibly binding to an enzyme and rendering it inactive. This typically occurs through the formation of a covalent bond between some group on the enzyme molecule and the inhibitor. We shall discuss this type of inhibition in Chapter 9. Also, some inhibitors can bind so tightly to the enzyme that they are for all practical purposes permanently bound (i.e., their dissociation rates are very slow). These inhibitors, which form a special class known as tight binding inhibitors, are treated separately, in Chapter 8. In their most commonly encountered form, however, inhibitors are molecules that bind reversibly to enzymes with rapid association and dissociation rates. Molecules that behave in this way, known as classical reversible inhibitors, serve as the focus of our attention in this chapter.

Much of the basic and applied use of reversible inhibitors relies on their ability to bind specifically and with reasonably high affinity to a target enzyme. The relative potency of a reversible inhibitor is measured by its binding capacity for the target enzyme, and this is typically quantified by measuring the dissociation constant for the enzyme–inhibitor complex.

$$[E] + [I] \underset{K_d}{\rightleftharpoons} [EI]$$

$$K_d = \frac{[E][I]}{[EI]}$$

This dissociation constant is often referred to as the inhibitor constant and is given the special symbol K_i. The K_i value of a reversible enzyme inhibitor can be determined experimentally in a number of ways. One could apply the technique of equilibrium dialysis, for example, to determine the value of K_i of an inhibitor for an enzyme (see Segel, 1976, for a discussion of equilibrium dialysis methods).

In many cases the binding of an inhibitor to an enzyme will affect some spectroscopic features of enzyme or inhibitor or both, and these spectroscopic signals can be used to determine the equilibrium constant for the complex. A contemporary example is found in the recent work by Furfine and coworkers (1992) on the use of inhibitor fluorescence quenching to determine the K_i values of inhibitors of the aspartyl protease from the HIV virus. New instrumentation based on surface plasmon resonance technology (e.g., the

BIAcore system from Pharmacia Biosensor) allows one to measure binding interactions between ligands and macromolecules in real time (Chaiken et al., 1991; Karlsson, 1994). While this method has been mainly applied to determining the binding affinities for antigen–antibody and receptor–ligand interactions, the same technology holds great promise for the study of enzyme–ligand interactions as well. For example, this method has already been used to study the interactions between protein-based protease inhibitors and their enzyme targets (see, e.g., Ma et al., 1994). Although these and many other physicochemical methods have been applied to the determination of K_i values for enzyme inhibitors, the most common and straightforward means of assessing inhibitor binding consists of determining its effect on the catalytic activity of the enzyme. By measuring the diminution of initial velocity with increasing concentration of the inhibitor, one can find the relative concentrations of free enzyme and enzyme–inhibitor complex at any particular inhibitor concentration, and thus calculate the relevant equilibrium constants. For the remainder of this chapter, we shall focus on the determination of K_i values through initial velocity measurements of these types.

7.1 Equilibrium Treatment of Reversible Inhibition

To understand the molecular basis of reversible inhibition, it is useful to reflect upon the equilibria between the enzyme, its substrate, and the inhibitor that can occur in solution. Figure 7.1 provides a generalized scheme for the potential interactions between these molecules. In this scheme, K_S is the equilibrium constant for formation of the ES complex from the free enzyme and free substrate, K_i is the dissociation constant for the EI complex, and k_P is the forward rate constant for product formation from the ES or ESI complexes. The factor α reflects the effect of the inhibitor on the affinity of the enzyme for its substrate, and the effect of the substrate on the affinity of the enzyme for the inhibitor. The factor β reflects the modification of the rate of product formation by the enzyme that is caused by the inhibitor. An inhibitor that completely blocks enzyme activity will drive the value of β to zero. An inhibitor that only partially blocks product formation will yield a value of β between 0 and 1. An enzyme activator, on the other hand, will provide a value of β greater than 1.

The question is often asked: Why is the constant α the same for modification of K_S and K_i? The answer is that this constant must be the same for both on thermodynamic grounds. To illustrate, let us consider the following set of coupled reactions:

$$E + S \underset{K_S}{\rightleftharpoons} ES \qquad \Delta G = RT\ln(K_S) \tag{7.1}$$

$$ES + I \underset{\alpha K_i}{\rightleftharpoons} ESI \qquad \Delta G = RT\ln(\alpha K_i) \tag{7.2}$$

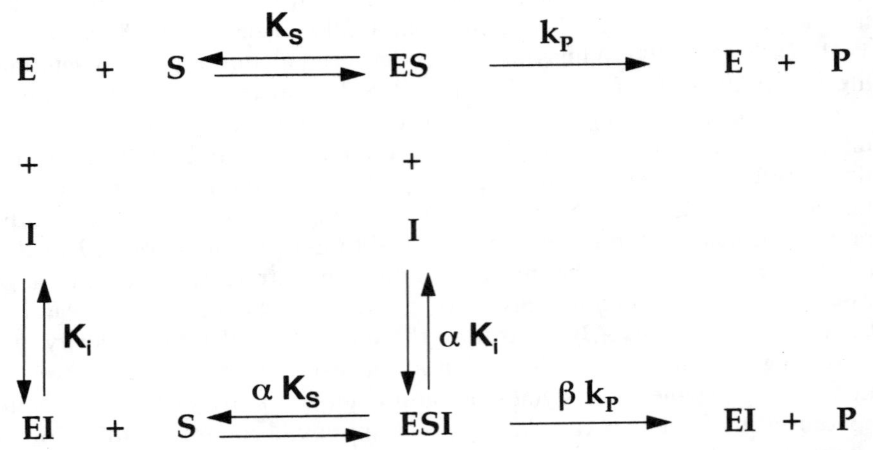

$$
\begin{array}{ccccccc}
E + S & \underset{}{\overset{K_S}{\rightleftharpoons}} & ES & \xrightarrow{k_P} & E + P \\
+ & & + & & \\
I & & I & & \\
\updownarrow K_i & & \updownarrow \alpha K_i & & \\
EI + S & \overset{\alpha K_S}{\rightleftharpoons} & ESI & \xrightarrow{\beta k_P} & EI + P
\end{array}
$$

Figure 7.1 Equilibrium scheme for enzyme turnover in the presence and absence of an inhibitor.

The net reaction of these two is:

$$E + S + I \rightleftharpoons ESI \qquad \Delta G = RT \ln(\alpha K_i K_S) \tag{7.3}$$

Now consider two other coupled reactions:

$$E + I \underset{K_i}{\rightleftharpoons} EI \qquad \Delta G = RT \ln(K_i) \tag{7.4}$$

$$EI + S \underset{aK_s}{\rightleftharpoons} ESI \qquad \Delta G = RT \ln(aK_S) \tag{7.5}$$

The net reaction here is:

$$E + S + I \rightleftharpoons ESI \qquad \Delta G = RT \ln(aK_S K_i) \tag{7.6}$$

Both sets of coupled reactions yield the same overall net reaction. Since, as we reviewed in Chapter 2, ΔG is a path-independent function, it follows that Equations 7.3 and 7.6 have the same value of ΔG. Therefore:

$$
\begin{aligned}
RT \ln(\alpha K_i K_S) &= RT \ln(aK_S K_i) \\
\therefore \quad \alpha(K_i K_S) &= a(K_i K_S) \\
\therefore \quad \alpha &= a
\end{aligned}
\tag{7.7}
$$

Thus, the value of α is indeed the same for the modification of K_S by inhibitor and the modification of K_i by substrate.

The values of α and β provide information on the degree of modification that one ligand (i.e., substrate and inhibitor) has on the binding of the other ligand, and define different modes of inhibitor interaction with the enzyme.

7.2 Modes of Reversible Inhibition

7.2.1 Competitive Inhibition

Competitive inhibition refers to the case of the inhibitor binding directly to the substrate binding site of the enzyme (or in close enough proximity to effectively occlude the substrate binding site), setting up a competition between the two ligands for the same binding site on the enzyme. Competitive inhibitors therefore diminish the initial velocity of the catalytic reaction by reducing the proportion of enzyme molecules that are available to bind substrate. In a sense, a competitive inhibitor reduces the effective concentration of active enzyme molecules in the sample. When the concentration of inhibitor is such that less than 100% of the enzyme molecules in solution are bound to inhibitor, one will observe residual activity due to the population of free enzyme. The molecules of free enzyme in this population will turn over at the same rate as in the absence of the inhibitor. Hence, the presence of a competitive inhibitor decreases the apparent affinity of the enzyme molecules for substrate but does not affect the maximum rate of enzyme turnover. In other words, competitive inhibitors affect the apparent K_m value of the enzyme for its substrate but not the V_{max} of the enzyme. For complete competitive inhibition, inhibitor binding blocks entirely the binding of substrate to the active site of an enzyme molecule. Hence, referring to the equilibria in Figure 7.1, complete competitive inhibition is characterized by $\alpha = \infty$ and $\beta = 0$. Figure 7.2A illustrates, in cartoon fashion, the interactions of a competitive inhibitor with an enzyme. Because competitive inhibitors must bind to the substrate binding site of the enzyme, these inhibitor generally share some structural commonality with the substrate or transition state of the reaction; that is, competitive inhibitors are structural mimics of the reaction substrate or transition state. Since these inhibitors compete with the substrate molecule for enzyme binding and vice versa, one of the hallmarks of competitive inhibition is that the inhibitor-bound enzyme will display increased relative activity at higher substrate concentrations. In other words, competitive inhibition can be overcome by increasing substrate concentration.

7.2.2 Noncompetitive Inhibition

Sometimes inhibitors bind with equal affinity to both the free enzyme (E) and the enzyme–substrate complex; thus complete noncompetitive inhibition is characterized by $\alpha = 1$ and $\beta = 0$. Inhibitors of this type bind to the enzyme at a site that is distinct from the substrate binding site and inhibit enzyme turnover by any of a number of different mechanisms. Noncompetitive inhibition cannot be overcome by increasing substrate concentration, since the two ligands do not compete for the same binding site on the enzyme. Thus the apparent effect of a noncompetitive inhibitor is to decrease the V_{max} of the enzyme without affecting the apparent K_m. Figure 7.2B illustrates the interactions between a noncompetitive inhibitor and its enzyme target.

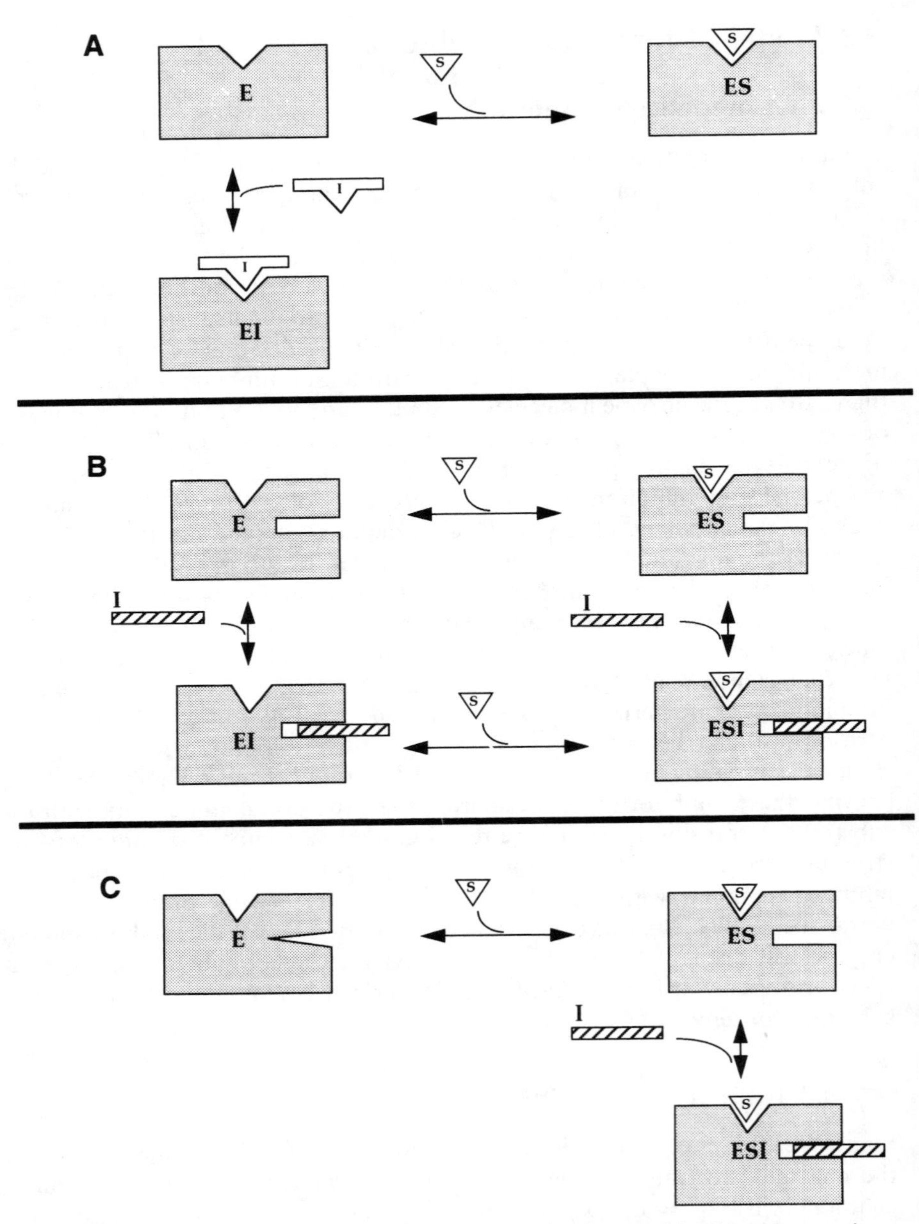

Figure 7.2 Cartoon representations of the three major forms of inhibitor interactions with enzymes. (A) competitive inhibition, (B) noncompetitive inhibition, and (C) uncompetitive inhibition.

7.2.3 Uncompetitive Inhibitors

Uncompetitive inhibitors bind to the enzyme at a site distinct from the substrate binding site, but in this case the inhibitor binds exclusively to the ES complex rather than to the free enzyme. The apparent effect of an uncompetitive inhibitor is to decrease V_{max} and to actually decrease K_m (i.e., increase the affinity of the enzyme for its substrate). Therefore, complete uncompetitive inhibitors are characterized by $\alpha < 1$ and $\beta = 0$ (Figure 7.2C).

Note that a truly uncompetitive inhibitor would have *no* affinity for the free enzyme; hence the value of K_i would be infinite. The inhibitor would, however, have a measurable affinity for the ES complex, so that αK_i would be finite. Obviously this situation is not well described by the equilibria in Figure 7.1. For this reason many authors choose to distinguish between the dissociation constants for [E] and [ES] by giving them separate symbols, such as K_{iE} and K_{iES}, K_i and K_I, K_{i1} and K_{i2}. Only rarely, however, does the inhibitor have no affinity whatsoever for the free enzyme. Rather, for uncompetitive inhibitors it is usually the case that $K_{iE} \gg K_{iES}$. Thus we can still apply the scheme in Figure 7.1 with the condition that for uncompetitive inhibitors $\alpha \ll 1$.

7.2.4 Mixed Inhibitors

The three types of inhibitor described thus far are the most commonly encountered forms of inhibition and will remain the focus of our attention for the rest of this chapter. There are a number of examples, however, of inhibitors that behave with a mixture of the characteristics of these other inhibitor types. Such molecules referred to as *mixed inhibitors*, generally affect both V_{max} and K_m, but in varying ways, depending on the details of their interactions with the enzyme molecule. Because mixed inhibitors will have different affinities for the free enzyme and the ES complex, we cannot describe these cases by a single inhibitor dissociation constant, K_i. Rather, in these cases one must determine the values of both K_i and αK_i to fully characterize the inhibition pattern observed (in some literature, the two dissociation constants are given different symbols, as discussed earlier). In Section 7.3 we shall describe graphical methods for diagnosing the mode by which an inhibitor interacts with an enzyme, including the patterns seen for mixed inhibitors. Readers who encounter this form of inhibition in their research should consult the more thorough treatment of these inhibitors in the text by Segel (1975).

7.2.5 Partial Inhibitors

Until now we have assumed that inhibitor binding to an enzyme molecule completely blocks subsequent product formation by that molecule. Referring to the scheme in Figure 7.1, this is equivalent to saying that $\beta = 0$ in these cases. In some situations, however, the enzyme can still turn over with the inhibitor bound, albeit at a far reduced rate compared to the uninhibited enzyme. Such situations, which manifest *partial inhibition*, are characterized by $0 < \beta < 1$. The distinguishing feature of a partial inhibitor is that the activity of the enzyme cannot be driven to zero even at very high concentrations of the inhibitor. This form of inhibition is relatively rare, however, and we shall not discuss it further. A more complete description of partial inhibitors has been presented elsewhere (Segel, 1975).

7.3 Graphic Determination of Inhibitor Type

7.3.1 Competitive Inhibitors

A number of graphic methods have been described for determining the mode of inhibition of a particular molecule. Of these, the double-reciprocal or Lineweaver–Burk plot is the most straightforward means of diagnosing inhibitor type. Recall from Chapter 5 that a double-reciprocal plot graphs the value of 1/velocity as a function of 1/[substrate] to yield, in most cases, a straight line. As well shall see, overlaying the double-reciprocal lines for an enzyme reaction carried out at several fixed inhibitor concentrations will yield a pattern of lines that is characteristic of a particular inhibitor type. The double-reciprocal plot was introduced in the days prior to the widespread use of computer-based curve-fitting methods, as a means of easily estimating the kinetic values K_m and V_{max} from the linear fits of the data in these plots. As described in Chapter 5, however, there are systematic weighting errors associated with the data manipulations that must be performed in constructing such plots.

To avoid weighting errors, and still use these reciprocal plots qualitatively to diagnose inhibitor type, we make the following recommendation. To diagnose inhibitor type, measure the initial velocity as a function of substrate concentration at several, fixed concentrations of the inhibitor of interest. Plot the data in terms of velocity as a function of substrate concentration for each inhibitor concentration, and fit these data to the Henri–Michaelis–Menten equation (Equation 5.16). Determine the values of K_m^{app} (i.e., the *apparent* value of K_m at different inhibitor concentrations) and V_{max}^{app} directly from the nonlinear least-squares best fits of the untransformed data. Finally, plug these values of K_m^{app} and V_{max}^{app} into the reciprocal equation (Equation 5.19) to obtain a linear function, and plot this linear function for each inhibitor concentration on the same double-reciprocal plot. In this way the double-reciprocal plots can be used to determine inhibitor type from the patterns of lines that result from varying inhibitor concentrations, but without introducing systematic errors that could compromise the interpretation.

Let's walk thorough an example of this to illustrate the method, and to determine the expected pattern for a competitive inhibitor. Let us say that we

Table 7.1 Hypothetical Velocity as a Function of Substrate Concentration at Three Fixed Concentrations of a Competitive Inhibitor

	Velocity (arbitrary units)		
[S] (μM)	[I] = 0	[I] = 10 μM	[I] = 25 μM
1	9.09	3.23	1.69
2	16.67	6.25	3.23
4	28.57	11.77	6.25
6	37.50	16.67	9.09
8	44.44	21.05	11.77
10	50.00	25.00	14.29
20	66.67	40.00	25.00
30	75.00	50.00	33.33
40	80.00	57.14	40.00
50	83.33	62.50	45.46

measure the initial velocity of our enzymatic reaction as a function of substrate concentration at 0, 10, and 25 μM concentrations of an inhibitor and obtain the results shown in Table 7.1. If we were to plot these data and fit them to Equation 5.16, we would obtain a graph such as that illustrated in Figure 7.3A. From the fits of the data, we would obtain the following *apparent* values of the kinetic constants:

$[I] = 0\ \mu M \quad V_{max} = 100 \quad K_m = 10.00\ \mu M$

$[I] = 10\ \mu M \quad V_{max}^{app} = 100 \quad K_m^{app} = 30.00\ \mu M$

$[I] = 25\ \mu M \quad V_{max}^{app} = 100 \quad K_m^{app} = 60.00\ \mu M$

If we plug these values of V_{max}^{app} and K_m^{app} into Equation 5.19 and plot the resulting linear functions, we obtain a graph such as that shown in Figure 7.3B.

The pattern of straight lines with interesecting y intercepts, as seen in Figure 7.3B, is the characteristic signature of a competitive inhibitor. The lines intersect at their y intercepts because a competitive inhibitor does not affect the apparent value of V_{max}, which, as we saw in Chapter 5, is defined by the reciprocal of the y intercept in a double-reciprocal plot. The slopes of the lines, which are given by K_m^{app}/V_{max}^{app}, vary among the lines because of the effect imposed on K_m by the inhibitor. The degree of perturbation of K_m will vary with the inhibitor concentration and will depend also on the value of K_i for the particular inhibitor. The influence of these factors on the initial velocity is given by:

$$v = \frac{V_{max}[S]}{[S] + K_m(1 + [I]/K_i)} \qquad (7.8)$$

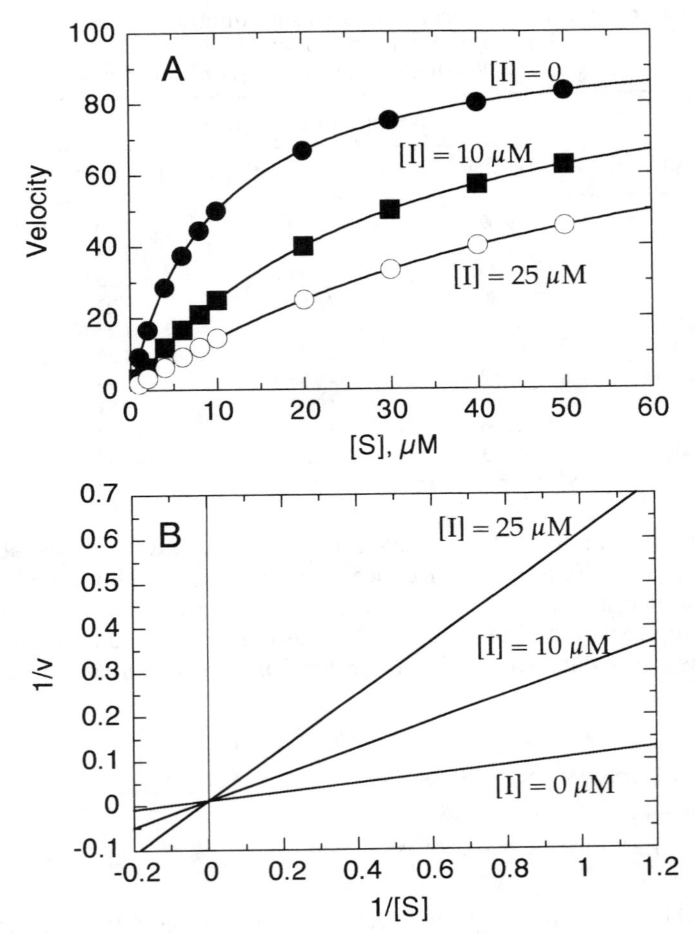

Figure 7.3 Untransformed (A) and double-reciprocal (B) plots for the effects of a competitive inhibitor on the velocity of an enzyme-catalyzed reaction. The lines in (B) are obtained by applying Equation 5.16 to the data (A) and using the apparent values of the kinetic constants in conjunction with Equation 5.19. See text for further details.

or, taking the reciprocal of this equation, we obtain:

$$\frac{1}{v} = \frac{1}{V_{\text{max}}} + \frac{1}{[\text{S}]} \frac{K_{\text{m}}}{V_{\text{max}}} \left(1 + \frac{[\text{I}]}{K_{\text{i}}}\right) \tag{7.9}$$

Now, comparing Equation 7.9 and 5.19, we see that the slopes of the double-reciprocal lines at inhibitor concentrations of 0 and i differ by the factor

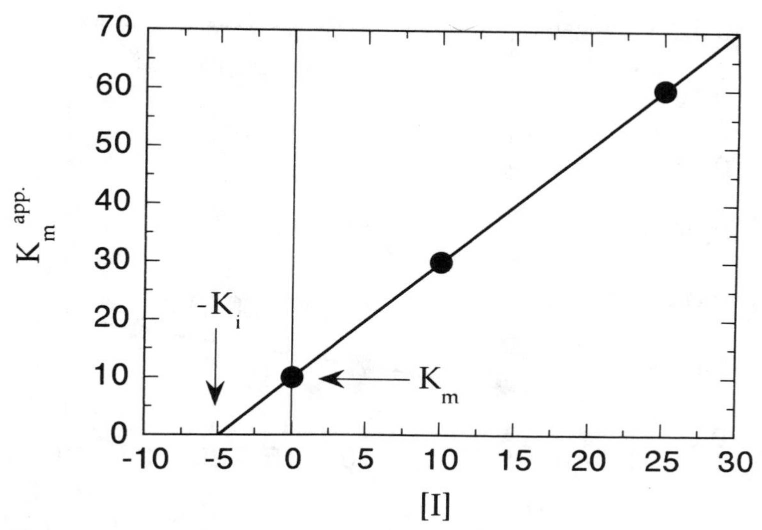

Figure 7.4 Secondary plot of K_m^{app} as a function of inhibitor concentration [I] for a competitive inhibitor. The value of the inhibitor constant K_i can be determined from the negative value of the x intercept of this type of plot.

$(1 + [I]/K_i)$. Thus, the ratio of these slope values is:

$$\frac{\text{slope}_i}{\text{slope}_0} = 1 + \frac{[I]}{K_i} \tag{7.10}$$

or, rearranging:

$$K_i = \frac{[I]}{\text{slope}_i/\text{slope}_0 - 1} \tag{7.11}$$

Thus, in principle, one could measure the velocity as a function of substrate concentration in the absence of inhibitor, and at a single, fixed values of [I] and use Equation 7.11 to determine the K_i of the inhibitor from the double-reciprocal plots. This method can be potentially misleading, however, because it relies on a single inhibitor concentration for the determination of K_i.

A more common approach to determining the K_i value of a competitive inhibitor is to replot the kinetic data obtained in plots such as Figure 7.3A as the apparent K_m value as a function of inhibitor concentration. The x intercept of such a "secondary plot" is equal to the negative value of the K_i, as illustrated in Figure 7.4, using the data from Table 7.1.

A third method for determining the K_i value of a competitive inhibitor was suggested by Dixon (1953). In this method one measures the initial velocity of the reaction as a function of inhibitor concentration at two or more fixed concentrations of substrate. The data are then plotted as $1/v$ as a function of

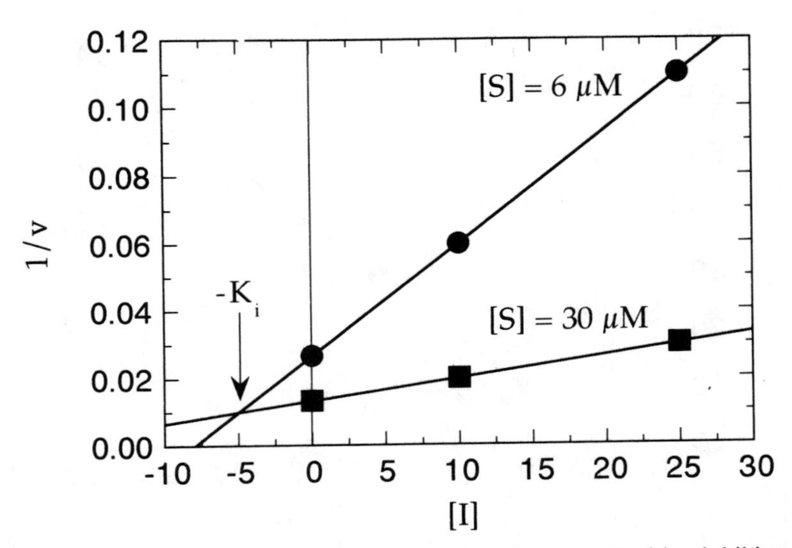

Figure 7.5 Dixon plot ($1/v$ as a function of [I]) for a competitive inhibitor at two different substrate concentrations The K_i value for this type of inhibitor is determined from the negative of the x-axis value at the point of intersection of the two lines.

[I] for each substrate concentration, and the value of $-K_i$ is determined from the x-axis value at which the lines interesect, as illustrated in Figure 7.5. The Dixon plot ($1/v$ as a function of [I]) is useful in determining the K_i values for other inhibitor types as well, as we shall see later in this chapter.

7.3.2 Noncompetitive Inhibitors

We have stated that a noncompetitive inhibitor affects the apparent V_{max} of the reaction but not K_m. The velocity equation for noncompetitive inhibition is

$$v = \frac{\dfrac{V_{max}}{1 + [\mathrm{I}]/K_i}[\mathrm{S}]}{K_m + [\mathrm{S}]} \tag{7.12}$$

or, in reciprocal form,

$$\frac{1}{v} = \left(\frac{1}{V_{max}} + \frac{K_m}{V_{max}}\frac{1}{[\mathrm{S}]}\right)\left(1 + \frac{[\mathrm{I}]}{K_i}\right) \tag{7.13}$$

From the form of Equation 7.13 we see that the double-reciprocal plots for varying concentrations of a noncompetitive inhibitor will yield a nest of lines that intersect at the x intercept of the graph, as illustrated in Figure 7.6. The point of intersection of these lines occurs at $-1/K_m$, while the y intercepts of the individual lines have values equal to $1/V_{max}(1 + [\mathrm{I}]/K_i)$.

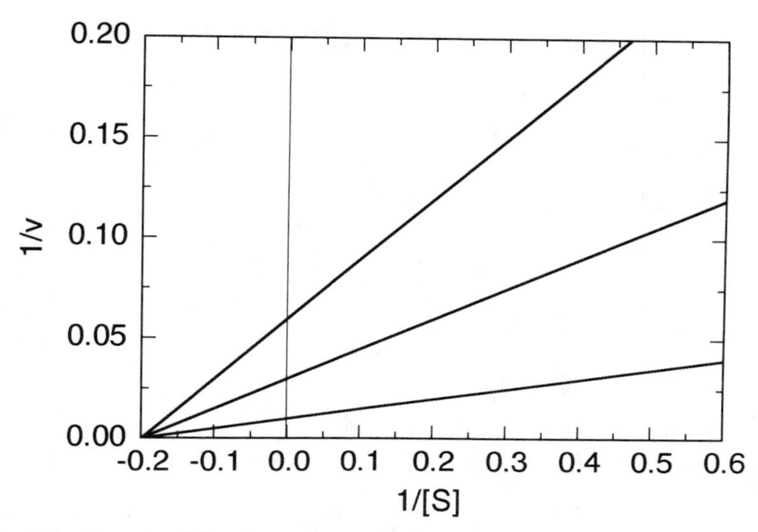

Figure 7.6 Pattern of lines in the double-reciprocal plot of a noncompetitive inhibitor Note the intersection of the lines at the $1/v = 0$, and $1/[S] < 0$.

To determine the K_i value of a noncompetitive inhibitor, researchers most commonly rely on the Dixon plot of $1/v$ as a function of [I] at a fixed substrate concentration (usually the studies are done at saturating substrate concentrations where $v \approx V_{max}$). The relationship between $1/v$ and [I] for a noncompetitive inhibitor is given by:

$$\frac{1}{v} = \frac{\dfrac{1 + K_m}{[S]}}{V_{max} K_i}[I] + \frac{1}{V_{max}}\left(1 + \frac{K_m}{[S]}\right) \tag{7.14}$$

If we divide the y intercept by the slope for this linear relationship, we obtain:

$$\frac{\dfrac{1}{V_{max}}\left(1 + \dfrac{K_m}{[S]}\right)}{\left(1 + \dfrac{K_m}{[S]}\right) \bigg/ V_{max} K_i} = \frac{K_i \dfrac{1}{V_{max}}\left(1 + \dfrac{K_m}{[S]}\right)}{\dfrac{1}{V_{max}}\left(1 + \dfrac{K_m}{[S]}\right)} = K_i \tag{7.15}$$

Thus, the K_i of a noncompetitive inhibitor can be obtained from a Dixon plot by dividing the y-intercept value by the slope. Recall, however, that the equation for any straight line can be defined as $y = mX + b$, where m is the slope and b is the y intercept. If we set $y = 0$ and rearrange, we find that $b/m = -x$. In other words, the y intercept divided by the slope of any linear function is equivalent to the negative of the x-intercept value. Therefore, we can very simply determine the K_i value of a noncompetitive inhibitor as the

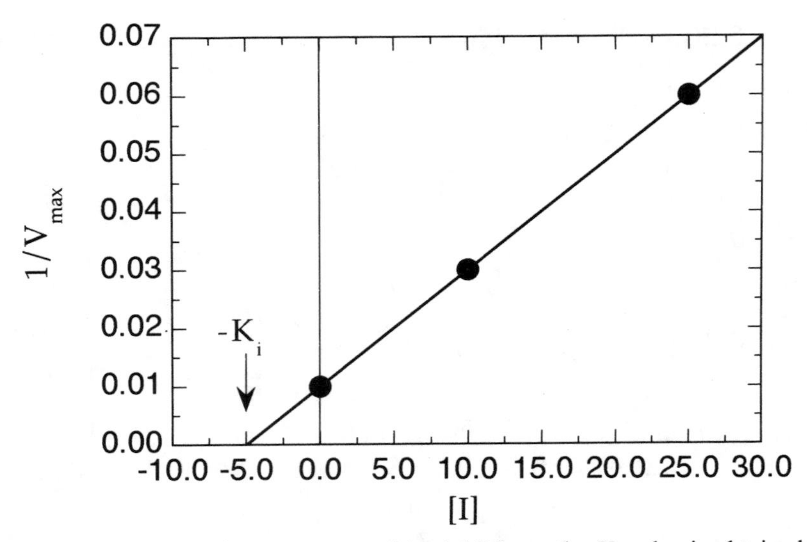

Figure 7.7 Dixon plot for a noncompetitive inhibitors; the K_i value is obtained from the negative value of the x intercept of the line.

negative of the x-intercept value from a Dixon plot. Figure 7.7 is a plot of this type for a noncompetitive inhibitor.

7.3.3 Uncompetitive Inhibitors

Both V_{max} and K_m are affected by the presence of an uncompetitive inhibitor. The form of the velocity equation therefore contains the K_i term in both the numerator and denominator:

$$v = \frac{\dfrac{V_{max}}{1 + [I]/K_i}[S]}{\dfrac{K_m}{1 + [I]/K_i} + [S]} \tag{7.16}$$

or in reciprocal form, we obtain:

$$\frac{1}{v} = \frac{\dfrac{K_m}{1 + [I]/K_i}}{\dfrac{V_{max}}{1 + [I]/K_i}} \frac{1}{[S]} + \frac{1}{\dfrac{V_{max}}{1 + [I]/K_i}} \tag{7.17}$$

Note that the term $1 + [I]/K_i$ appears in both the numerator and denominator of the slope term. We can therefore cancel these sums to obtain the simpler

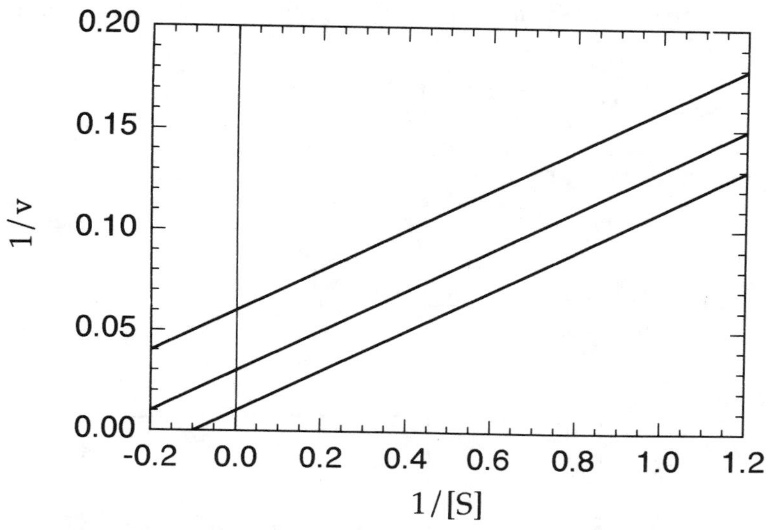

Figure 7.8 Pattern of lines in the double-reciprocal plot of an uncompetitive inhibitor.

equation:

$$\frac{1}{v} = \frac{K_m}{V_{max}} \frac{1}{[S]} + \frac{1}{\dfrac{V_{max}}{1 + [I]/K_i}} \tag{7.18}$$

We see from Equation 7.18 that the slope of the double-reciprocal plot is independent of inhibitor concentration and that the y intercept increases steadily with increasing inhibitor. Thus, the overlaid double-reciprocal plot for an uncompetitive inhibitor at varying concentrations appears as a series of parallel lines that intersect the y axis at different values, as illustrated in Figure 7.8.

For an uncompetitive inhibitor the x intercept of a Dixon plot will be equal to $-K_i(1 + K_m/[S])$. At first glance this relationship may not appear to be particularly convenient. If, however, one is working at saturating conditions, where $[S] \gg K_m$, the value of $K_m/[S]$ becomes very small and can be assumed to be zero. Under these conditions, the x intercept of the Dixon plot will be equal to $-K_i$. Thus, *under conditions of saturating substrate*, one can determine the value of K_i directly from the x intercept of a Dixon plot, as described earlier for the case of noncompetitive inhibition.

7.3.4 Mixed Inhibitors

We have seen that for mixed inhibitors we must take into consideration the inhibitor dissociation constant for both the free enzyme and the enzyme–

substrate complex. The form of the velocity equation in this case will be:

$$v = \frac{V_{\max}[S]}{[S]\left(1 + \frac{[I]}{\alpha K_i}\right) + K_m\left(1 + \frac{[I]}{K_i}\right)} \tag{7.19}$$

The reciprocal form of this equation (after some canceling of terms) has the form:

$$\frac{1}{v} = \left(1 + \frac{[I]}{K_i}\right)\frac{K_m}{V_{\max}}\frac{1}{[S]} + \frac{1 + \frac{[I]}{\alpha K_i}}{V_{\max}} \tag{7.20}$$

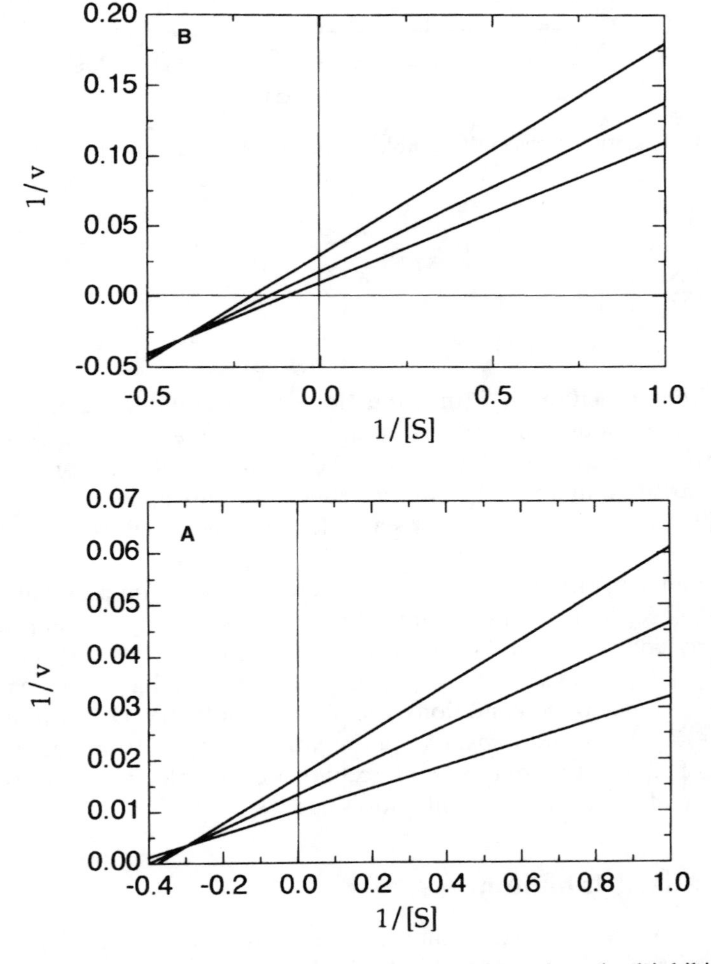

Figure 7.9 Patterns of lines in the double-reciprocal plots for mixed inhibitors when $\alpha > 1$ (A) and when $\alpha < 1$ (B).

As described by Equation 7.20, both the slope and y intercept of the double-reciprocal plot will be affected by the presence of inhibitor. The pattern of lines seen when the plots for varying inhibitor concentrations are overlaid will depend on the value of α for a mixed inhibitor. When $\alpha > 1$, the lines will intersect at a value of $1/[S]$ less than zero and a value of $1/v$ of greater than zero (Figure 7.9A). If, on the other hand, $\alpha < 1$, the lines will intersect below the x and y axes, at negative values of $1/[S]$ and $1/v$ (Figure 7.9B). To obtain the values of K_i and αK_i, one must construct two secondary plots. The first of

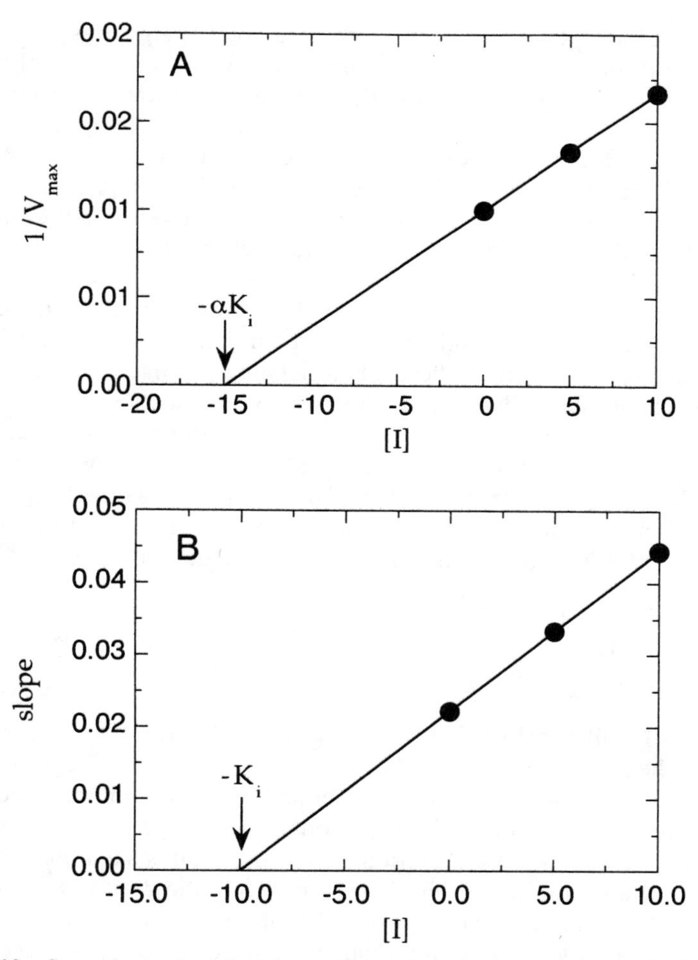

Figure 7.10 Secondary plots for the determination of the inhibitor constants for a mixed inhibitor. (A) $1/V_{max}$ versus [I]; the value of $-\alpha K_i$ is determined from the x intercept of the line. (B) The value of $-K_i$ is determined from the x intercept of a plot of the slope of the lines from the double-reciprocal (Lineweaver–Burk) plot as a function of [I].

these is a Dixon plot of $1/V_{max}$ (i.e., at saturating substrate concentration) as a function of [I], from which the value of $-\alpha K_i$ can be determined as the x intercept (Figure 7.10A). In the second plot, one graphs the slope of the double-reciprocal line (from the Lineweaver–Burk plot) as a function of [I]. For this plot, the x intercept will be equal to $-K_i$ (Figure 7.10B). Combining the information from these two plots, one can readily determine the value of both inhibitor dissociation constants.

7.4 Dose–Response Curves of Enzyme Inhibition

In many biological assays one can measure a specific signal as a function of the concentration of some exogenous substance. A plot of the signal obtained as a function of the concentration of the exogenous substance is referred to as a *dose–response plot*, and the function that describes the change in signal as a function of the concentration of substance is known as the *dose–response curve.* These plots can be used, for example, to follow the binding of ligands to receptors in vitro, to describe the effects of media additives on cell proliferation and growth, and to keep track of other cellular events that occur downstream of the initial ligand binding event. Usually the signal is plotted on a linear y-axis scale, and the concentration is plotted on a logarithmic x-axis scale. The resulting plot gives a sigmoidal or S-shaped curve from which the concentration of substance required to effect a half-maximal change in the signal being monitored can easily be determine. The same plots also can be used to follow the effects of an inhibitor on the initial velocity of an enzymatic reaction at a fixed concentration of substrate. The concentration of inhibitor required to achieve a half-maximal degree of inhibition is referred to as the IC_{50} value (for Inhibitor Concentration giving 50% inhibition), and the equation describing the effect of inhibitor concentration on reaction velocity is as follows:

$$\frac{v_i}{v_0} = \frac{1}{1 + \dfrac{[I]}{IC_{50}}} \tag{7.21}$$

where v_i is the initial velocity in the presence of inhibitor at concentration [I] and v_0 is the initial velocity in the absence of inhibitor. The ratio v_i/v_0 is referred to as the *fractional activity* remaining at a given inhibitor concentration. Figure 7.11, a typical dose–response plot for an enzyme inhibitor, illustrates the graphical determination of the IC_{50} from such a plot (today, the IC_{50} value of an inhibitor is determined by fitting the data in a plot such as Figure 7.11 to Equation 7.21). Note that the inhibitor concentration is typically varied over several logs to obtain good estimates of the IC_{50} values.

Dose–response plots are very widely used for comparing the relative inhibitor potencies of multiple compounds for the same enzyme, under well-controlled conditions. The method is popular because it permits the analyst to

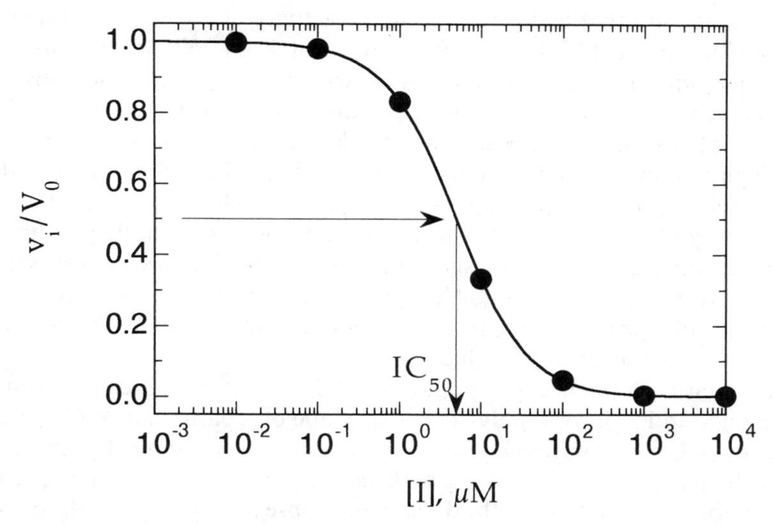

Figure 7.11 Dose–response plot of enzyme fractional activity as a function of inhibitor concentration. Note that the inhibitor concentration is plotted on a log scale. The value of the IC_{50} for the inhibitor can be determined graphically, as illustrated.

determine the IC_{50} by making measurements over a broad range of inhibitor concentrations, at a single fixed substrate concentration. This is very convenient when many compounds of unknown and varying inhibitory potency are to be screened. In the pharmaceutical industry, for example, one may wish to screen several thousand potential inhibitory substances to find those that have some potency against a particular target enzyme. These compounds are likely to span a wide range of IC_{50} values. Thus, one would set up a standard screening protocol in which the initial velocity of an enzymatic reaction is measured over five or more logs of inhibitor concentrations. In this way the IC_{50} values of many of the compounds could be determined without any prior knowledge of the range of concentrations required to effect potent inhibition of the enzyme.

The IC_{50} value is a practical readout of the relative effects on enzyme activity of different substances under a specific set of solution conditions. In many instances, it is the net effect of the inhibitor on enzyme activity, rather than its true dissociation constant for the enzyme, that is the ultimate criterion by which the effectiveness of a compound is judged. In some situations, a K_i value cannot be rigorously determined because of lack of knowledge or control over the assay conditions; many times, in these cases, the only measure of relative inhibitor potency is an IC_{50} value. For example, consider the task of determining the relative effectiveness of a series of inhibitors for a target enzyme in a cellular assay. Often, in these cases, the inhibitor is added to the cell medium and the effects of inhibition are measured indirectly by a readout of biological activity that is dependent on the activity of the target enzyme. In a cellular situation like this, one often does not know either the substrate concentration within the cell or the relative amounts of enzyme and substrate (recall that in vitro we set up our steady state conditions so that $[S] \gg [E]$, but this is not necessarily the case in the cell). Also, in these situations, one does not truly know the effective concentration of inhibitor *within* the cell that is causing the degree of inhibition being measured. This is because the cell membrane may block the transport of the bulk of the added inhibitor into the cell, and because cellular metabolism may diminish the effective concentration of inhibitor that reaches the target enzyme. Because of these uncontrollable factors in the cellular environment, often it is necessary to report the effectiveness of an inhibitor as an IC_{50} value.

Despite their convenience and popularity, IC_{50} value measurements can be misleading if used inappropriately. The IC_{50} value of a particular inhibitor can change with changing solution conditions, so it is very important to report the details of the assay conditions along with the IC_{50} value. For example, in the case of competitive inhibition, the IC_{50} value observed for an inhibitor will depend on the concentration of substrate present in the assay, relative to the K_m of that substrate. This is illustrated in Figure 7.12 for a competitive inhibitor under conditions of $[S] = K_m$ and $[S] = 10 \times K_m$. Thus, in comparing a series of competitive inhibitors, one must ensure that the IC_{50} values are measured at the same substrate concentration. For the same reasons, it is not rigorously correct to compare the relative potencies of inhibitors of different types by use of IC_{50} values. The IC_{50} values of a noncompetitive and

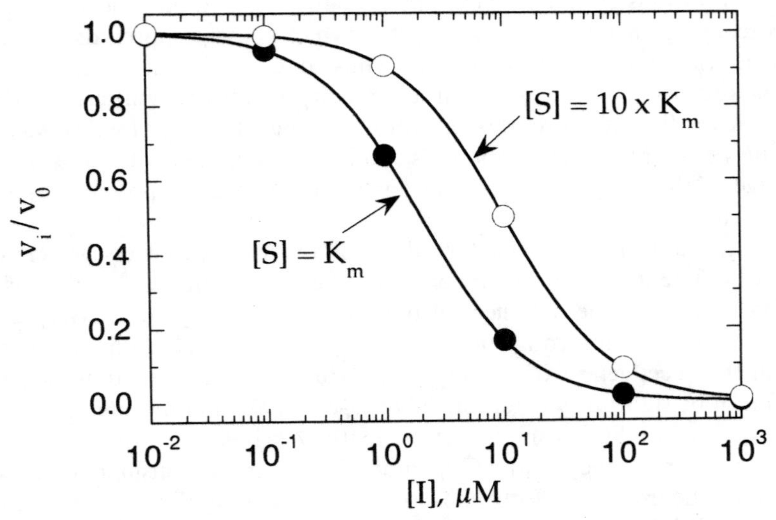

Figure 7.12 Effect of substrate concentration on the IC_{50} value of a competitive inhibitor.

a competitive inhibitor will vary with substrate concentration, but in different ways. Hence, the relative effectiveness observed in vitro under a particular set of solution conditions may not be the same relative effectiveness observed in vivo, where the conditions are quite different. Whenever possible, therefore, the K_i values should be used to compare the inhibitory potency of different compounds.

It is possible to take advantage of the convenience of IC_{50} measurements and still report inhibitor potency in terms of true K_i values when the mode of inhibition for a series of compounds is known, as well as the values of [S] and K_m. The relationship between the K_i, [S], K_m, and IC_{50} values can be derived from the velocity equations already presented. The derivations have been described in detail by Cheng and Prusoff (1973) for competitive, noncompetitive, and uncompetitive inhibitors. The reader is referred to the original paper for the derivations. Here we shall simply present the final forms of the relationships:

For competitive inhibitors:

$$K_i = \frac{IC_{50}}{1 + \frac{[S]}{K_m}} \tag{7.22}$$

For noncompetitive inhibitors:

$$K_i = IC_{50} \qquad \text{(assuming that } \alpha = 1\text{)} \tag{7.23}$$

For uncompetitive inhibitors:

$$K_i = \frac{IC_{50}}{1 + \frac{K_m}{[S]}} \qquad \text{(if [S]} \gg K_m\text{, then } K_i \approx IC_{50}\text{)} \tag{7.24}$$

Equations 7.22–7.24, known as the Cheng and Prusoff relationships, can be conveniently used to convert IC_{50} values to K_i values. To ensure that the correct relationship can be applied, however, it is critical, to know the mode of inhibition of the compounds being tested. It might thus seem that there is no great advantage to the use of the Cheng and Prusoff relationships if the mode of inhibition for each compound must be determined by Lineweaver–Burk analysis anyway. In many cases, however, one will wish to measure the relative inhibitory potency of a series of structurally related compounds. If these compounds represent small structural perturbations from a common parent molecule, it is often safe to assume that all the derivative molecules share the same mode of inhibition as the parent. In such a situation, one could determine the mode of inhibition for the parent molecule only and then apply the appropriate Cheng and Prusoff relationship to the rest of the molecular series.

There is, of course, the possibility of an inadvertent change in the mode of inhibition as a result of the structural perturbations. This is usually not a great danger if the perturbations are minor, and one can spot-check by performing Lineweaver–Burk analysis on a subgroup of compounds representing a wide range of perturbations within the series. This is a common strategy used in the development of structure–activity relationships for the determination of the key structural components in the inhibitory mechanism shared by a series of related molecules, as described in Section 7.5. The reader should note, however, that many scientists consider the K_i values derived by application of the Cheng and Prusoff relationships to be less accurate than those obtained by the more traditional methods described earlier. There is lower confidence in the former results partly because the effects of the inhibitor are examined at only a single, fixed substrate concentration. Nevertheless, because of their convenience, the Cheng and Prusoff relationships are commonly used for high throughput inhibitor screening.

At the beginning of this chapter we mentioned that some inhibitors do not block completely the ability of the enzyme to turnover when bound to the inhibitor. These partial inhibitors will not display the same dose–response curves as full inhibitors since, for these compounds, one can never drive the reaction velocity to zero, even at very high inhibitor concentrations. Rather, the dose–response curve for a partial inhibitor will be best fit by a more generalized form of Equation 7.21, given by:

$$y = \frac{y_{max} - y_{min}}{1 + \dfrac{[I]}{IC_{50}}} + y_{min} \qquad (7.25)$$

where y is the fractional activity of the enzyme in the presence of inhibitor at concentration [I], y_{max} is the maximum value of y that is observed at zero inhibitor concentration (for fractional activity, this is 1.0), and y_{min} is the minimum value of y that can be obtained at high inhibitor concentrations.

Unlike the case of full inhibitors, the dose–response curve for a partial inhibitor will reach a minimum, nonzero value of v_i/v_0 at high values of [I]. In Figure 7.13A, for example, the value of β for our inhibitor is 0.05, so that even at very high inhibitor concentrations, the enzyme still displays 5% of its uninhibited velocity. When behavior of this type is observed, one must be very careful to ensure that the lack of complete inhibition is not an experimental artifact. For example, in densitometry measurements one often observes some finite background density that is difficult to completely subtract out and can give the appearance of partial inhibition when, in fact, full inhibition is taking place. A more diagnostic signature of partial inhibition can be obtained by presenting the data as a Dixon plot. While all the full inhibitors discussed thus far yielded linear fits in Dixon plots, partial inhibitors typically display hyperbolic fits of the data in these plots (Figure 7.13B). In these cases one can extract the values of α, K_i, and β for the inhibitor, depending on the mode of partial inhibition that is taking place. These analyses are, however, beyond the

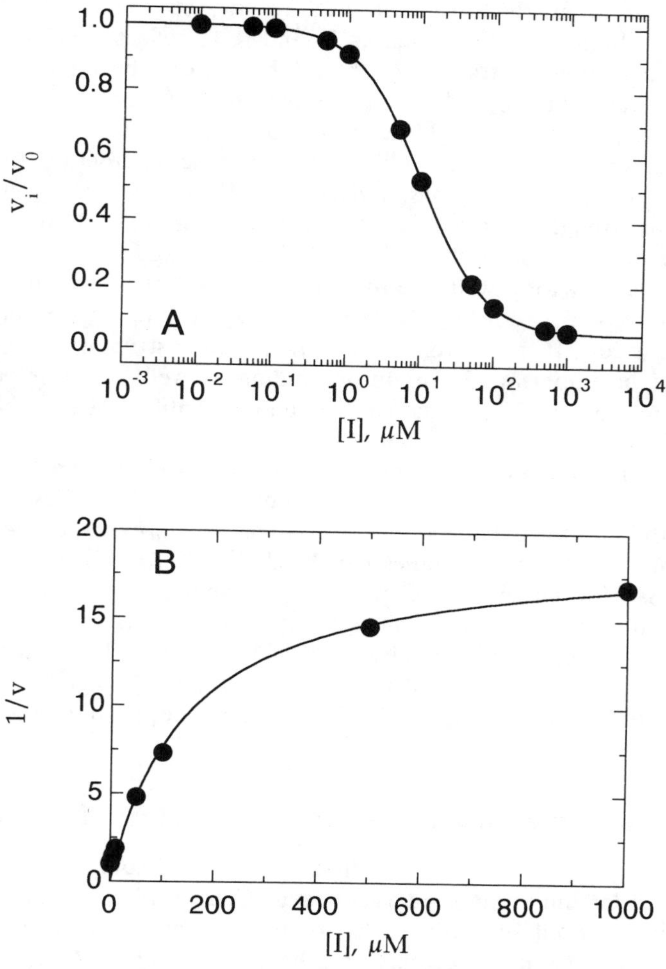

Figure 7.13 Dose–response (A) and Dixon (B) plots for a partial inhibitor. Note that in (A) the value of v_i/v_0 reaches a nonzero plateau at high inhibitor concentrations. The hyperbolic nature of the Dixon plot (B) is characteristic of partial inhibition.

scope of the present text. The reader who encounters this relatively unusual form of enzyme inhibition is referred to a text by Segel (1975) for a more comprehensive discussion of the data analysis.

7.5 Structure–Activity Relationships and Inhibitor Design

Modern attempts to identify inhibitors of specific enzymes have largely focused on elucidating the stereochemical and physicochemical features of inhibitory molecules that allow them to bind well to the enzyme (Suckling, 1991). Measures of inhibitor potency, such as K_i and IC_{50}, reflect the change in free energy that accompanies transfer of the inhibitor from the solvated aqueous state to the bound state in the enzyme binding pocket (i.e., the ΔG of binding). We have already discussed physicochemical forces that are important in protein structure and ligand binding: hydrophobic interactions, hydrogen bonding, electrostatic interactions, and van der Waals forces. The same forces determine the strength of interaction between an inhibitor and an enzyme.

Likewise, we have seen that the shape or topology of a substrate will determine its ability to fit well into the binding pocket of an enzyme, based on the structural complementarity between the enzyme binding pocket and the substrate molecule. It stands to reason that the same structural complementarity should be important in inhibitor binding as well, and this is what is observed empirically. Thus if one can somehow identify a reasonably potent inhibitor of an enzyme, one can begin to make analogues of that molecule with varied structural and physicochemical properties to determine the effect of these changes on inhibitor potency. Attempts to correlate these structural changes with inhibitor potency are referred to as *structure–activity relationship* (SAR) studies. Today SAR studies can be divided into two major strategic categories: SAR in the absence of structural information on the target enzyme and SAR that utilizes structural information about the enzyme obtained from x-ray crystallographic or multidimensional NMR studies. The latter category also is referred to as rational or structure-based inhibitor design. Sections 7.5.1 and 7.5.2 introduce some of the techniques used for both these strategies.

7.5.1 SAR in the Absence of Enzyme Structural Information

Any SAR study begins with the identification of a lead compound that shows some potency for inhibiting the target enzyme. This lead compound might be identified by random screening of a compound library, such as a natural products library, or it might be based on the known structures of the substrate or product of a particular enzymatic reaction. With a lead compound in hand, analysts subject the substance to small structural perturbations and test these analogues for inhibitor potency. This most basic form of SAR study has been conducted in one way or another since the nineteenth century. The goals of these studies are to determine what structural changes will lead to improved inhibitor potency and to identify the *pharmacophore*, the minimal structure required for inhibition. Once identified, the *pharmacophore* serves as a template for further inhibitor design.

Consider the enzyme dihydrofolate reductase (DHFR), which catalyzes a key step in the biosynthesis of deoxythymidine. Inhibition of this enzyme blocks DNA replication and thus acts to inhibit cell growth and proliferation. DHFR inhibitors are therefore potentially useful therapeutic agents for the control of aberrant cell growth in cancer, and as antibiotics for the control of bacterial growth. Early attempts to identify inhibitors of this enzyme were based on synthesizing analogues of the substrate dihydrofolate. Figure 7.14 illustrates the chemical structures of dihydrofolate and two classes of DHFR inhibitors, the pterdines, exemplified by methotrexate, and the 5-substituted 2,4-diaminopyrimidines. Methotrexate was identified as a potent inhibitor of DHFR because of its striking structural similarity to the substrate dihydrofolate.

Next, the question of what portions of the methotrexate molecule were critical for DHFR inhibition was addressed by synthesizing various structural analogues of methotrexate. From these studies it was determined that the critical pharmacophore (i.e., the minimal structural component required for inhibition) was the 2,4-diaminopyrimidine ring system. This discovery led to the development of the second class of inhibitors illustrated in Figure 7.14, the 5-substituted 2,4-diaminopyrimidines, of which trimethoprim is a well-known example. Methotrexate is now a prescribed drug for the treatment of human cancers. Trimethoprim also is a prescribed drug, but its use is in the control of bacterial infections.

An unexpected outcome of the studies on these inhibitors was the finding that the pteridines, such as methotrexate, are potent inhibitors of both mammalian and bacterial DHFR, while trimethoprim and its analogues are much better inhibitors of the bacterial enzymes. The K_i values for trimethoprim for *E. coli* and human lymphoblast DHFR are 1.35 and 170,000 nM, respectively (Li and Poe, 1988), a selectivity for the bacterial enzyme of roughly 126,000-fold! The reason for this spectacular species selectivity was not clear until the crystal structures of mammalian and bacterial DHFR were solved. (See Mathews et al., 1985, for a very clear and interesting account of these crystallographic studies and their interpretation.)

Today the enzymologist attempts to develop higher potency inhibitors not simple by random replacement of structural components on a molecule, but rather by systematic and rational changes in stereochemical and physicochemical properties of the substituents. Some properties to be changed are obvious from the structure of the lead compound. If, for example, the lead inhibitor is seen to contain a carboxylic acid group, one immediately wonders whether acid–base-type interactions with a group on the enzyme are involved in binding. One might substitute the carboxylate moiety with an ester, for example, to determine the importance of the carboxylate in binding. More general properties of chemical substituent can be examined as well. These studies call for quantitative measures of the different physicochemical properties to be considered. Among the relevant general properties of chemical substituents, steric bulk, hydrophobicity, and electrophilicity/nucleophilicity are generally agrees to be important factors, and chemists have therefore developed quantitative measures of these parameters.

Several measures have been suggested to quantify steric bulk or molecular volume. One of the earliest attempts at this was the Taft steric parameter E_S, which was defined as the logarithm of the rate of acid-catalyzed hydrolysis of

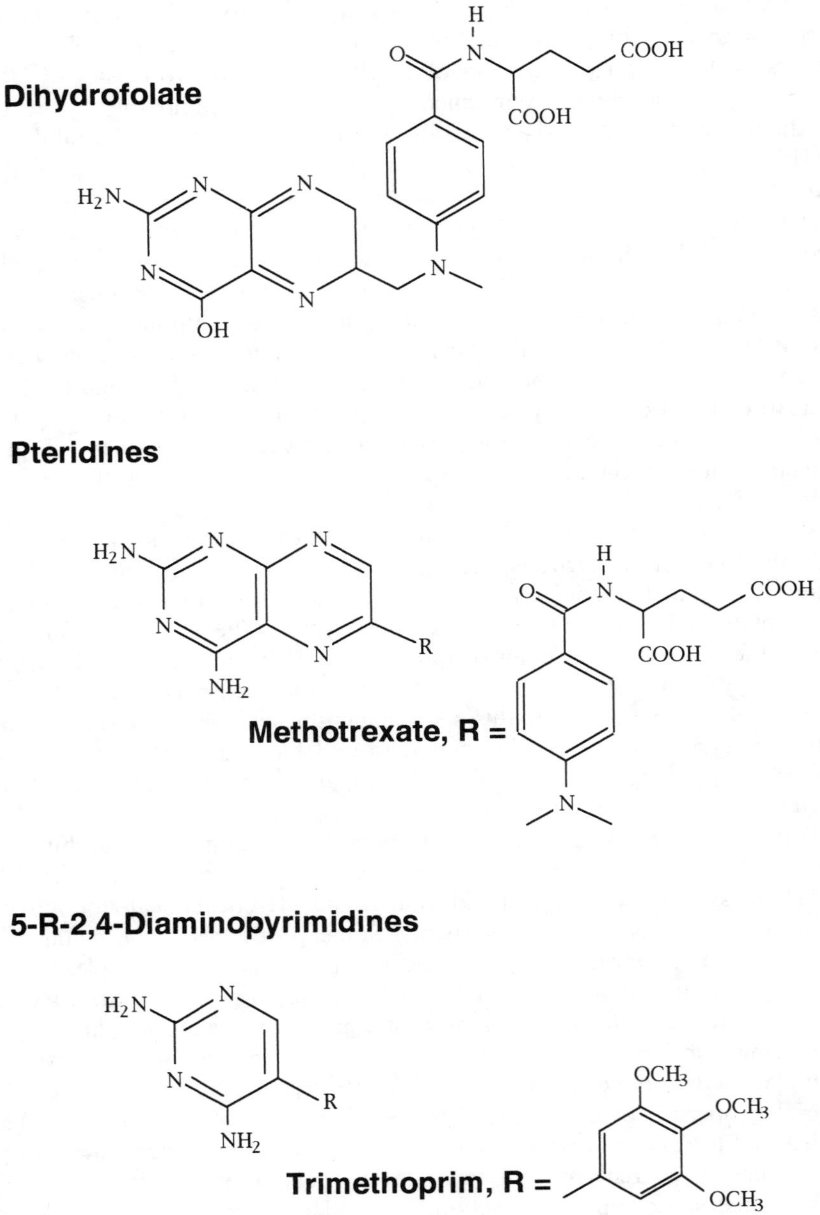

Dihydrofolate

H_2N N N N N OH H N O COOH COOH

Pteridines

H_2N N N N N NH_2 R

Methotrexate, R = H N O COOH COOH N

5-R-2,4-Diaminopyrimidines

H_2N N N NH_2 R

Trimethoprim, R = OCH_3 OCH_3 OCH_3

Figure 7.14 Chemical structures of the substrate (dihydrofolate) and two types of inhibitor of the enzyme dihydrofolate reductase (DHFR).

a carboxymethyl-substituted molecule relative to the rate for the methyl acetate analogue (Taft, 1953; Nogrady, 1985):

$$E_S = \log(k_{XCOOCH_3}) - \log(k_{CH_3COOCH_3}) \tag{7.26}$$

A more geometric measure of steric bulk is provided by the Verloop steric parameter, which basically measures the bond angles and bond lengths of the substituent group (Nogrady, 1985). More recently, chemists have used molar refractivity as a measure of molecular volume of substituents (Pauling and Pressman, 1945; Hansch and Klein, 1986). The molar refractivity MR is defined as follows:

$$MR = \frac{n^2 - 1}{n^2 + 1}\frac{MW}{d} \tag{7.27}$$

where n is the index of refraction, MW is the molecular weight, and d is the density of the substituent under consideration. Since n does not vary widely among organic molecules, MR is mainly a measure of molecular volume.

The relative hydrophobicity of molecular substituents is most commonly measured by their partition coefficient between a polar and nonpolar solvent. For this purpose, chemists have made water and octanol the solvents of choice. The molecule is dissolved in a 1:1 mixture of the two solvents, and its concentration in each solvent is measured at equilibrium. The partition coefficient is then calculated as the equilibrium constant:

$$P = \frac{[I]_{octanol}}{[I]_{water}} \tag{7.28}$$

In measuring the relative hydrophobicity of different substituents, their effect on the partition coefficient of benzene in octanol/water is used as a standard. The hydrophobic parameter π is used for this purpose, and is defined as (Hansch and Klein, 1986):

$$\pi = \log(P_x) - \log(P_H) \tag{7.29}$$

where P_x is the partition coefficient for a monosubstituted benzene with substituent x, and P_H is the partition coefficient of benzene itself.

The most widely used index of electronic effects in inhibitor design is the Hammett σ constant. Originally developed to correlate quantitatively the relationship between the electron-donating or -accepting nature of a parasubstituent on the ionization constant for benzoic acid in water (Hammett, 1970; Nogrady, 1985), this index is defined as follows:

$$\sigma = \log(K_x) - \log(K_H) \tag{7.30}$$

where K_x is the ionization constant for the parasubstituted benzoic acid with substituent x, and K_H is the ionization constant for benzoic acid. Groups that are electron acceptors (e.g., COOH, NO_2, NR_3^+) withdraw electron density from the ring system, hence stabilize the ionized form of the acid; such groups have positive values of σ. Electron-donating groups (e.g., OH, OCH_3, NH_2) have the opposite effect on the ionization constant and thus have negative values of σ. Values of σ for a very large number of organic substituent have been tabulated by several authors. One of the most comprehensive lists of σ values can be found in the text by Martin (1978).

The ability to quantify these various physicochemical properties has led to attempts to express the inhibitor potency of molecules as a mathematical function of these parameters. This strategy of *quantitative structure–activity relationships* (QSAR), was first championed by Hansch and his coworkers (Hansch, 1969; Hansch and Leo, 1979). In a typical QSAR study, a series of analogues of a lead inhibitor is prepared with substituents that systematically vary the parameters described earlier. The experimentally determined potencies of these compounds are then fit to varying linear and nonlinear weighted sums of the parameter indices to obtain the best correlation by regression analysis. Equation 7.31–7.33 illustrate forms typically used in QSAR work.

$$\log\left(\frac{1}{K_i}\right) = a\pi + b\sigma + cMR + d \tag{7.31}$$

$$\log\left(\frac{1}{K_i}\right) = a\pi^2 + b\sigma + cMR + d \tag{7.32}$$

$$\log\left(\frac{1}{K_i}\right) = a\pi^2 + b\sigma + c\log(\beta MR) + d \tag{7.33}$$

Equation 7.31 is a simple linear relationship, while Equations 7.32 and 7.33 have nonlinear components. In these equations the values *a*, *b*, *c*, *d*, and β are proportionality constants, determined from the regression analysis. In developing a mathematical expression for the correlation relationship here, one hopes to predict the inhibitory potency of further compounds, prior to their synthesis, based on the equation established from the QSAR. In practice, the predictive power of these QSAR equations varies dramatically. When such predictions fail, it is usually because additional factors that influence inhibitor potency were not quantitatively included in the functional expression. In some cases these additional factors are neither well understood nor easily quantified.

As a simple example of QSAR, let us again consider the inhibition of bacterial DHFR by pteridines and 5-substituted 2,4-diaminopyrimidines. Coats et al. (1984) have studied the QSARs of both classes of compounds for their ability to inhibit the DHFR from the bacterium *Lactobacillus casei*. They have measured the IC_{50} values for 25 pteridine analogues with different R substituents, and also for 33 5-substituted 2,4-diaminopyrimidine analogues with different R groups (see basic types, Figure 7.14). From these data they determined the QSAR equations given by Equations 7.34 and 7.35 for the pteridines and 5-R-2,4-diaminopyrimidines, respectively:

$$\log\left(\frac{1}{IC_{50}}\right) = 0.23\pi - 0.004\pi^2 + 0.77I + 3.39 \tag{7.34}$$

$$\log\left(\frac{1}{IC_{50}}\right) = 0.38\pi - 0.007\pi^2 + 0.66I + 2.15 \tag{7.35}$$

In these equations, the π parameters refers to the hydrophobicity of the R group, and the index I is an empirical parameter related to the presence of an —N—C— or —C—N— bridge between the parent ring system and an aromatic ring on the substituent (Coats et al., 1984). The relationships between the IC_{50} value calculated from these equations and the experimentally determined IC_{50} values are illustrated in Figure 7.15. Again, one must keep in mind that these correlations are for the molecules used to establish the QSAR equations. The value of the equations in predicting the inhibitor potency of other molecules will depend on how significantly other unaccounted for factors influence potency. Nevertheless, QSAR provides a means of rationalizing the observed potencies of structurally related compounds in terms of familiar physiochemical properties. An up-to-date volume by Kubinyi (1993) provides a detailed and practical introduction to the field of QSAR. This text should be consulted as a starting point for those interested in a more in-depth treatment of the subject.

Another approach to designing potent inhibitors of enzymes is to consider the probable structure of the transition state of the chemical reaction catalyzed by the enzyme. As described in Chapter 4, the catalytic efficiency of enzymes is due largely to their ability to achieve transition state stabilization. If this stabilization is equated with binding energy, a stable analogue that mimics the structure of the transition state should bind to an enzyme some 10^{10}–10^{15} times greater than the corresponding ground state substrate molecule (Wolfenden, 1972). Since typical substrate K_m values are in the millimolar-to-nanomolar range, this suggests that a true transition state mimic would bind to its target enzyme with a K_i value between 10^{-13} and 10^{-24} M! With such incredibly tight binding affinities, such inhibitors would behave practically as irreversible enzyme inactivators.

The foregoing approach to inhibitor design has been hindered by the great difficulty of obtaining information on transition state structure by traditional physical methods. Because they are so short-lived, the transition state species of most enzymatic reactions are present under steady state conditions at very low concentrations (i.e., femtomolar or less). Hence, attempts to obtain structural information on these species from spectroscopic or crystallographic methods have been largely unsuccessful. Information on transiton state structure can, however, be gleanded from analysis of kinetic isotope effects on enzyme catalysis, as recently reviewed by Schramm et al. (1994). As discussed in Chapter 6, kinetic isotope effects are observed because of the changes in vibrational frequencies for the reactant and transition state species that

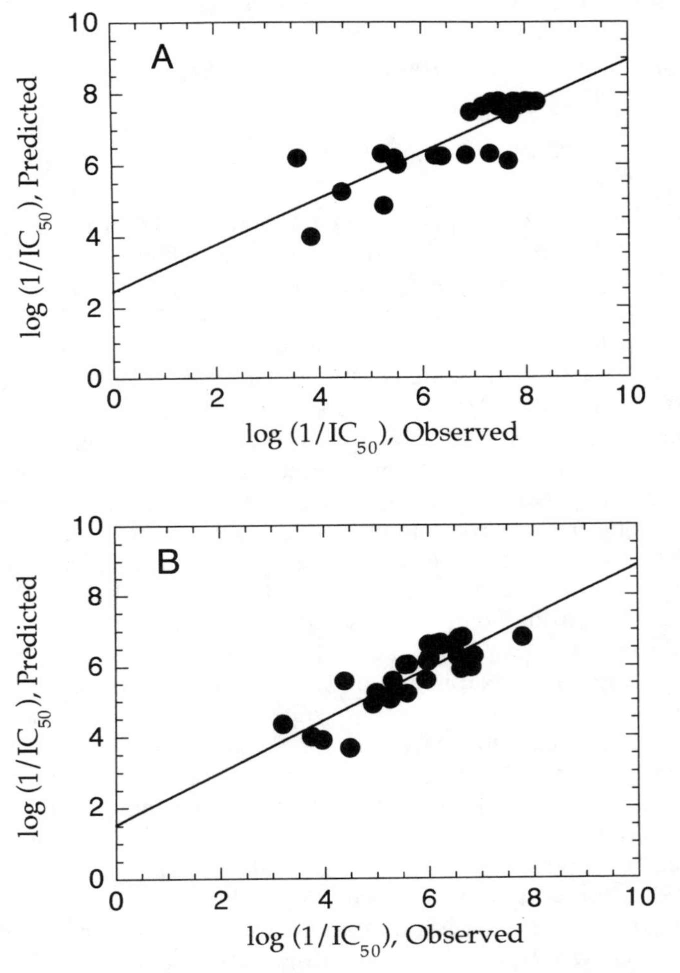

Figure 7.15 QSAR correlation plots for the potencies of pteridines (A) and 5-substituted 2,4-diaminopyrimidines; (B) as inhibitors of the dihydrofolate reductase from *L. casei*. [Data from Coats et al. (1984).]

accompany heavy isotope incorporation. By synthesizing substrate analogues with heavy isotopes at specific locations, one can determine the kinetic isotope effects imparted by each replacement. From this type of information, one can use vibrational normal mode calculations to identify the vibrational modes that are most strongly perturbed in the transformation from reactant to transition state of the substrate, hence to map out the structural changes that have occurred in the molecule. The information thus obtained can then be used to design molecules that mimic the structure of the reaction transition state.

This approach has recently been applied to the design of transition state analogues of the aspartyl protease renin (Blundell et al., 1987).

In vivo, renin is responsible for the proteolytic processing of angiotensinogen to angiotensin I, the progenitor of the vasoconstrictor peptide angiotensin II. The substrate is hydrolyzed at a Leu-Val peptide bond, and the hydrolysis reaction is proposed to utilize an active site water molecule as the attacking nucleophile to produce the tetrahedral transition state illustrated in Figure 7.16A. The peptide sequence of the renin substrate angiotensinogen is shown in Figure 7.16C. In their first attempt at an inhibitor of renin Blundell and coworkers replaced the P1 carbonyl group by a methylene linkage, yielding the reduced isostere (—CH_2—NH—) containing peptide inhibitor **1** (Figure 7.16C). These investigators next noted that the pepstatins are naturally occurring protease inhibitors that contain the unusual amino acid statine (Figure 7.16B), which in turn contains a —CH(OH)— moiety that resembles the proposed transition state of renin. Pepstatin is a poor renin inhibitor, but reasoning that the statine group was a better transition state mimic than the reduced peptide isostere (—CH_2—NH—), Blundell et al incorporated this structure into their peptide inhibitor to produce **2** (Figure 7.16C). The

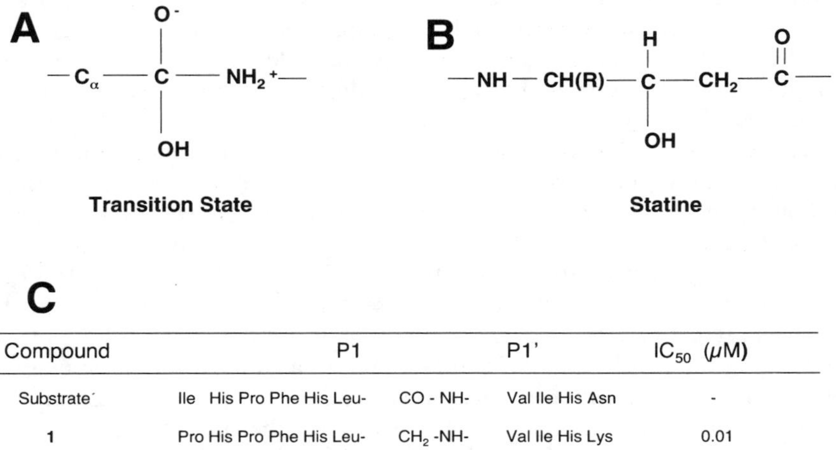

A

O^- (top), —C_α—C—NH_2^+—, OH (bottom)

Transition State

B

H (top), —NH—CH(R)—C—CH_2—C(=O)—, OH (bottom)

Statine

C

Compound	P1		P1'	IC_{50} (μM)
Substrate	Ile His Pro Phe His Leu-	CO - NH-	Val Ile His Asn	-
1	Pro His Pro Phe His Leu-	CH_2 -NH-	Val Ile His Lys	0.01
2	Boc His Pro Phe His Leu-	CH(OH)-CH_2-CO-NH-Leu Phe		0.01
3	Boc His Pro Phe His Leu-	CH(OH)-CH_2-	Val Ile His	0.0007

Figure 7.16 (A) Proposed structure of the tetrahedral transition state of the renin proteolysis reaction. (B) Chemical structure of the statine moiety of natural protease inhibitors, such as pepstatin. (C) Structures and IC_{50} values for the peptide substrate and inhibitors of renin, incorporating various forms of transition state analogues. [Data from Blundell et al. (1987).]

closest analogue to the true transition state of the reaction would be one incorporating a —CH(OH)—NH— group at P1—P1′ of the peptide. Since, however, synthesis of this transition state analogue was hampered by the instability of the resulting compound, Blundell's group instead synthesized a closely related analogue containing a hydroxyethylene moiety (—CH(OH)—CH_2—), **3**, which proved to be an extremely potent inhibitor of the enzyme (Figure 7.16C).

To date, the de novo design of transition state analogues as enzyme inhibitors has been applied to only a limited number of enzymes by a handful of laboratories. With improvements in the computational methods associated with this strategy, however, more widespread use of this approach is likely to be seen in the future.

7.5.2 Inhibitor Design Based on Enzyme Structure

In the search for potent enzyme inhibitors, knowledge of the three-dimensional structure of the inhibitor binding site on the enzyme provides the ultimate guide to designing new compounds. The structures of enzyme active sites can be obtained in atomic detail from x-ray crystallography and multidimensional NMR spectroscopy. A detailed discussion of these methods is beyond the scope of the present text. Our discussion will focus instead on the use of the structural details obtained from these techniques. For the reader interested in learning about protein crystallography and NMR spectroscopy, many excellent review articles and texts deal with these subjects. (See McRee, 1993, and Fesik, 1991, for good introductions to protein crystallography and NMR spectroscopy, respectively.)

The crystal or NMR structure of an enzyme with an inhibitor bound provides structural details at the atomic level on the interactions between the inhibitor and the enzyme that promotes binding. Hydrogen bonding, salt bridge formation, other electrostatic interactions, and hydrophobic interactions can be readily inferred from inspection of a high resolution structure. Figure 7.17 provides simplified schematic representations of the binding interactions between DHFR and its substrate dihydrofolate and the inhibitor methotrexate, illustrating the involvement of common amino acid residues in the binding of both ligands. These structural diagrams also indicate that the orientation and hydrogen bonding patterns are not identical for the substrate and the ligand. Nevertheless, the major forces involved in binding of both ligands to the enzyme are hydrogen bonds between amino acid residues of the active site and the 2,4-diaminopyrimidine ring of the ligands. Visual inspection of these crystal structures by means of molecular graphics methods suggested that this ring constituted the critical pharmacophore and led to the design of the prototypical 5-substituted 2,4-diaminopyrimidine, trimethoprim (Marshall and Cramer, 1988). As we have seen, these structural inferences are consistent with the SAR and QSAR studies of DHFR inhibitors.

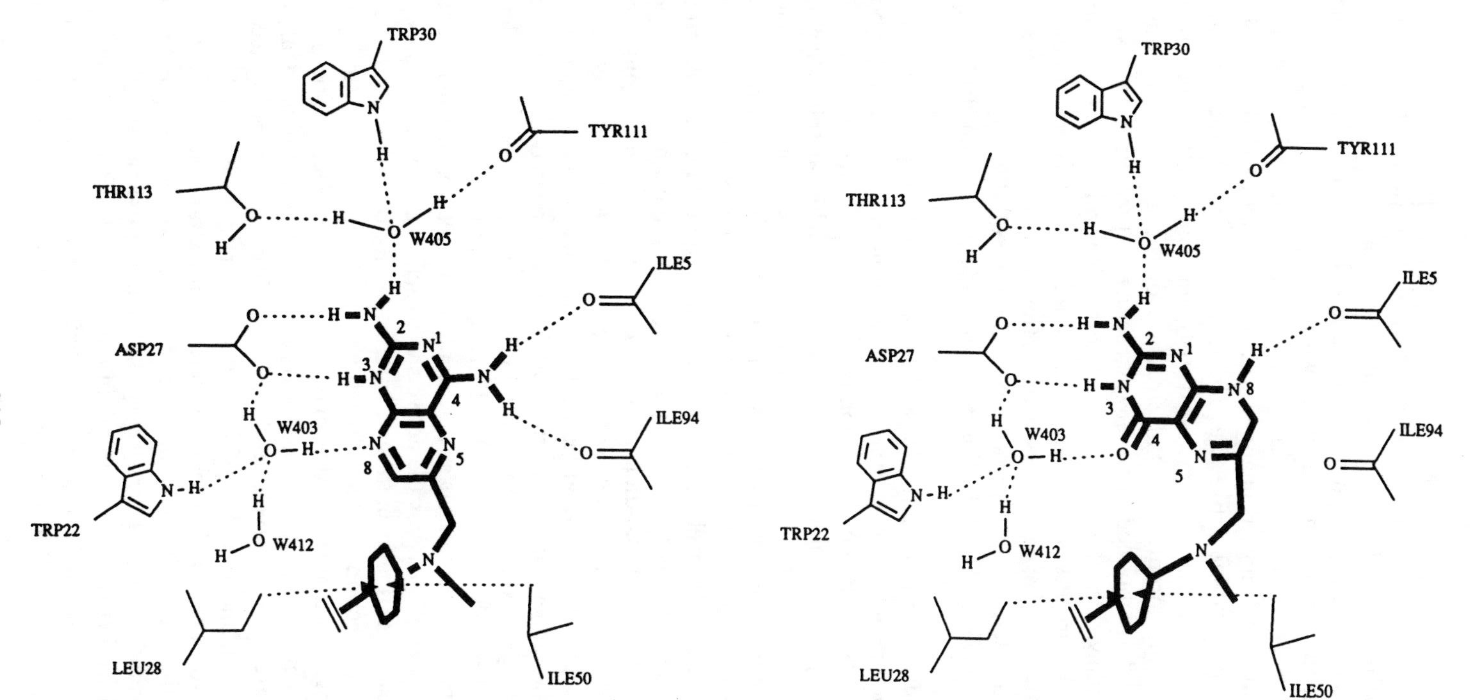

Figure 7.17 Interactions of the dihydrofolate reductase active site with the inhibitor methotrexate (left) and the substrate dihydrofolate (right). [Reprinted from Klebe (1994) with permission from Academic Press Limited.]

The example of trimethoprim suggests a straightforward, if tedious, means of utilizing structural information in the design of new enzyme inhibitors; namely, the iterative design, synthesis, and crystallization of inhibitor–enzyme complexes. In this approach, one starts with the crystal structure of the free enzyme or of the enzyme–lead inhibitor complex. Based on inspection of the crystal structure, one suggests changes in chemical structure of the inhibitor to better engage the enzyme active site. The new compound is then synthesized and tested for inhibitory potency. Next, to determine whether the predicted interactions in fact occur, a crystal structure of the enzyme with this new inhibitor bound is obtained. This new structure is then used to search for additional changes to the inhibitor structure that might further improve potency, and the process is continued until an inhibitor of sufficient potency is obtained. This iterative structure-based inhibitor design method was used recently in the design and synthesis of inhibitors of thymidylate synthase reported by Appelt et al. (1991); this paper provides a good illustration of the method.

Thus, the first step to structure-based inhibitor design is to obtain a crystal or NMR structure of the target enzyme, with or without a lead inhibitor bound to it. In some cases, the determination of a crystal or NMR structure of the target protein proves problematic because of the technical difficulties associated with crystallographic and NMR methods. If the structure of a closely related enzyme has been reported, however, one can still attempt to model the three-dimensional structure of the target enzyme by means of homology modeling (Lesk and Boswell, 1992). In homology modeling one attempts to build a model of the target enzyme by superimposing the amino acid residues of the target onto the three-dimensional structure of a homologous protein whose structure has been solved. For this method to work, the target enzyme and its homologue must share at least 30% amino acid sequence identity. The accuracy of the model obtained in this way is directly related to the degree of sequence identity between the two proteins; the greater the sequence identity, the greater the accuracy of the modeled structure will be.

With the modeled or actual structure of the target enzyme active site in hand, the next step is to assess the active site structure in a meaningful way, to permit the use of this information to predict inhibitor binding motifs. The simple visual inspection of such structures can be augmented today with computer programs that allow the analyst to map the electrostatic potential surface of the active site, identify and localize specific types of functional group within the active site (potential acid–base groups, hydrogen-bonding acceptors or donators, etc.), and the like. When the active site has been well described, one attempts to design inhibitors with stereochemical and functional complementarity to the active site structure. Again, these activities are greatly aided by high powered computer programs that make possible the probing of complementarity between a potential inhibitor and the enzyme active site. Assessment of the stereochemical complementarity of a potential inhibitor is aided by the use of molecular dynamics simulation programs by means of which the most energetically favorable conformations of inhibitory molecules can be assessed to determine whether they will adapt a conformation that is complementary to the enzyme active site.

New programs allow one to perform free energy perturbation calculations in which the structure of a bound inhibitor is slowly mutated and the difference in calculated free energy of binding between the starting and final structures determined (Marshall and Cramer, 1988). In this way, one can search for structural perturbations that will increase the affinity of an inhibitor for the enzyme active site. The complementarity of functional groups can be probed by computational methods as well. For example, the computer program GRID (Goodford, 1985) can be used to search the structure of an enzyme active site for areas that are likely to interact strongly with a particular functional group probe.

A recent example of the use of such programs comes from the studies by von Itzstein et al. (1993) aimed at designing potent inhibitors of the sialidase enzyme from influenza virus. This group started with a visual inspection of the enzyme active site obtained from a series of crystals for the enzyme to which various sialic acid analogues were bound. This visual inspection of the cocrystal structure for the enzyme bound to the unsaturated sialic acid analogue Neu5Ac2en suggested that replacement of the 4-hydroxyl group of the substrate by an amino group might be useful. A GRID calculation was performed with a protonated primary amine group as the probe, and a "hot spot" (i.e., an area of likely strong interaction) was identified within the enzyme active site. The results of this calculation and visual inspection suggested that replacement of the 4-hydroxyl group with a amino group would lead to much tighter binding because a salt bridge would form between the amino group and the side chain carboxylate of Glu 119 of the enzyme. Further evaluation of the computational data suggested that replacement of the 4-hydroxyl group with a guanidinyl group would even further enhance inhibitor binding by engaging both Glu 119 and Glu 227 through lateral binding of the two terminal nitrogens on this functional group. Based on these results, the 4-amino and 4-guanidino derivatives of Neu5Ac2en were synthesized and, as expected, found to be potent inhibitors of the enzyme, with K_i values of 50 and 0.2 nM, respectively. The crystal structures of the enzyme bound to each of these new inhibitors was then determined, and the predicted modes of inhibitor interactions with the enzyme were by and large confirmed.

The design of new enzyme inhibitors, both by structure-based design methods and in the absence of enzyme structural information, is a large and growing field. We have only briefly introduced this complex and exciting area in this chapter. There are many excellent sources for additional information on strategies for inhibitor design. These include several texts devoted entirely to this subject (e.g., Sandler and Smith, 1994). Also, most modern medicinal chemistry textbooks contain sections on SAR and inhibitor design (see, e.g., Nogrady, 1985; Dean, 1987). Finally, a number of primary journals commonly feature papers in the field of inhibitor design and SAR. These include *Journal of Medicinal Chemistry* (ACS), *Journal of Enzyme Inhibitors*, *Bioorganic and Medicinal Chemistry Letters*, and *Journal of Computer-Aided Molecular Design*. These sources, and the specific references at the end of this chapter, will provide good starting points for the reader interested in exploring these subjects in greater depth.

7.6 Summary

In this chapter we described the modes by which an inhibitor can bind to an enzyme molecule and thus render it inactive. Graphical methods were introduced for the diagnosis of the mode of inhibitor interaction with the enzyme on the basis of the effects of that inhibitor on the apparent values of the kinetic constants K_m and V_{max}. Having thus identified the mode of inhibitor interaction, we described methods for quantifying the inhibitor potency in terms of K_i, the dissociation constant for the enzyme–inhibitor complex.

Also in this chapter, we introduced some of the physicochemical determinants of enzyme–inhibitor interactions and saw how these could be systematically varied for the design of more potent inhibitors. Finally we introduced the concept of structure-based inhibitor design in which the crystal or NMR structure of the target enzyme is used to aid the design of new inhibitory molecules in an iterative process of enzyme–inhibitor structure determination, new inhibitor design and synthesis, and quantitation of new inhibitor potency.

References and Further Reading

Appelt, K., Bacquet, R. J., Bartlett, C. A., Booth, C. L. J., Freer, S. T., Fuhry, M. A. M., et al. (1991) *J. Med. Chem.* **34**, 1925.

Blundell, T. L., Cooper, J., Foundling, S. I., Jones, D. M., Atrash, B., and Szelke, M. (1987) *Biochemistry*, **26**, 5586.

Chaiken, I., Rose, S., and Karlsson, R. (1991) *Anal. Biochem.* **201**, 197.

Cheng, Y.-C., and Prusoff, W. H. (1973) *Biochem. Pharmacol.* **22**, 3099.

Cleland, W. W. (1979) *Methods Enzymol.* **63**, 103.

Coats, E. A., Genther, C. S., and Smith, C. C. (1984) In *QSAR in Design of Bioactive Compounds*, M. Kuchar, ed., J. R. Prous Science, Barcelona, Spain, pp. 71–85.

Dean, P. M. (1987) *Molecular Foundations of Drug–Receptor Interactions*, Cambridge University Press, New York.

Dixon, M. (1953) *Biochem. J.* **55**, 170.

Fesik, S. W. (1991) *J. Med. Chem.* **34**, 2937.

Furfine, E. S., D'Souza, E., Ingold, K. J., Leban, J. J., Spectro, T., and Porter, D. J. T. (1992) *Biochemistry*, **31**, 7886.

Goodford, P. J. (1985) *J. Med. Chem.* **28**, 849.

Hammett, L. P. (1970) *Physical Organic Chemistry*, McGraw-Hill, New York.

Hansch, C. (1969) *Acc. Chem. Res.* **2**, 232.

Hansch, C., and Klein, T. E. (1986) *Acc. Chem. Res.* **19**, 392.

Hansch, C., and Leo, A. (1979) *Substituent Constants for Correlation Analysis in Chemistry and Biology*, Wiley, New York.

Karlsson, R. (1994) *Anal. Biochem.* **221**, 142.

Klebe, G. (1994) *J. Mol. Biol.* **237**, 212.

Kubinyi, H. (1993) *QSAR: Hansch Analysis and Related Approaches*, VCH Publishers, New York.

Lesk, A. M., and Boswell, D. R. (1992) *Curr. Opin. Struct. Biol.* **2**, 242.

Li, R.-L., and Poe, M. (1988) *J. Med. Chem.* **31**, 366.

Ma, H., Yang, H. Q., Takano, E., Hatanaka, M., and Maki, M. (1994) *J. Biol. Chem.* **269**, 24430.

Marshall, G. R., and Cramer, R. D., III (1988) *Trends Pharmacol. Sci.* **9**, 285.

Martin, Y. C. (1978) *Quantitative Drug Design*, Dekker, New York.

Matthews, D. A., Bolin, J. T., Burridge, J. M., Filman, D. J., Volz, K. W., and Kraut, J. (1985) *J. Biol. Chem.* **260**, 392.

McRee, D. E. (1993) *Practical Protein Crystallography*, Academic Press, San Diego, CA.

Nogrady, T. (1985) *Medicinal Chemistry: A Biochemical Approach*, Oxford University Press, New York.

Pauling, L., and Pressman, D. (1945) *J. Am. Chem. Soc.* **75**, 4538.

Sandler, M., and Smith, H. J. (1994) *Design of Enzyme Inhibitors as Drugs*, Vols 1 and 2, Oxford University Press, New York.

Schramm, V. L., Horenstein, B. A., and Kline, P. C. (1994) *J. Biol. Chem.* **269**, 18259.

Segel, I. H. (1975) *Enzyme Kinetics*, Wiley, New York.

Segel, I. H. (1976) *Biochemical Calculations*, 2nd ed., Wiley, New York.

Suckling, C. J. (1991) *Experientia*, **47**, 1139.

Taft, R. W. (1953) *J. Am. Chem. Soc.* **75**, 4538.

Von Itzstein, M., Wu, W.-Y., Kok, G. B., Pegg, M. S., Dyason, J. C., Jin, B., Phan, T. V., Smythe, M. L., White, H. F., Oliver, S. W., Colman, P. M., Varghese, J. N., Ryan, D. M., Woods, J. M., Bethell, R. C., Hotham, V. J., Cameron, J. M., and Penn, C. R. (1993) *Nature*, **363**, 418.

Wolfenden, R. (1972) *Acc. Chem. Res.* **5**, 10.

CHAPTER 8

Tight Binding Inhibitors

In Chapter 7 we discussed reversible inhibitors of enzymes that bind and are released at rates that are rapid in comparison to the rate of enzyme turnover and have overall dissociation constants that are large in comparison to the total concentration of enzyme present. The interactions of these inhibitors with their target enzymes can be analyzed by the Henri–Michaelis–Menten type equations discussed in Chapters 5 and 7, because the free inhibitor concentration can be assumed to be well modeled by the total concentration of added inhibitor; that is, since $[E_t]$ is much smaller than K_i, the concentration of EI complex is very small compared to $[I_t]$. This assumption, however, is not valid for all inhibitors. Some inhibitors bind to their target enzyme with such high affinity that the population of free inhibitor molecules is significantly depleted by formation of the enzyme–inhibitor complex. For these *tight binding inhibitors*, the steady state approximations we have used thus far are no longer valid; in fact, it has been suggested that whenever the K_i of an inhibitor is less than 1000-fold greater than the total enzyme concentration, these assumptions should be abandoned (Goldstein, 1944; Dixon and Webb, 1979). In this chapter we shall describe alternative methods for data analysis in the case of tight binding inhibitors that allow one to characterize the type of inhibition mechanisms involved and to quantify correctly the dissociation constant for the enzyme–inhibitor complex.

8.1 Identifying Tight Binding Inhibition

In this chapter we shall consider the steady state approach to studying tight binding inhibitors. Such work requires assay conditions that permit all the equilibra involving the inhibitor, substrate, and enzyme to be reached rapidly with respect to measurement of the steady state velocities. Many tight binding inhibitors, however, are restricted in their action by a time-dependent component. We deal with time-dependent inhibition explicitly in Chapter 9. For our present discussion, we shall assume that either the establishment of equilibrium is rapid or sufficient time has been allowed before the initiation of reaction by the substrate for the inhibitor and enzyme to establish equilibrium (i.e., a preincubation of the enzyme with the inhibitor has been incorporated into the experimental design).

The simplest determination that tight binding inhibition is occurring comes from measurement of the dose–response curve for inhibition (see Chapter 7). An IC_{50} value obtained from this treatment of the data that is similar to the concentration of total enzyme in the sample (i.e., within a factor of 10) is a good indication that the inhibitor may be of the tight binding type. A more defining feature of tight binding inhibitors is the variation of the IC_{50} value observed for these inhibitors with total enzyme concentration at a fixed substrate concentration. This is true because a tight binding inhibitor interacts with the enzyme in nearly stoichiometric fashion. Hence, the higher the concentration of enzyme present, the higher the concentration of inhibitor required to reach half-maximal saturation of the inhibitor binding sites (Figure 8.1A). Several authors have derived equations similar in form to Equation 8.1, which demonstrates that the IC_{50} value of a tight binding inhibitor will track linearly with the total concentration of enzyme $[E_t]$ (Myers, 1952; Cha et al.,

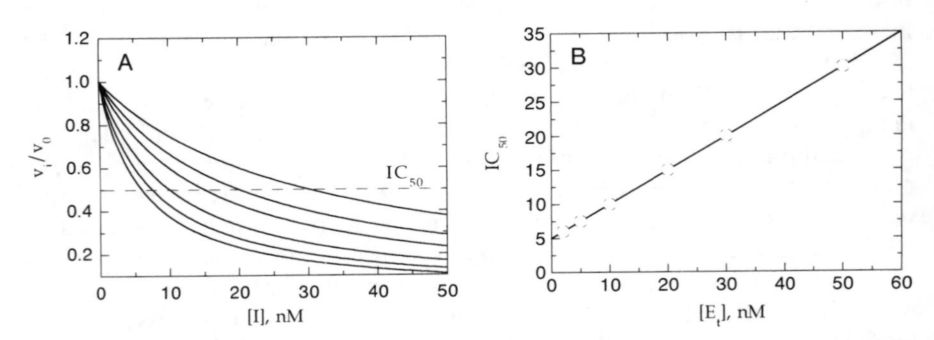

Figure 8.1 (A) Dose–response plot of fractional velocity as a function of tight binding inhibitor concentration at different enzyme concentrations. Note that this plot is the same as the dose–response plots introduced in Chapter 7, except that here the x axis is plotted on a linear, rather than a logarithmic, scale. (B) Plot of IC_{50} value obtained from the curves in (A) as a function of enzyme concentration.

1975; Greco and Hakala, 1979; Williams and Morrison, 1979):

$$IC_{50} = \tfrac{1}{2}[E_t] + K_i^{app} \qquad (8.1)$$

Thus, a plot of IC_{50} as a function of $[E_t]$ (at a single, fixed substrate concentration) is expected to yield a straight line with slope of 0.5 and y intercept equal to K_i^{app}, as illustrated in Figure 8.1B. The value K_i^{app} is related to the true K_i by factors involving the substrate concentration and K_m, depending on the mode of interaction between the inhibitor and the enzyme.

8.2 Distinguishing Inhibitor Type for Tight Binding Inhibitors

Morrison (Morrison, 1969; Williams and Morrison, 1979) has provided in-depth mathematical treatments of the effects of tight binding inhibitors on the initial velocities of enzymatic reactions. These studies revealed, among other things, that the classical double-reciprocal plots used to distinguish inhibitor type for simple enzyme inhibitors fail in the case of tight binding inhibitors. For example, based on the work just cited by Morrison and coworkers, the double-reciprocal plot for a tight binding competitive inhibitor would give the pattern of lines illustrated in Figure 8.2. The data at very high substrate

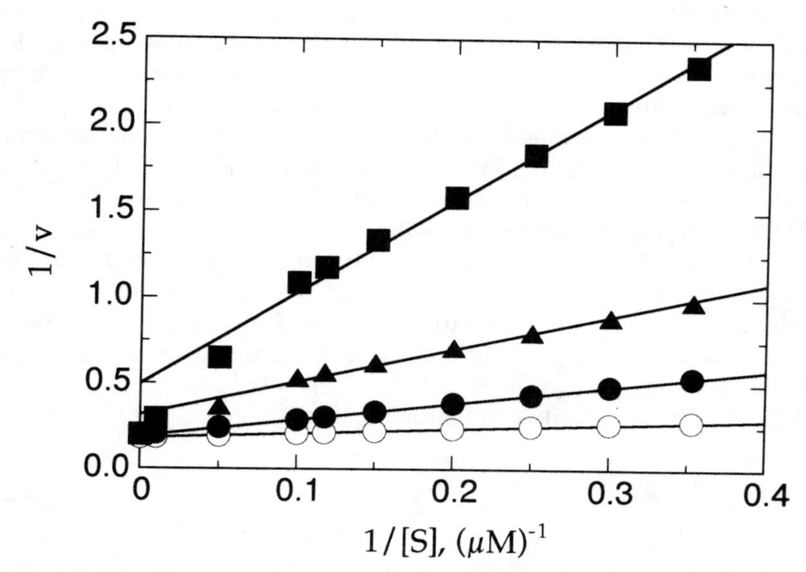

Figure 8.2 Double-reciprocal plot for a tight binding competitive inhibitor: the pattern of lines is similar to that expected for a classical noncompetitive inhibitor (see Chapter 7).

concentrations curve downward in this plot, and the curves at different inhibitor concentrations converge at the y axis. Note, however, that this curvature is apparent only at very high substrate concentrations and in the presence of high inhibitor concentrations. This subtlety in the data analysis is easy to miss if care is not taken to include such extreme conditions, or if these conditions are not experimentally attainable. Hence, if the few data points in the very high substrate region are ignored, it is tempting to fit the data in Figure 8.2 to a series of linear functions, as has been done in this illustration. The pattern that emerges from this treatment of the data is a series of lines that intersect at or near the x axis, to the left of the y axis. This is the expected result for a classical noncompetitive inhibitor (see Chapter 7), and we can generally state that *tight binding inhibitors display double-reciprocal plots that appear similar to the classical pattern for noncompetitive inhibitors, regardless of their true mode of interaction with the enzyme.*

As one might imagine, this point has led to a number of misinterpretations of kinetic data for inhibitors in the literature. For example, the naturally occurring inhibitors of ribonuclease are nanomolar inhibitors of this enzyme. Initial evaluation of the inhibitor type by double-reciprocal plots indicated that these inhibitors acted through classical noncompetitive inhibition. It was not until Turner et al. (1983) performed a careful examination of these inhibitors, over a broad range of inhibitor and substrate concentrations, and properly evaluated the data (as discussed later), that these proteins were recognized to be tight binding *competitive* inhibitors.

How then can one determine the true mode of interaction between an enzyme and a tight binding inhibitor? Several graphical approaches have been suggested. One of the most straightforward is to determine the IC_{50} values for the inhibitor at a fixed enzyme concentration, but at a number of different substrate concentrations. As with simple reversible inhibitors, the IC_{50} of a tight binding inhibitor depends on the K_i of the inhibitor, the substrate concentration, and substrate K_m in different ways, depending on the mode of inhibition. For tight binding inhibitors we must additionally take into consideration the enzyme concentration in the sample, since this will affect the measured IC_{50}, as discussed earlier. The appropriate relationships between these factors and the IC_{50} for different types of tight binding inhibitor have been derived several times in the literature (Cha, 1975; Williams and Morrison, 1979; Copeland et al., 1995). Rather than working through these derivations again, we shall simply present the final form of the relationships.

For tight binding competitive inhibitors:

$$IC_{50} = K_i\left(1 + \frac{[S]}{K_m}\right) + \tfrac{1}{2}[E_t] \qquad (8.2)$$

For tight binding noncompetitive inhibitors:

$$IC_{50} = K_i + \tfrac{1}{2}[E_t] \qquad (8.3)$$

For tight binding uncompetitive inhibitors:

$$\mathrm{IC}_{50} = K_i\left(1 + \frac{K_m}{[\mathrm{S}]}\right) + \tfrac{1}{2}[\mathrm{E_t}] \tag{8.4}$$

For tight binding mixed inhibitors:

$$\mathrm{IC}_{50} = \frac{[\mathrm{S}] + K_m}{K_m/K_i + [\mathrm{S}]/\alpha K_i} + \tfrac{1}{2}[\mathrm{E_t}] \tag{8.5}$$

From the form of these equations, we see that a plot of the IC_{50} value as a function of substrate concentrations will yield quite different patterns, depending on the inhibitor type. For a tight binding competitive inhibitor, the IC_{50} value will increase linearly with increasing substrate concentration (Figure 8.3A). For an uncompetitive inhibitor, a plot of IC_{50} value as a function of substrate concentration will curve downward sharply, while for a noncompetitive inhibitor the IC_{50} will be independent of substrate concentration (Figure 8.3A). In the case of a mixed inhibitor, the plot of IC_{50} versus [S] will curve either upward or downward, depending on whether α is greater than or less than 1.0 (Figure 8.3B).

In an alternative graphical method for determining the inhibitor type and obtaining an estimate of the inhibitor K_i, the fractional velocity of the enzyme reaction is plotted as a function of inhibitor concentration at some fixed substrate concentration (Dixon, 1972). The data can be fit to Equation 7.21 to yield a curvilinear fit, as shown in Figure 8.4A. (Note that this is the same as the dose–response plots discussed in Chapter 7, except here the x axis is plotted on a linear, rather than a logarithmic, scale.) A line is drawn from the v value at $[\mathrm{I}] = 0$ (referred to here as the starting point) through the point on the curve where $v = v_0/2$ ($n = 2$) and extended to the x axis. A second line is drawn from the starting point through the point on the curve where $v = v_0/3$

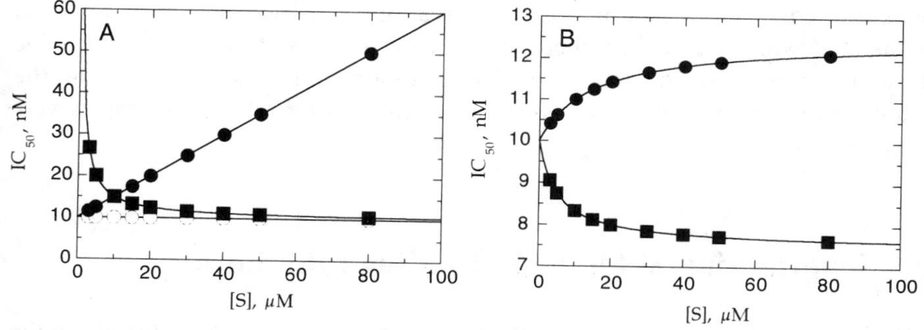

Figure 8.3 (A) The effects of substrate concentration on the IC_{50} values of competitive (solid circles), noncompetitive (open circles), and uncompetitive (solid squares) tight binding inhibitors. (B) The effects of substrate concentration on the IC_{50} values of mixed tight binding inhibitors when $\alpha > 1$ (squares) and when $\alpha < 1$ (circles).

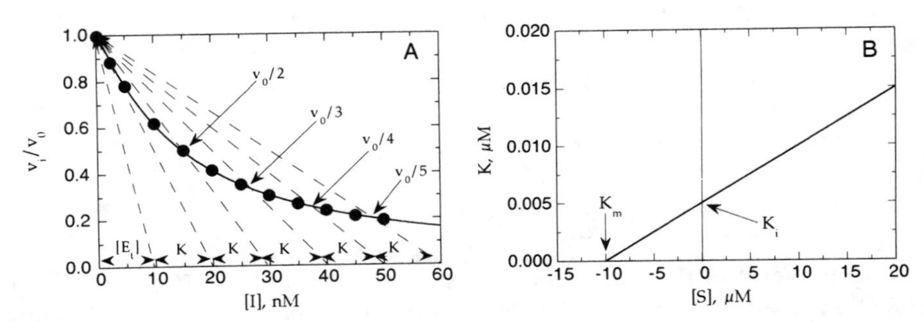

Figure 8.4 (A) Determination of "K" by the graphical method of Dixon (1972): dashed lines connect the starting point ($v_i/v_0 = 1$, $[I] = 0$) with points on the curve where $v_i = v_0/n$ ($n = 2, 3, 4$, and 5). Additional lines are drawn for apparent $n = 1$ and apparent $n = 0$, based on the x-axis spacing value "K," determined from the $n = 2$–5 lines (see text for further details). (B) Secondary plot of the "K" as a function of substrate concentration for a tight binding competitive inhibitor. Graphical determinations of K_i and K_m are obtained from the values of the y and x intercepts of the plot, respectively, as shown.

($n = 3$), and, in a similar fashion, additional lines are drawn from the starting point through other points on the curve where $v = v_0/n$ (n is an integer). The nest of lines thus drawn will intersect the x axis at a constant spacing, which is defined as K.

Knowing the value of K from the nest of lines, one can draw additional lines from the x axis to the origin at spacings of K on the x axis, for apparent values of $n = 1$ and $n = 0$. From this treatment, the line corresponding to $n = 0$ will intersect the x axis at a displacement from the origin that is equal to the total enzyme concentration $[E_t]$. Dixon goes on to show that in the case of a noncompetitive inhibitor, the spacing value K is equal to the inhibitor K_i, and a plot of K as a function of substrate concentration will be a horizontal line; that is, the value of K for a noncompetitive inhibitor is independent of substrate concentration. For a competitive inhibitor, however, the measured value of K will increase with increasing substrate concentration. A replot of K as a function of substrate concentration yields estimates of the K_i of the inhibitor and the K_m of the substrate from the y and x intercepts, respectively (Figure 8.4B).

8.3 Determining K_i for Tight Binding Inhibitors

The literature describes several methods for determining the K_i value of a tight binding enzyme inhibitor. We have already discussed the graphical method of Dixon (1972), which allows one to simultaneously distinguish inhibitor type and calculate the K_i. A more mathematical treatment of tight binding inhibitors, presented by Morrison (1969), led to a generalized equation to describe the fractional velocity of an enzymatic reaction as a function of inhibitor concentration, at fixed concentrations of enzyme and substrate:

$$\frac{v_i}{v_0} = 1 - \frac{([E_t] + [I] + K_i^{app}) - \sqrt{([E_t] + [I] + K_i^{app})^2 - 4[E_t][I]}}{2[E_t]} \tag{8.6}$$

The form of K_i^{app} in this equation varies with inhibitor type. The explicit forms of this parameter for the different inhibitor types, which follow, are similar to those presented in Equations 8.2–8.5 for the IC_{50} values.

For competitive inhibitors:

$$K_i^{app} = K_i\left(1 + \frac{[S]}{K_m}\right) \tag{8.7}$$

For noncompetitive inhibitors:

$$K_i^{app} = K_i \tag{8.8}$$

For uncompetitive inhibitors:

$$K_i^{app} = K_i\left(1 + \frac{K_m}{[S]}\right) \tag{8.9}$$

For mixed inhibitors:

$$K_i^{app} = \frac{[S] + K_m}{K_m/K_i + [S]/\alpha K_i} \tag{8.10}$$

Prior to the widespread use of computer-based routines for curve fitting, the direct use of the Morrison equation was inconvenient for extracting inhibitor constants from experimental data. To overcome this limitation, Henderson (1972) presented the derivation of a linearized form of the Morrison equation that allowed graphical determination of K_i and $[E_t]$ from measurements of the fractional velocity as a function of inhibitor concentration at a fixed substrate concentration. The generalized form of the Henderson equation is as follows:

$$\frac{[I]}{1 - v_i/v_0} = K_i^{app}\left(\frac{v_0}{v_i}\right) + [E_t] \tag{8.11}$$

where K_i^{app} has the same forms as presented in Equations 8.7–8.10 for the various inhibitor types.

Inspection reveals that Equation 8.11 is a linear equation. Hence, if one were to plot $[I]/(1 - v_i/v_0)$ as a function of v_0/v_i (i.e., the reciprocal of the fractional velocity), the data could be fit to a straight line with slope equal to K_i^{app} and y intercept equal to $[E_t]$, as illustrated in Figure 8.5. Note that the Henderson method yields a straight-line plot regardless of the inhibitor type. The slope of the lines for such plots will, however, vary with substrate concentration in

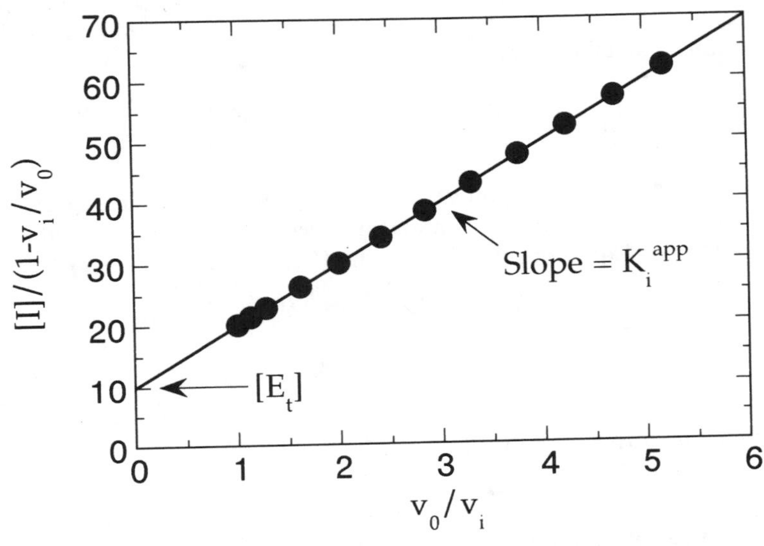

Figure 8.5 Henderson plot for a tight binding inhibitor.

different ways depending on the inhibitor type. The variation observed is similar to that presented in Figure 8.3 for the variation in IC_{50} value for different tight binding inhibitors as a function of substrate concentration. Thus, the Henderson plots also can be used to distinguish among the varying inhibitor binding mechanisms.

While linearized Henderson plots are convenient in the absence of a computer curve-fitting program, the data treatment does introduce some degree of systematic error. (See Henderson, 1973, for a discussion of the statistical treatment of such data.) Today, with the availability of robust curve-fitting routines on laboratory computers, it is no longer necessary to resort to linearized treatments of data such as the Henderson plots. The direct fitting of fractional velocity versus inhibitor concentration data to the Morrison equation (Equation 8.6) is thus much more desirable, and is strongly recommended.

Figure 8.6 illustrates the direct fitting of fractional velocity versus inhibitor concentration data to Equation 8.6. Such data analysis would call for predetermination of the K_m value for the substrate (as described in Chapter 5) and knowledge of the substrate concentration in the assays. Then the data, such as the points in Figure 8.6, would be fit to the Morrison equation, allowing both K_i^{app} and $[E_t]$ to be simultaneous determined as fitting parameters. Measurements of this type at several different substrate concentrations would allow determination of the mode of inhibition, and thus the experimentally measured K_i^{app} values could be converted to true K_i values.

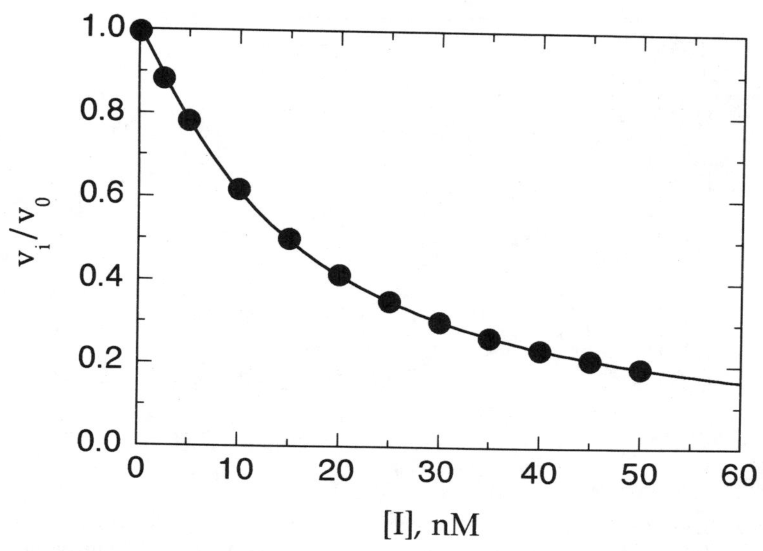

Figure 8.6 Plot of fractional velocity as a function of inhibitor concentration for a tight binding inhibitor. The solid curve drawn through the data points represents the best fit to the Morrison equation (Equation 8.6).

8.4 Use of Tight Binding Inhibitors to Determine Active Enzyme Concentration

In many experimental strategies one wishes to know the concentration of enzyme in a sample for subsequent data analysis. This approach applies not only to kinetic data, but also to other biochemical and biophysical studies with enzymes. The literature gives numerous methods for determining total protein concentration in a sample on the basis of spectroscopic, colorimetric, and other analytical techniques (see Copeland, 1994, for some examples). All these methods, however, measure bulk protein concentration rather than the concentration of the target enzyme in particular. Also, these methods do not necessarily distinguish between active enzyme molecules and molecules of denatured enzyme. In many of the applications one is likely to encounter, it is the concentration of active enzyme molecules that is most relevant. The availability of a tight binding inhibitor of the target enzyme provides a convenient means of accurately determining the concentration of active enzyme in the sample, even in the presence of denatured enzyme or other nonenzymatic proteins (Williams et al., 1979).

Referring back to Equation 8.6, if we set up an experiment in which $[E_t]$ and [I] are both much greater than K_i^{app}, we can largely ignore the K_i^{app} term

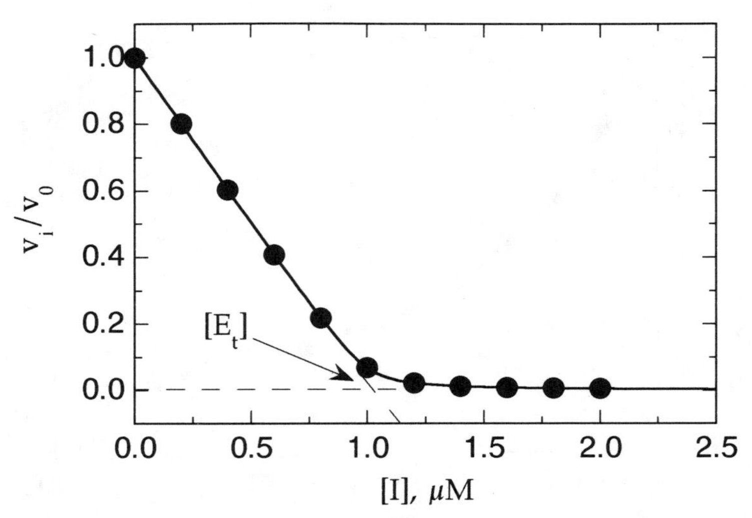

Figure 8.7 Determination of active enzyme concentration by titration with a tight binding inhibitor: $[E_t] = 1.0\,\mu M$, $K_i = 5\,nM$ (i.e., $[E_t]/K_i = 200$). The solid curve drawn through the data is the best fit to the Morrison equation (Equation 8.6). The dashed lines were drawn by linear least-squares fits of the data at inhibitor concentrations that were low (0–0.6 μM) and high (1.4–2.0 μM), respectively. The active enzyme concentration is determined from the x-axis value at the intersection of the two straight lines.

in this equation Under these conditions, the fractional velocity of the enzymatic reaction will fall off quasi-linearly with increasing inhibitor concentration until $[I] = [E_t]$. At this point the fractional velocity will approach zero and remain there at higher inhibitor concentrations. In this case, a plot of fractional velocity as a function of inhibitor concentration will resemble Figure 8.7, when fit to the Morrison equation. The data in Figure 8.7 were generated for the following hypothetical situation: K_i of inhibitor, 5 nM; active enzyme concentration of the sample, 1.0 μM (i.e., $[E_t]/K_i = 200$). The data at lower inhibitor concentration can be fit to a straight line that is extended to the x axis (short dashed line in Figure 8.7), and the data points at higher inhibitor concentrations can be fit to a straight horizontal line at $v_i/v_0 = 0$ (longer dashed line in Figure 8.7). The two lines thus drawn will intersect at a point on the x axis where $[I] = [E_t]$. Note, however, that this treatment only works when $[E_t]$ is much greater than K_i. When $[E_t]$ is less than about $200\,K_i$, the data are not well described by two intersecting straight lines. In such cases the data can be fit directly to Equation 8.6 to determine $[E_t]$ as described earlier.

This type of treatment is quite convenient for determining the active enzyme concentration of a stock enzyme solution (i.e., at high enzyme concentration) that will be diluted into a final reaction mixture for experimentation. For example, one might wish to store an enzyme sample at a nominal enzyme concentration of 100 μM in a solution containing 1 mg/mL gelatin for stability purposes (see discussion in Chapter 6). The presence of the gelatin would preclude accurate determination of enzyme concentration by one of the traditional colorimetric protein assays; moreover, *active* enzyme concentration could not be determined by means of such assays. Given a nanomolar inhibitor of the target enzyme, one could dilute a sample of the stock enzyme to some convenient concentration for an enzymatic assay that was still much greater than the K_i (e.g., 1 μM). Treatment of the fractional velocity versus inhibitor concentration as described here would thus lead to determination of the true concentration of active enzyme in the working solution, and from this one could back-calculate to arrive at the true concentration of active enzyme in the enzyme stock. This is a routine strategy in many enzymology laboratories, and numerous examples of its application can be found in the literature.

8.5 Summary

In this chapter we have described a special case of enzyme inhibition in which the dissociation constant of the inhibitors is similar to the total concentration of enzyme in the sample. These inhibitors offer a special challenge to the enzymologist because they cannot be analyzed by the traditional methods described in Chapter 7. We have seen that tight binding inhibitors yield double-reciprocal plots that appear to suggest noncompetitive inhibition, regardless of the true mode of interaction between the enzyme and the inhibitor. Thus, whenever noncompetitive inhibition is diagnosed through the use of double-reciprocal plots, the data should be reevaluated to ensure that tight binding inhibition is not occurring. Methods for determining the true mode of inhibition and the K_i for these tight binding inhibitors have been described in this chapter.

Tight binding inhibitors are an important class of molecules in many industrial enzyme applications. Many contemporary therapeutic enzyme inhibitors, for example, act as tight binders. Recent examples include inhibitors of dihydrofolate reductase (as anticancer drugs), inhibitors of the HIV aspartyl protease (as anti-AIDS drugs), and inhibitors of metalloproteases (as potential cartilage protectants). Many of the naturally occuring enzyme inhibitors, which play a role in metabolic homeostasis, are tight binding inhibitors of their target enzymes. Thus tight binding inhibitors are an important and commonly encountered class of enzyme inhibitors. The need for special treatment of enzyme kinetics in the presence of these inhibitors must not be overlooked.

References and Further Reading

Bieth, J. (1974) In *Proteinase Inhibitors, Bayer-Symposium V*, Springer-Verlag, New York, pp. 463–469.

Cha, S. (1975) *Biochem. Pharmacol.* **24**, 2177.

Cha, S. (1976) *Biochem. Pharmacol.* **25**, 2695.

Cha, S., Agarwal, R. P., and Parks, R. E., Jr. (1975) *Biochem. Pharmacol.* **24**, 2187.

Copeland, R. A. (1994) *Methods of Protein Analysis: A Practical Guide to Laboratory Protocols*, Chapman & Hall, New York.

Copeland, R. A., Lombardo, D., Giannaras, J., and DeCicco, C. P. (1995) *Bioorg. Med. Chem. Lett.* **5**, 1947.

Dixon, M. (1972) *Biochem. J.* **129**, 197.

Dixon, M., and Webb, E. C. (1979) *Enzymes*, 3rd ed., Academic Press, New York.

Goldstein, A. (1944) *J. Gen. Physiol.* **27**, 529.

Greco, W. R., and Hakala, M. T. (1979) *J. Biol. Chem.* **254**, 12104.

Henderson, P. J. F. (1972) *Biochem. J.* **127**, 321.

Henderson, P. J. F. (1973) *Biochem. J.* **135**, 101.

Morrison, J. F. (1969) *Biochim. Biophys. Acta*, **185**, 269.

Myers, D. K. (1952) *Biochem. J.* **51**, 303.

Turner, P. M., Lerea, K. M., and Kull, F. J. (1983) *Biochem. Biophys. Res. Commun.* **114**, 1154.

Williams, J. W., and Morrison, J. F. (1979) *Methods Enzymol.* **63**, 437.

William, J. W., Morrison, J. F., and Duggleby, R. G. (1979) *Biochemistry*, **18**, 2567.

CHAPTER

9

Time-Dependent Inhibition

All the inhibitors we have encountered thus far have established their binding equilibrium with the enzyme on a time scale that is rapid with respect to the turnover rate of the enzyme-catalyzed reaction. In Chapter 8 we noted that many tight binding inhibitors establish this equilibrium on a slower time scale, but in our discussion we eliminated this complication by pretreating the enzyme with the inhibitor long enough to ensure that equilibrium had been fully reached before steady state turnover was initiated by addition of substrate. In this chapter we shall explicitly deal with inhibitors that bind slowly to the enzyme on the time scale of enzymatic turnover, and thus display a change in initial velocity with time. These inhibitors, that is, act as *slow binding* or *time-dependent* inhibitors of the enzyme.

We can distinguish four different modes of interaction between an inhibitor and an enzyme that would result in slow binding kinetics. The equilibria involved in these processes are represented in Figure 9.1. Figure 9.1A shows the equilibrium associated with the uninhibited turnover of the enzyme, as we discussed in Chapter 5: k_1, the rate constant associated with substrate binding to the enzyme to form the ES complex, is sometimes refered to as k_{on} (for substrate coming *on* to the enzyme). The constant k_2 in Figure 9.1A is the dissociation or off rate constant for the ES complex dissociating back to free enzyme and free substrate, and k_{cat} is the catalytic rate constant as defined in Chapter 5.

In the remaining schemes of Figure 9.1 (B–D), the equilibrium described by Scheme A occurs as a competing reaction (as we saw in connection with simple reversible enzyme inhibitors in Chapter 7).

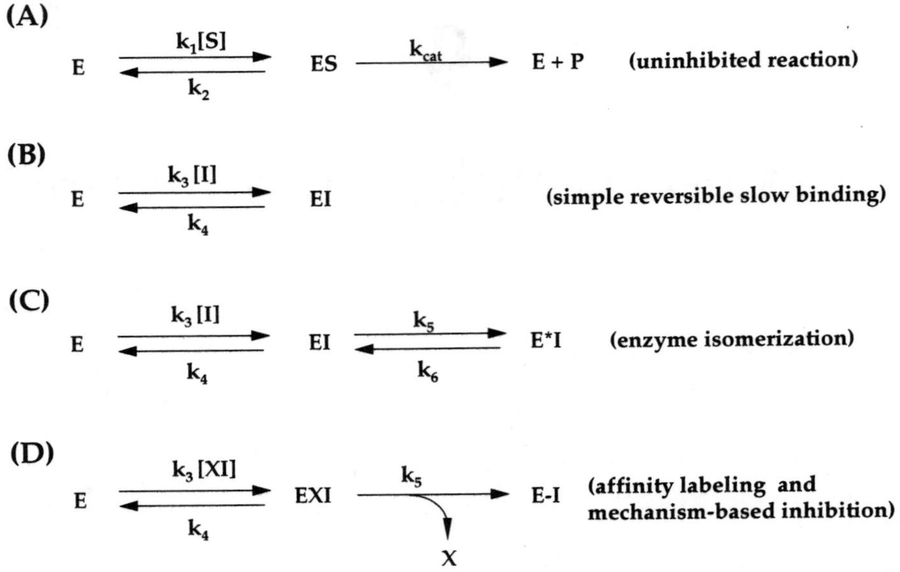

(A)

$$\mathrm{E} \underset{k_2}{\overset{k_1[\mathrm{S}]}{\rightleftharpoons}} \mathrm{ES} \xrightarrow{k_{cat}} \mathrm{E + P} \quad \textbf{(uninhibited reaction)}$$

(B)

$$\mathrm{E} \underset{k_4}{\overset{k_3[\mathrm{I}]}{\rightleftharpoons}} \mathrm{EI} \quad \textbf{(simple reversible slow binding)}$$

(C)

$$\mathrm{E} \underset{k_4}{\overset{k_3[\mathrm{I}]}{\rightleftharpoons}} \mathrm{EI} \underset{k_6}{\overset{k_5}{\rightleftharpoons}} \mathrm{E^*I} \quad \textbf{(enzyme isomerization)}$$

(D)

$$\mathrm{E} \underset{k_4}{\overset{k_3[\mathrm{XI}]}{\rightleftharpoons}} \mathrm{EXI} \xrightarrow[\downarrow \mathrm{X}]{k_5} \mathrm{E\text{-}I} \quad \textbf{(affinity labeling and mechanism-based inhibition)}$$

Figure 9.1 Schemes for time-dependent enzyme inhibition. Scheme A, which describes the turnover of the enzyme in the absence of inhibitor, is a competing reaction for all the other schemes. Scheme B illustrates the equilibrium for a simple reversible inhibition process that leads to time-dependent inhibition because of the low values of k_3 and k_4 relative to enzyme turnover. In Scheme C, an initial binding of the inhibitor to the enzyme leads to formation of the EI complex, which undergoes an isomerization of the enzyme to form the new complex E*I. Scheme D represents the reactions associated with irreversible enzyme inactivation due to covalent bond formation between the enzyme and some reactive group on the inhibitor, leading to the covalent adduct E–I. Inhibitors that conform to Scheme D may act as affinity labels of the enzyme, or they may be mechanism-based inhibitors.

Scheme B illustrates the case of the inhibitor binding to the enzyme in a simple bimolecular reaction, similar to what we discussed in Chapters 7 and 8. Here, however, the association and dissociation rate constants (k_3 and k_4, respectively) are such that the equilibrium is established slowly. As with rapid binding inhibitors, the equilibrium dissociation constant K_i is given here by:

$$K_i = \frac{k_4}{k_3} = \frac{[\mathrm{E}][\mathrm{I}]}{[\mathrm{EI}]} \tag{9.1}$$

Morrison and Walsh (1988) have pointed out that even when k_3 is diffusion limited, if K_i is low and [I] is varied in the region of K_i, both k_3[I] and k_4 will be low in value. Hence, under these circumstances onset of inhibition would be slow even though the magnitude of k_3 is that expected for a rapid reaction. This is why most tight binding inhibitors display time-dependent inhibition. If the observed time dependence is due to an inherently slow rate of binding, the inhibitor is said to be a *slow binding* inhibitor, and its dissociation constant is given by Equation 9.1. If, on the other hand, the inhibitor is also a tight binder, it is said to be a *slow, tight binding* inhibitor, and the depletion of the free enzyme and free inhibitor concentrations due to formation of the EI complex also must be taken into account:

$$K_i = \frac{([E_t] - [EI])([I] - [EI])}{EI} \qquad (9.2)$$

In Scheme C, the enzyme encounters the inhibitor and establishes a binding equilibrium that is defined by the on and off rate constants k_3 and k_4, just as in scheme B. In Scheme C, however, the binding of the inhibitor induces in the enzyme a conformational transition, or isomerization, that leads to a new enzyme–inhibitor complex E*I; the forward and reverse rate constants for the equilibrium between these two inhibitor-bound conformations of the enzyme are given by k_5 and k_6, respectively. The dissociation constant for the initial EI complex is still given by K_i (i.e., k_4/k_3), but a second dissociation constant for the second enzyme conformation, K_i^* must be considered as well. This second dissociation constant is given by:

$$K_i^* = \frac{K_i k_6}{k_5 + k_6} = \frac{[E][I]}{[EI] + [E^*I]} \qquad (9.3)$$

To observe a slow onset of inhibition, K_i^* must be much less than K_i. Hence, in this situation, the isomerization of the enzyme leads to much tighter binding between the eznyme and the inhibitor. As with scheme B, if the inhibitor is of the slow, tight binding variety, the diminution of free enzyme and free inhibitor must be explicitly accounted for in the expressions for both K_i and K_i^* (see Morrison and Walsh, 1988).

Note that to observe slow binding kinetics it is not sufficient for the conversion of EI to E*I alone to be slow. The reverse reaction must be slow as well. In fact, for the slow binding to be detected, the reverse rate constant (k_6) must be less than the forward isomerization rate (k_5). In the extreme case ($k_6 \ll k_5$), one does not observe a measurable return to the EI conformation and the enzyme isomerization step will appear to lead to irreversible inhibition. Under these conditions, k_6 can be considered to be insignificant, and the isomerization can be treated practically as an irreversible step dominated by the rate constant k_5.

Finally, in Scheme D we consider two modes of interaction of the inhibitor with the enzyme for which k_6 is truly equal to zero; that is, we are dealing with irreversible enzyme inactivation. We must make the distinction here between reversible and irreversible inhibition. In all the inhibitory schemes we have considered thus far, even in the case of slow tight binding inhibition, k_6 has been nonzero. This rate constant may be very small, and the inhibitors may act, for all practical purposes, as irreversible. With enough dilution of the EI complex and enough time, however, one can eventually recover an active free enzyme population. In the case of an irreversible inhibitor, the enzyme molecule that has bound the inhibitor is permanently incapacitated. No amount of time or dilution will result in a reactivation of the enzyme that has encountered inhibitors of these types. Such inhibitors hence are often referred to as *enzyme inactivators*.

The first example of irreversible inhibition is the process known as *affinity labeling* or *covalent modification* of the enzyme. In this case, the inhibitory compound binds to the enzyme and covalently modifies a catalytically essential residue or residues on the enzyme. The covalent modification involves some chemical alteration of the inhibitory molecule, but the process is based on chemistry that occurs at the modification site in the absence of any enzyme-catalyzed reaction. Affinity labels are useful not only as inhibitors of enzyme activity; they also have become valuable research tools. Some of these compounds are very selective for specific amino acid residues and can thus be used to identify key residues involved in the catalytic cycle of the enzyme (Lundblad, 1991; Copeland, 1994).

In the second form of irreversible inactivation we shall consider, *mechanism-based inhibition*, the inhibitory molecule binds to the enzyme active site and is recognized by the enzyme as a substrate analogue. The inhibitor is therefore chemically transformed through the catalytic mechanism of the enzyme to form an E–I complex that can no longer function catalytically. Many of these inhibitors inactivate the enzyme by forming an irreversible covalent E–I adduct. In other cases, the inhibitory molecule is subsequently released from the enzyme (a process referred to as noncovalent inactivation), but the enzyme has been permanently trapped in a form that can no longer support catalysis. Because they are chemically altered via the mechanism of enzymatic catalysis at the active site, mechanism-based inhibitors always act as competitive enzyme inactivators. These inhibitors have been referred to by a variety of names in the literature: suicide substrates, suicide enzyme inactivators, k_{cat} inhibitors, enzyme-activated irreversible inhibitors, Torjan horse inactivators, enzyme-induced inactivators, dynamic affinity labels, trap substrates, and so on (Silverman, 1988a).

In the discussion that follows we shall describe experimental methods for detecting the time dependence of slow binding inhibitors, and data analysis methods that allow one to distinguish among the different potential modes of interaction with the enzyme. We shall also discuss the appropriate determination of the inhibitor constants K_i and K_i^* for these inhibitors.

9.1 Progress Curves for Slow Binding Inhibitors

The progress curves for an enzyme reaction in the presence of a slow binding inhibitor will not display the simple linear product versus time relationship we have seen for simple reversible inhibitors. Rather, product formation over time

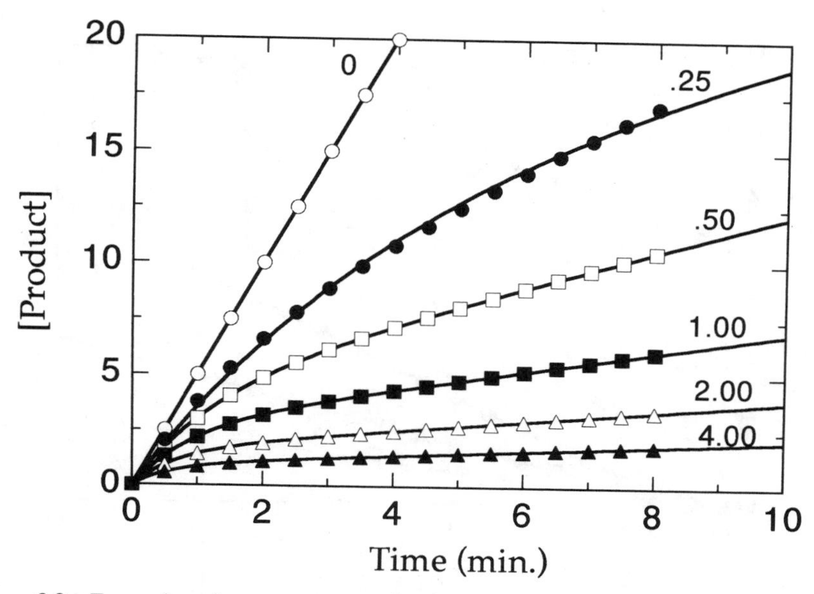

Figure 9.2 Examples of progress curves in the presence of varying concentrations of a time-dependent enzyme inhibitor for a reaction initiated by adding enzyme to a mixture containing substrate and inhibitor. Curves are numbered to indicate the relative concentrations of inhibitor present. Note that over the entire time window used (here 10 min), the uninhibited enzyme displays a linear progress curve.

will be a curvilinear function because of the slow onset of inhibition for these compounds. Figure 9.2 illustrates typical progress curves for a slow binding inhibitor when the enzymatic reaction is initiated by addition of enzyme. Over a time period in which the uninhibited enzyme displays a simple linear progress curve, the data in the presence of the slow binding inhibitor will display a quasi-linear relationship with time in the early part of the curve, converting later to a different (slower) linear relationship between product and time. Note that it is critical to establish a time window covering the linear portion of the uninhibited reaction progress curve, during which one can observe the change in slope that occurs with inhibition. If the onset of inhibition is very slow, a long time window may be required to observe the changes illustrated in Figure 9.2. With long time windows, however, one runs the risk of reaching significant substrate depletion which would invalidate the subsequent data analysis. Thus it may be necessary to evaluate several combinations of enzyme, substrate, and inhibitor concentrations to find an appropriate range of each for conducting time-dependent measurements. With these cautions addressed, the progress curves at different inhibitor concentrations can be described by Equation 9.4:

$$[\mathrm{P}] = v_{\mathrm{s}}t + \frac{v_{\mathrm{i}} - v_{\mathrm{s}}}{k_{\mathrm{obs}}}[1 - \exp(-k_{\mathrm{obs}}t)] \tag{9.4}$$

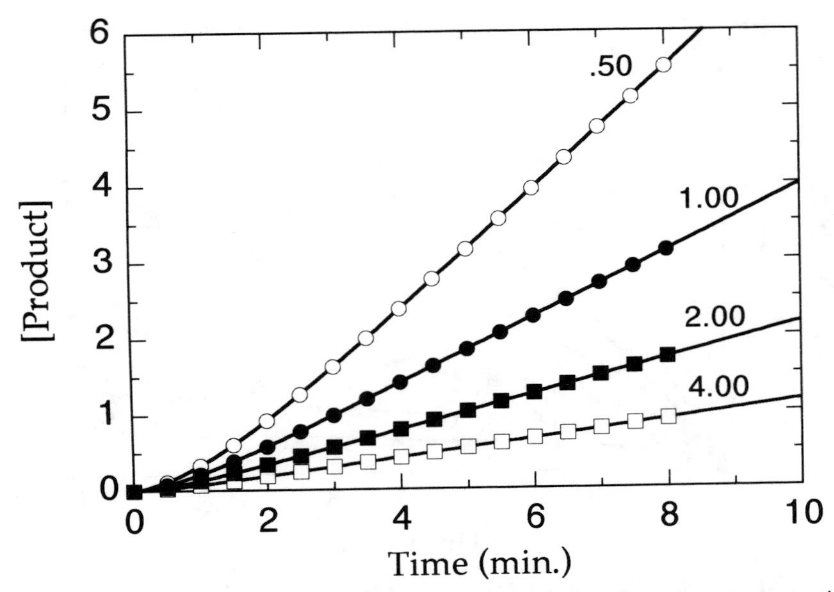

Figure 9.3 Examples of progress curves in the presence of varying concentrations of a time-dependent enzyme inhibitor for a reaction initiated by diluting an enzyme–inhibitor complex into the reaction buffer containing substrate. Curves are numbered to indicate the relative concentrations of inhibitor.

where v_i and v_s are the initial and steady state (i.e., final) velocities of the reaction in the presence of inhibitor, k_{obs} is the apparent first-order rate constant for the interconversion between v_i and v_s, and t is time. Morrison and Walsh (1988) have provided explicit mathematical expressions for v_i and v_s in the case of a competitive slow binding inhibitor, illustrating that v_i and v_s are functions of V_{max}, [S], K_m, and either K_i or K_i^* (for inhibitors that act according to Scheme C in Figure 9.1), respectively. For our purposes, it is sufficient to treat Equation 9.4 as an empirical equation that makes possible the extraction from the experimental data of values for v_i, v_s, and most importantly, k_{obs}. Note that v_i may or may not vary with inhibitor concentration, depending on the relative values of K_i and K_i^*, and the ratio of [I] to K_i (Morrison and Walsh, 1988). The value of v_s will be a finite, nonzero value as long as the inhibitor is not an irreversible enzyme inactivator. In the latter case, the value of v_s will eventually reach zero.

A second strategy for measuring progress curves for slow binding inhibitors is to preincubate the enzyme with the inhibitor for a long time period relative to the rate of inhibitor binding, and to then initiate the reaction by diluting the enzyme–inhibitor solution with a solution containing the substrate for the enzyme. During the preincubation period the equilibria between enzyme and inhibitor are established, and addition of substrate perturbs this equilibrium.

Because of the slow off rate of the inhibitor, the progress curve will display an initial shallow slope, which eventually turns over to the steady state velocity, as illustrated in Figure 9.3. The progress curves seen here also are well described by Equation 9.4, except that now the initial velocity is lower than the steady state velocity, whereas for data obtained by initiating the reaction with enzyme, the initial velocity is greater than the steady state velocity. To highlight this difference, some authors replace the term v_i in Equation 9.4 with v_r in the case of reactions initiated with substrate. Morrison and Walsh (1988) again provide an explicit mathematical form for v_r, which depends on the V_{max}, [S], K_m, [I], K_i, K_i^*, and the volume ratio between the preincubation enzyme–inhibitor solution and the final volume of the total reaction mixture. Again, for our purposes we can use Equaton 9.4 as an empirical equation, allowing v_i (or v_r), v_s, and k_{obs} to be adjustable parameters whose values are determined by non-linear curve fitting analysis.

Inhibitors that are very tight binding, as well as time dependent, almost always conform to Scheme C of Figure 9.1 (Morrison and Walsh, 1988). In this case the progress curves also will be influenced by the depletion of the free enzyme and free inhibitor populations that occurs. To account for these diminished populations, Equation 9.4 must be modified as follows:

$$[P] = v_s t + \frac{(v_i - v_s)(1 - \gamma)}{k_{obs}\gamma}[1 - \exp(-k_{obs}\gamma t)] \tag{9.5}$$

where γ is given by

$$\gamma = \frac{K_i^{*app} + [E_t] + [I_t] - Q}{K_i^{*app} + [E_t] + [I_t] + Q} \tag{9.6}$$

and

$$Q = [(K_i^{*app} + [I_t] - [E_t])^2 + 4(K_i^{*app}[E_t])]^{1/2} - (K_i^{*app} + [I_t] - [E_t]) \tag{9.7}$$

Further discussion of the data analysis for slow, very tight binding inhibitors can be found in the review by Morrison and Walsh (1988).

If inhibitor binding is very slow compared to the rate of uninhibited enzyme turnover, another convenient experimental strategy can be employed to determine k_{obs}. Essentially, the enzyme is preincubated with the inhibitor for different lengths of time before the steady state velocity of the reaction is measured. For example, if the steady state velocity of the reaction can be measured over a 30-second time window, but the inhibitor binding event occurs over the course of tens of minutes, the enzyme could be preincubated with the inhibitor between 0 and 120 minutes in 5-minute intervals, and the velocity of the reaction measured after each of the different preincubation times. Figure 9.4 illustrates the type of data this treatment would produce. For a fixed inhibitor concentration, the fractional velocity remaining after a given

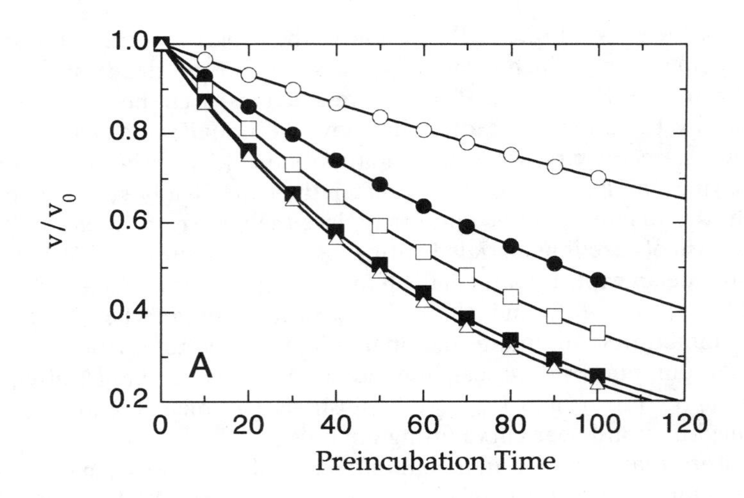

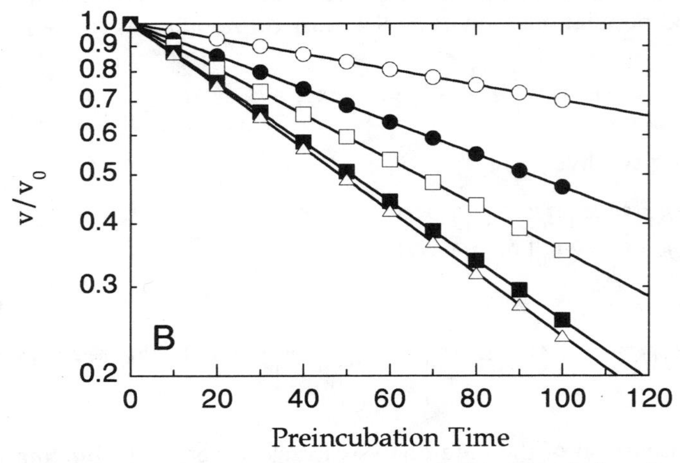

Figure 9.4 Preincubation time dependence of the fractional velocity of an enzyme-catalyzed reaction in the presence of varying concentrations of a slow binding inhibitor: data on a linear scale (A) and on a semilog scale (B).

preincubation time will fall off according to Equation 9.8:

$$\frac{v}{v_0} = \exp(-k_{\text{obs}}t) \tag{9.8}$$

Therefore, at a fixed inhibitor concentration, the fractional velocity will decay exponentially with preincubation time, as in Figure 9.4A. For convenience, we can recast Equation 9.8 by taking the logarithm of each side to obtain a linear function:

$$2.303 \log_{10}\left(\frac{v}{v_0}\right) = -k_{obs}t \quad (9.9)$$

Thus the value of k_{obs} at a fixed inhibitor concentration can be determined directly from the slope of a semilog plot of fractional velocity as a function of preincubation time, as in Figure 9.4B.

9.2 Distinguishing Between Slow Binding Schemes

To distinguish among the schemes illustrated in Figure 9.1, one must determine the effect of inhibitor concentration on the apparent first-order rate constant k_{obs}. We shall present the relationships between k_{obs} and [I] for these various schemes without deriving them explicitly. A full treatment of the derivation of these equations can be found in Morrison and Walsh (1988) and references therein.

9.2.1 Scheme B

For an inhibitor that binds according to Scheme B of Figure 9.1, the relationship between k_{obs} and [I] is given by Equation 9.10:

$$k_{obs} = k_4\left(1 + \frac{[\mathrm{I}]}{K_i^{app}}\right) \quad (9.10)$$

where K_i^{app} is the apparent K_i, which is related to the true K_i by different functions depending on the mode of inhibitor interaction with the enzyme (i.e., competitive, noncompetitive, uncompetitive, etc.; see Section 9.3). From Equation 9.10 we see that a plot of k_{obs} as a function of [I] should yield a straight line with slope equal to k_4/K_i^{app} and y intercept equal to k_4 (Figure 9.5). Thus from linear regression analysis of such data, one can simultaneously determine the values of k_4 and K_i^{app}.

9.2.2 Scheme C

For inhibitors corresponding to Scheme C of Figure 9.1, k_{obs} is related to [I] as follows:

$$k_{obs} = k_6 + \left[\frac{k_5[\mathrm{I}]}{K_i^{app} + [\mathrm{I}]}\right] \quad (9.11)$$

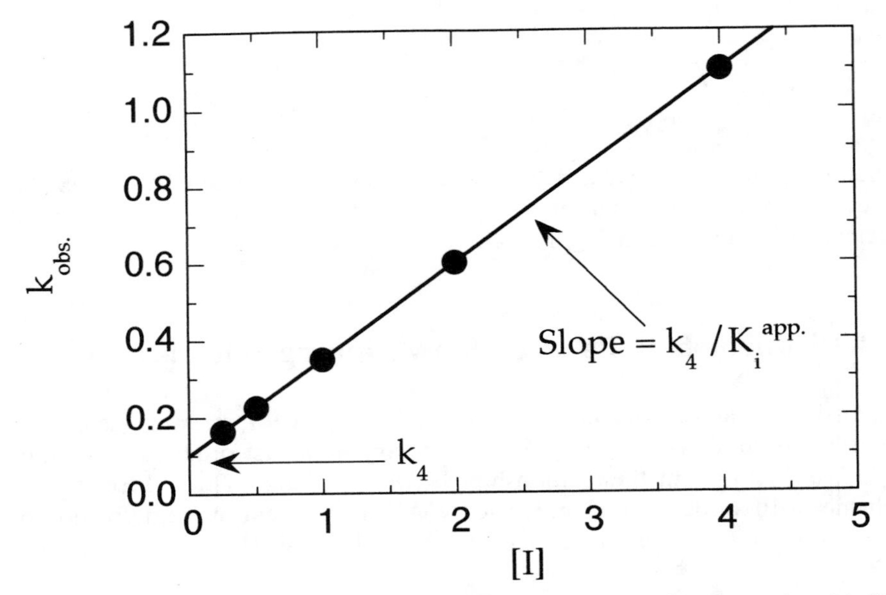

Figure 9.5 Plot of k_{obs} as a function of inhibitor concentration for a slow binding inhibitor that conforms to Scheme B of Figure 9.1.

which can be recast thus:

$$k_{obs} = k_6 \left[\frac{1 + \dfrac{[I]}{K_i^{*app}}}{1 + \dfrac{[I]}{K_i^{app}}} \right] \tag{9.12}$$

The form of Equations 9.11 and 9.12 predicts that k_{obs} will vary as a hyperbolic function of [I], as illustrated in Figure 9.6. The *y* intercept of the curve in this figure provides an estimate of the rate constant k_6, while the maximum value of k_{obs} expected at infinite inhibitor concentration according to Equation 9.11, is $k_6 + k_5$. Hence, by nonlinear curve fitting of the data to Equation 9.11 one can simultaneously determine the values of k_6, K_i^{app}, and K_i^{*app}.

Note that if K_i were much greater than K_i^*, the concentations of inhibitor required for slow binding inhibition would be much less than K_i. Under these circumstances, the steady state concentration of [EI] would be kinetically insignificant, and Equation 9.11 would thus reduce to:

$$k_{obs} = k_6 \left[1 + \frac{[I]}{K_i^{*app}} \right] \tag{9.13}$$

Thus for this situation a plot of k_{obs} as function of [I] would again yield a straight-line relationship, as we saw for inhibitors associated with Scheme B.

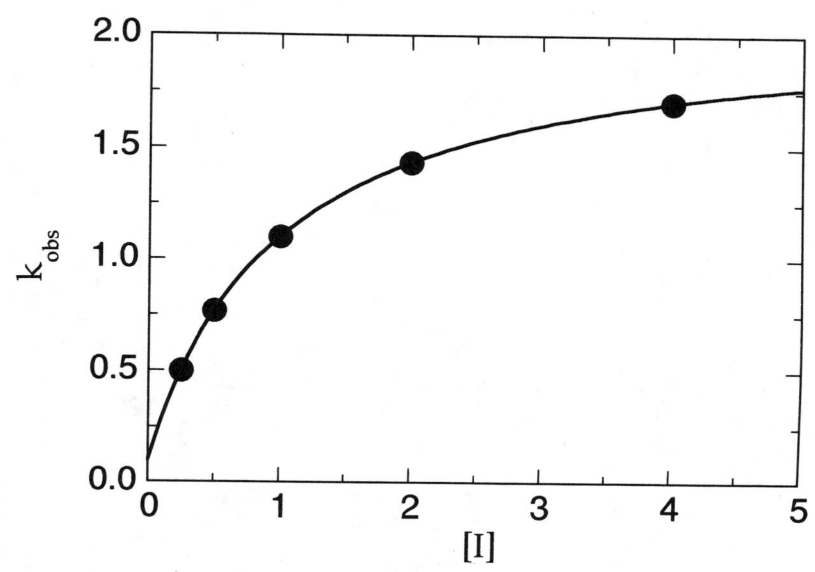

Figure 9.6 Plot of k_{obs} as a function of inhibitor concentration for a slow binding inhibitor that conforms to Scheme C of Figure 9.1.

In fact, when a straight-line relationship is observed in the plot of k_{obs} versus [I], one cannot readily distinguish between these two situations.

9.2.3 Scheme D

If the kinetic constant k_6 is very small in Scheme C, or zero as in Scheme D, the inhibitor acts, for all practical purposes, as an irreversible inactivator of the enzyme. In such cases, Equation 9.11 reduces to:

$$k_{obs} = \frac{k_5[\mathrm{I}]}{K_i^{app} + [\mathrm{I}]} \tag{9.14}$$

Here again, a plot of k_{obs} as a function of [I] will yield a hyperbolic curve (Figure 9.7A), but now the y intercept will be zero (reflecting the zero, or near-zero, value of k_6).

For irreversible inhibitors, the return to free E and free I from the EI complex is greatly perturbed by the irreversibility of the subsequent inactivation event (represented by k_5). For this reason, Tipton (1973) and Kitz and Wilson (1962) make the point that for irreversible inactivators, the term K_i no longer represents the simple dissociation constant for the EI complex. Rather, the term K_i^{app} in Equation 9.14 is defined as the apparent concentration of inhibitor required to reach half-maximal rate of inactivation of the enzyme.

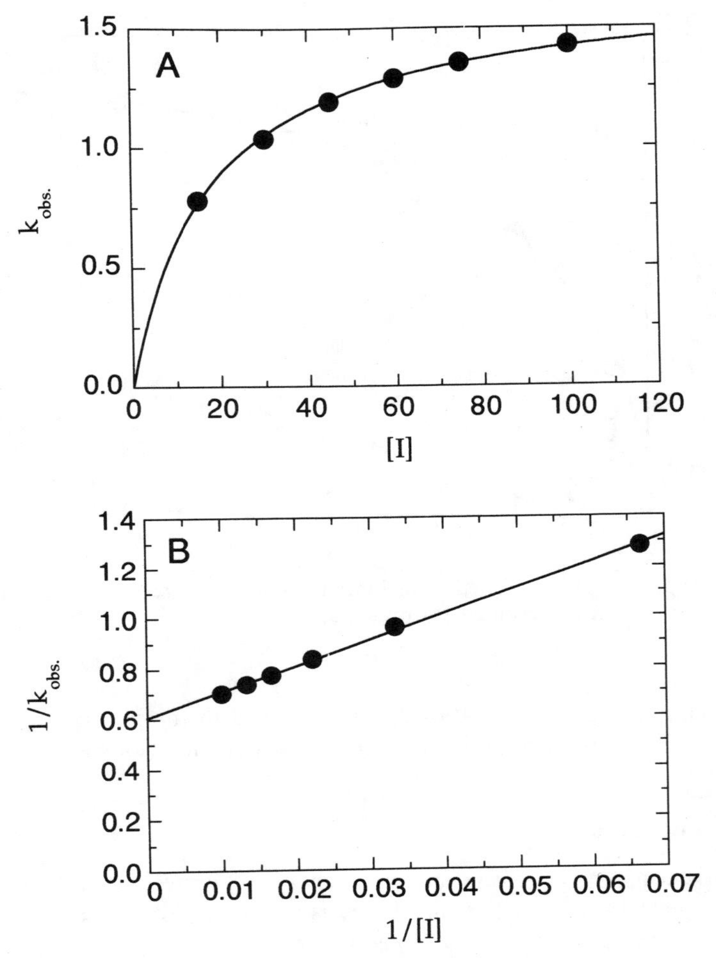

Figure 9.7 (A) Plot of k_{obs} as a function of inhibitor concentration for a slow binding inhibitor that conforms to Scheme D of Figure 9.1. (B) The data as in (A) presented as a double-reciprocal plot. The nonzero intercept indicates that the inactivation proceeds through a two-step mechanism: an initial binding step followed by a slower inactivation event.

Kitz and Wilson (1962) also replace k_5 in Equation 9.14 with k_{inact}, which they define as the maximal rate of enzyme inactivation. With these definitions, the parameters k_{inact} and K_i^{app} are reminiscent of the parameters V_{max} and K_m, respectively, from the Henri–Michaelis–Menten equation (Chapter 5). Just as the ratio k_{cat}/K_m is the best measure of the catalytic efficiency of an enzyme-catalyzed reaction, the best measure of inhibitory potency for an irreversible inhibitor is the second-order rate constant obtained from the ratio k_{inact}/K_i.

Similar to the Lineweaver–Burk plots encountered in Chapter 5, a double-reciprocal plot of $1/k_{obs}$ as a function of 1/[I] yields a straight-line relationship. Most irreversible inhibitors bind to the enzyme active site in a reversible manner (represented by K_i^{app}) before the slower inactivation event (represented by k_5) proceeds. Thus, as illustrated by Scheme D in Figure 9.1, the inactivation of the enzyme requires two sequential steps: a binding event and an inactivation event. Irreversible inhibitors that behave in this fashion display a linear relationship between $1/k_{obs}$ and 1/[I] that intersects the y axis at a value

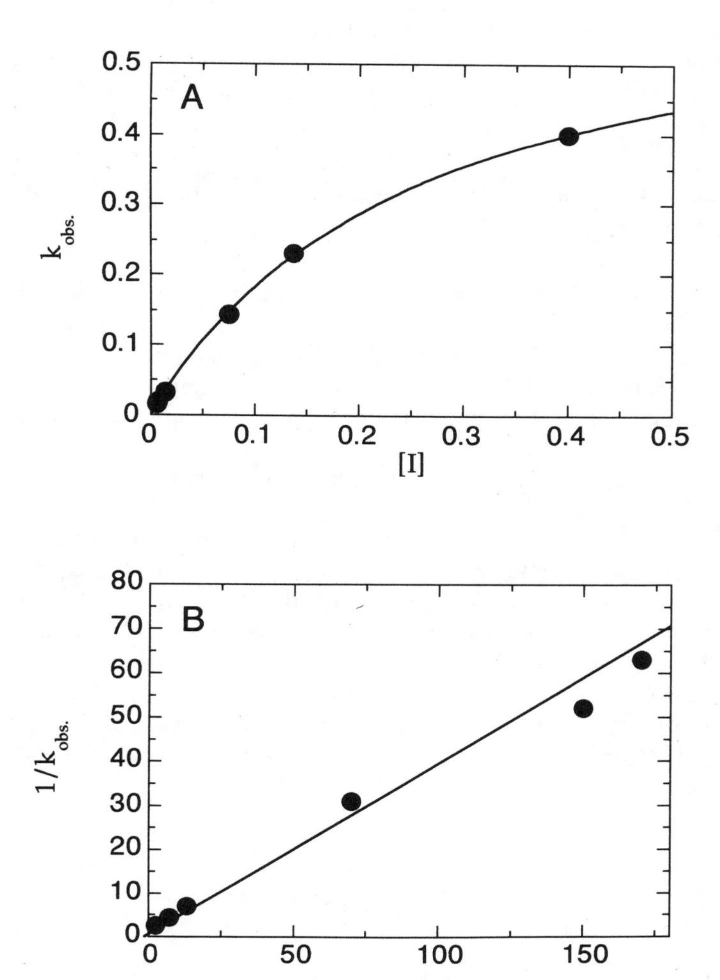

Figure 9.8 (A) Plot of k_{obs} as a function of inhibitor concentration for inhibition of acetylcholinesterase by methylsulfonyl fluoride. (B) The data in (A) as a double-reciprocal plot. [Data adapted from Kitz and Wilson (1962).]

greater than zero (Figure 9.7B). If, however, the formation of the reversible EI complex is kinetically insignificant relative to the rate of inactivation, the double-reciprocal plot will pass through the origin, reflecting a single-step inactivation process (Kitz and Wilson, 1962):

$$E + I \xrightarrow{k_{inact}} E - I$$

Although not as common as the two-step inactivation scheme shown in Figure 9.1D, this type of behavior is sometimes seen for small molecule affinity labels of enzymes. For example, Kitz and Wilson (1962) showed that the compound methylsulfonyl fluoride inactivates acetylcholinesterease by irreversible formation of a sulfonyl–enzyme adduct that appears to form in a single inactivation step (Figure 9.8).

9.3 Distinguishing Between Modes of Inhibitor Interaction with Enzyme

Morrison states that most slow binding enzyme inhibitors act as competitive inhibitors, binding at the enzyme active site (Morrison, 1982; Morrison and Walsh, 1988). Nevertheless it is possible, in principle at least, for slow binding inhibitors to interact with the enzyme by competitive, noncompetitive, uncompetitive, or mixed inhibition patterns. In the preceding equations, the relationships between K_i^{app} and K_i, and between K_i^{*app} and K_i^*, are the same as those presented in Chapters 7 and 8 for the relationships between K_i^{app} and K_i for the different modes of inhibition.

To distinguish the mode of inhibition that is taking place, hence to ensure the use of the appropriate relationships for K_i and K_i^* in the equations, one must determine the effects of varying substrate concentration on the value of k_{obs} at a fixed concentration of inhibitor. Tian and Tsou (1982, and references therein) have presented derivations of the relationships between k_{obs} and substrate concentration for competitive, noncompetitive, and uncompetitive irreversible inhibitors. (Similar patterns will be observed for slow binding inhibitors that conform to Scheme C as well.) More generalized forms of these relationships are given in Equations 9.15–9.17.

For competitive inhibition:

$$k_{obs} = \frac{k}{1 + [S]/K_m} \tag{9.15}$$

For noncompetitive inhibition:

$$k_{obs} = k \tag{9.16}$$

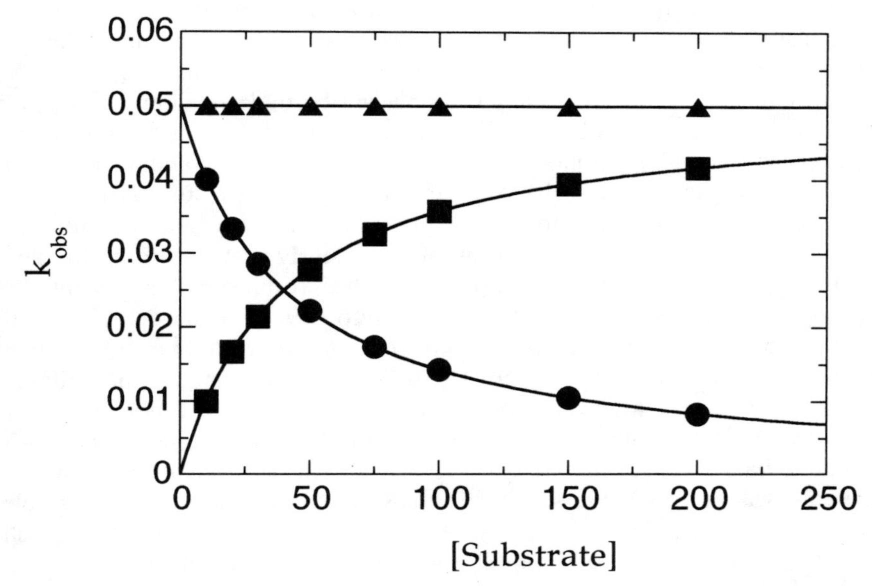

Figure 9.9 Expected dependence of k_{obs} on substrate concentration for time-dependent irreversible inhibitors that conform to competitive (circles), noncompetitive (triangles), and uncompetitive (squares) modes of interaction with the enzyme.

For uncompetitive inhibiton:

$$k_{obs} = \frac{k}{1 + K_m/[S]} \tag{9.17}$$

The constant k in these equations can be treated as an empirical variable for curve-fitting purposes (see Tian and Tsou, 1982, for the explicit form of k for irreversible inhibitors).

From the forms of Equations 9.15–9.17, we see that a competitive slow binding inhibitor will display a diminution in k_{obs} as the substrate concentration is raised. For a noncompetitive inhibitor, on the other hand, k_{obs} will not vary with substrate concentration, while for an uncompetitive inhibitor the value of k_{obs} will increase with increasing substrate concentration. These relationships between k_{obs} and substrate concentration are illustrated in Figure 9.9.

9.4 Determining Reversibility

It has been established that when an inhibitor conforming to Scheme C of Figure 9.1 has a very low value of k_6, it is difficult to differentiate this mode of inhibition from true irreversible inactivation according to Scheme D. To distinguish between these two possibilities, one must determine whether enzyme activity can be rescued by removal of unbound inhibitor from the enzyme solution. This is typically accomplished by dialysis, filter binding, or size exclusion chromatography (see Chapter 6 and references therein for details about these methods). Suppose, for example, that a slow binding inhibitor reduces the steady state velocity of an enzyme reaction by 50% at a concentration of 100 nM. If we prepare a 1 mL sample of enzyme with inhibitor present at 100 nM and dialyze this sample extensively against a liter of buffer, the final concentration of enzyme in the dialysis tubing will be essentially unchanged, but the concentration of free inhibitor will be reduced 1000-fold. If the inhibitor were binding reversibly to the enzyme we would have observed a postdialysis return of enzyme activity to close to the original uninhibited activity. For a very low value of k_6, it might take some time—hours or days—for the new equilibrium between free and bound inhibitor to establish itself after dialysis. If k_6 is nonzero, however, the expected reversal of inhibition eventually will occur. Of course, one must ensure that the enzyme itself is stable during these manipulations. Otherwise, it will be impossible to distinguish residual inhibition (due to the inhibitor), from enzyme inactivation (due to protein denaturation).

To distinguish covalent inactivation from noncovalent inhibition, one can look for the release of the original inhibitor molecule upon denaturation of the enzyme sample. Suppose that a slow binding inhibitor actually acted as a covalent affinity label of the target enzyme. If we were to denature the enzyme after inhibition and then separate the denatured protein from the rest of the solution (see Chapter 6), the inhibitory molecule would remain with the denatured protein, as a result of the covalent linkage between the inhibitor and the enzyme. If, on the other hand, the inhibitor were noncovalently associated with the inhibitor, it would be released into the solution upon denaturation of the enzyme.

An illustration of this type of experiment comes from work recently performed in our laboratory on an inhibitor of the inducible isoform of the enzyme prostaglandin G/H synthase (PGHS2). A compound we were investigating, DuP697, displayed the kinetic features of a competitive, slow binding, irreversible enzyme inactivator (Copeland et al., 1994). A plot of k_{obs} as a function of DuP697 concentration displayed a hyperbolic fit that passed through the origin of the plot. Extensive dialysis of the inhibited enzyme against buffer did not result in a return of enzymatic activity, suggesting either that the inhibitor was covalently associating with the enzyme or that the value of k_6 was extremely small.

To determine which way DuP697 was interacting with the enzyme, we treated a micromolar solution of the enzyme with a substoichiometric concentration of the inhibitor and allowed the resulting solution to equilibrate for a long time period, relative to the rate of enzyme inactivation. The enzyme was then denatured and precipitated by addition of 4 volumes of a 1 : 1 mixture of methanol/acetonitrile. The denatured protein solution was centrifuged through a 30 kDa cutoff filter, and the filtrate from this treatment was dried under nitrogen and redissolved in a small volume of dimethyl sulfoxide or acetonitrile. The amount of DuP697 released from the enzyme by this treatment was then determined by reversed phase HPLC and by measuring the ability of the redissolved filtrate to effect inhibition of fresh samples of the enzyme (Copeland et al., 1994, 1995). Upon finding that 97% of the DuP697 added to the starting enzyme sample was recovered in this way, we concluded that DuP697 is not a covalent modifier of the enzyme, but rather conforms to Scheme C of Figure 9.1 with an extremely small value for k_6.

9.5 Examples of Slow Binding Enzyme Inhibitors

The literature is filled with examples of slow binding, slow, tight binding, affinity label, and mechanism-based inhibitors of important enzymes. Extensive examples of slow binding inhibitors were presented in the review by Morrison and Walsh (1988). Silverman has devoted a two-volume text to the subject of mechanism-based enzyme inactivators (1988a), as well as an extensive review article (1988b) to their potential uses in medicine. (See also Trzaskos et al., 1995, for an interesting more recent example of mechanism-based inactivation of lanosterol 14α-methyl demethylase in the design of new cholesterol-lowering therapies.) Rather than providing an exhaustive review of the literature, we shall present only two examples of enzyme systems that have proved amenable to time-dependent inhibition: the serine proteases and prostaglandin G/H synthase. These examples should suffice to illustrate the importance of this class of enzyme inhibitors.

9.5.1 Serine Proteases

As we saw in Chapter 4, the serine proteases hydrolyze peptide bonds through the formation of a tetrahedral transition state involving a peptide carbonyl carbon of the substrate and an active site serine residue as the attacking nucleophile. Several groups have taken advantage of the ability of boron to adopt a tetrahedral ligand sphere in preparing transition state analogues as inhibitors of serine proteases. Kettner and Shervi (1984) have used this strategy to prepare selective inhibitors of the serine proteases chymotrypsin and leukocyte elastase based on α-aminoboronate peptide analogues (Figure 9.10). They found that succinamide methyl esters that incorporate aminoboronate analogues of phenylalanine and valine were highly selective inhibitors of chymotrypsin and leukocyte elastase, respectively. A selective inhibitor of leukocyte elastase could have potential therapeutic value in the treatment of a number of inflammatory diseases of the respiratory system (e.g., cystic fibrosis, asthma). Kinetic studies of R-Pro-boroPhe-OH binding to chymotrypsin and R-Pro-boroVal-OH binding to leukocyte elastase revealed that both inhibitors function as competitive slow binding inhibitors that conform to Scheme C of

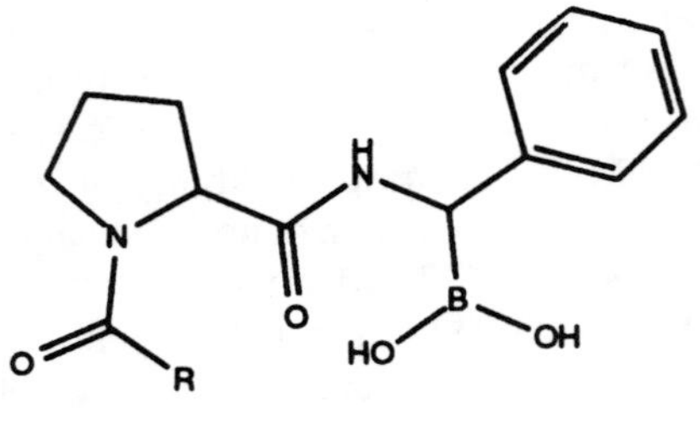

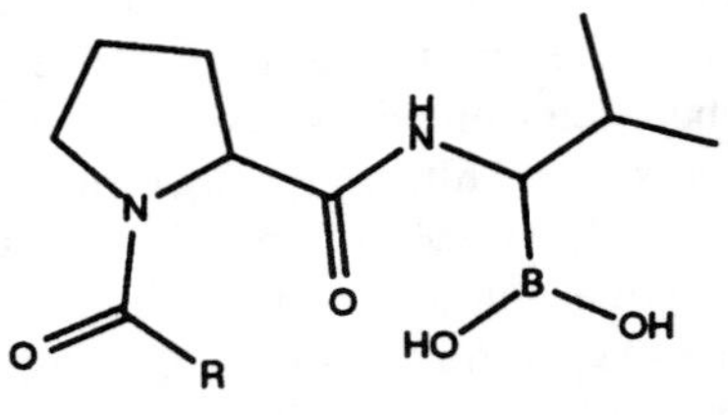

R-Pro-boroPhe-OH R-Pro-boroVal-OH

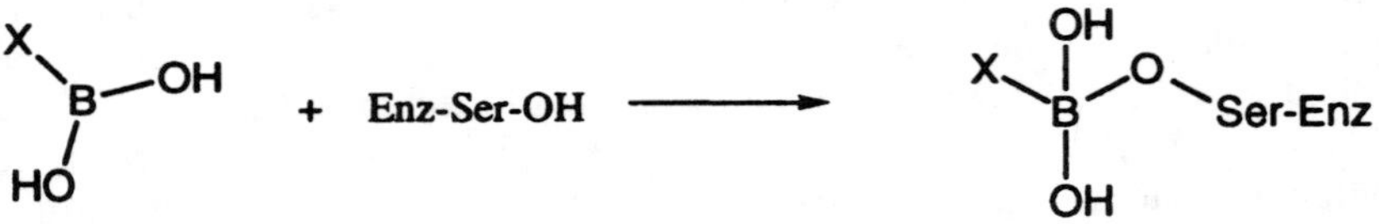

Figure 9.10 Examples of slow binding α-aminoboronate peptide inhibitors of serine proteases. These inhibitors form tetrahedral adducts with the active site serine of the proteases. See Kettner and Shervi (1984) for further details.

Figure 9.1. For chymotrypsin inhibition by R-Pro-boroPhe-OH, these workers determined values of K_i and K_i^* of 3.4 and 0.16 nM, respectively. Likewise, for leukocyte elastase inhibition by R-Pro-boroVal-OH, the values of K_i and K_i^* were found to be 15 and 0.57 nM, respectively. Interestingly, Kettner and Shervi also found R-Pro-boroPhe-OH to be a nanomolar inhibitor of the serine protease cathepsin G, but in this case no slow binding behavior was observed. They suggest that the slow binding behavior of these inhibitors reflects the formation of an initial tetrahedral adduct with the active site serine, followed by a conformational rearrangement of the enzyme to optimize binding (presumably this conformational readjustment does not occur in the case of cathepsin G).

9.5.2 Prostaglandin G/H Synthase

Prostaglandins are mediators of many of the physiological effects associated with inflammation that lead to such symptoms as pain, swelling, and fever. The biosynthesis of these mediators is rate-limited by the conversion of arachidonic acid to prostaglandin GH_2 by the enzyme prostaglandin G/H synthase (PGHS). One of the most widely used drugs today for the treatment of the pain, swelling, and fever associated with inflammation is aspirin, a compound first isolated from the bark of a certain type of willow tree that had been used for centuries as a folk treatment for pain and fever (Weissman, 1991).

In the early 1970s Vane and his coworkers showed that aspirin elicits its anti-inflammatory effects by inhibition of prostaglandin biosynthesis (Vane, 1971). It was subsequently found that aspirin functions as an affinity label for the enzyme PGHS, covalently inhibiting the enzyme by acetylating an active site serine (Ser 530). The acetylation of this residue irreversibly blocks the binding of arachidonic acid to the enzyme active site. Chronic aspirin use, however, may lead to stomach pain and ulceration, and renal failure, as a result of the breakdown of the mucosal linings of the stomach, intestines, and kidneys. For years, scientists and physicians have searched for anti-inflammatory drugs that could be taken over time without severe side effects. From their efforts a broad class of drugs, known as nonsteroidal anti-inflammatory drugs (NSAIDs) has emerged.

A large and highly prescribed class of NSAIDs are the carboxylic acid containing compounds, typified by the drugs flurbiprofen and indomethacin (Figure 9.11). These compounds have been shown to act as time-dependent inactivators of PGHS, conforming to Scheme C of Figure 9.1. However, the value of k_6 is so low that for all practical purposes, these compounds function as irreversible inactivators and can be treated kinetically as such. Rome and Lands (1975), who have studied the time dependence of these inhibitors in detail, noted a common structural feature, a carboxylic acid group. Reasoning that some acid–base chemistry at the enzyme active site might account for the time-dependent inhibitory effects observed, these workers prepared the methyl esters of eight carboxylate-containing PGHS inhibitors. The results of these studies are summarized in Table 9.1.

Rome and Lands found that for the most part, binding of the inhibitor (reflected in K_i) was not significantly affected by esterification, but in all cases the time dependence (reflected in k_{inact}) was completely lost.

A structural rationale for the foregoing results may now be available. Picot et al. (1994) have recently reported the crystal structure of PGHS with the carboxylate inhibitor flurbiprofen bound at the active site. They found that the carboxylate moiety of this inhibitor engages in formation of a salt bridge with Arg 120 in the arachidonic acid binding cavity. The formation of this salt bridge, by displacement of nearby amino acid residues, may be the rate-limiting step in the time-dependent inactivation of the enzyme by these inhibitors. Figure 9.11 provides a cartoon version of the proposed interactions between the active site arginine and the carboxylate group of NSAIDs. Note that residue Ser 530 is in close proximity to the bound inhibitor in the crystal structure. This serine is the site of covalent modification and irreversible inactivation by aspirin.

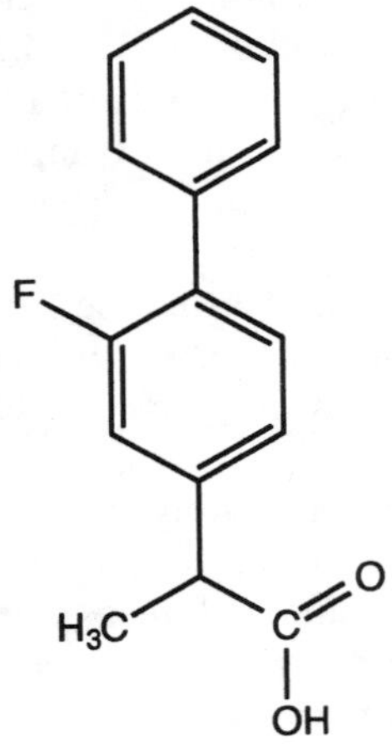

Flurbiprofen

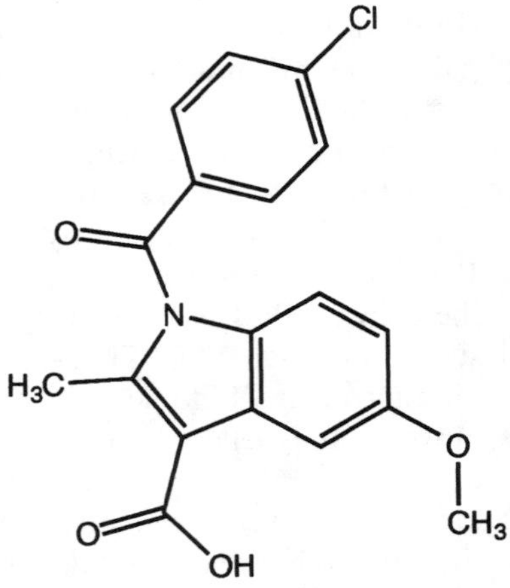

Indomethacin

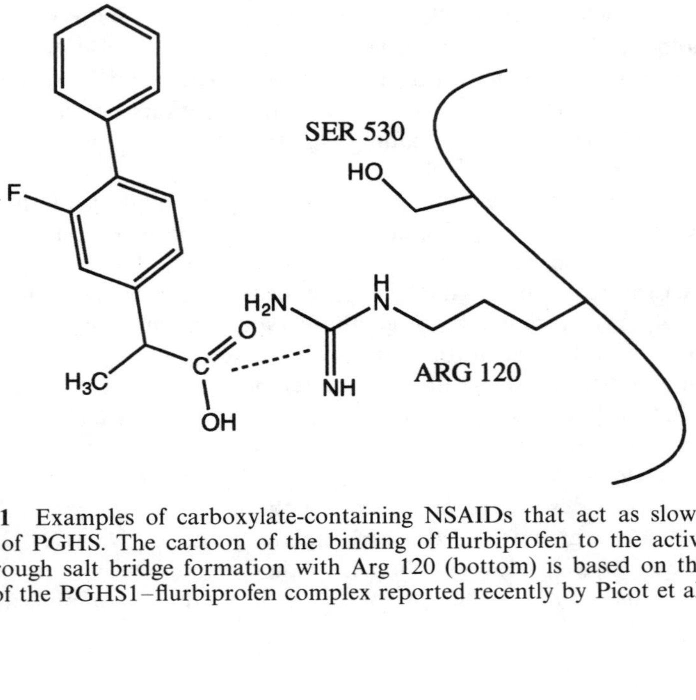

Figure 9.11 Examples of carboxylate-containing NSAIDs that act as slow binding inhibitors of PGHS. The cartoon of the binding of flurbiprofen to the active site of PGHS through salt bridge formation with Arg 120 (bottom) is based on the crystal structure of the PGHS1–flurbiprofen complex reported recently by Picot et al. (1994).

Table 9.1 Time-Dependent Kinetic Parameters for Carboxylate Inhibitors of PGHS and Their Methyl Esters

	K_i (μM)		k_{inact} ($\mu M^{-1}\ min^{-1}$)	
Inhibitor	Free Acid	Ester	Free Acid	Ester
Indomethacin	100	1	0.04	0
Aspirin	14,000	16,000	0.0003	0
Flurbiprofen	1	0.5	1.1	0
Ibuprofen	3	6	0	0
Meclofenamic acid	4	1	0.4	0
Mefenamic acid	1	3	0	0
BL-2338	1	5	0.08	0
BL-2365	14	9	0	0

Source: Data from Rome and Lands (1975).

Recently Penning and coworkers (Tang et al., 1995) developed affinity labels of PGHS by preparing bromoacetamido analogues of the existing NSAIDs indomethacin and mefenamic acid. The bromoacetamido group is attacked by an active site nucleophile to form a covalent adduct that leads to irreversible inactivation of the enzyme (Figure 9.12). Under strong acidic conditions, the amine-containing NSAID moiety is cleaved off, leaving behind a carboxymethylated version of the active site nucleophile. These affinity labels can thus be used as mechanistic probes of the enzyme active site. By incorporating a radiolabel into the methylene carbon of the bromoacetamido group, one can obtain selective radiolabel incorporation into the enzyme at the attacking nucleophile.

PGHS performs two catalytic conversions of its substrate, arachidonic acid: a cyclooxygenase step (in which two equivalents of molecular oxygen are added) and a peroxidase step (in which the incorporated peroxide moiety is converted to the final alcohol of prostaglandin GH_2). The classical NSAID inhibitors block enzyme turnover by inhibiting selectively the cyclooxygenase

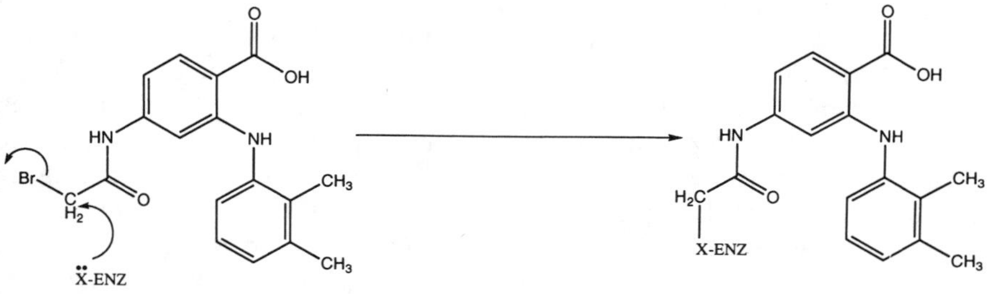

Figure 9.12 Affinity labeling of PGHS by the bromoacetamido analogue of the NSAID 2,3-dimethylanthranilic acid. [Adapted from Tang et al. (1995).]

step of the reaction. This observation has raised the question of whether the two enzymatic reaction steps involve the same set of active site residues or use distinct catalytic centers for each reaction. Tang et al. (1995) demonstrated that the bromoacetamido affinity labels bind to PGHS in a 2:1 stoichiometry and, unlike their NSAID analogues, abolish both the cyclooxygenase and peroxidase activities of the enzyme. Interestingly, they found that pretreatment of the enzyme with aspirin or mefenamic acid reduces the stoichiometry of the affinity label incorporation to 1:1. Furthermore, if the mefenamic acid saturated enzyme is treated with the affinity label and subsequently dialyzed to remove mefenamic acid, the version of the enzyme that results retains its cyclooxygenase activity but is devoid of peroxidase activity (Tang et al., 1995). These findings support the hypothesis that the catalytic centers for cyclooxygenase and peroxidase activities are distinct in PGHS. These affinity labels, and the peroxidase-deficient enzyme they provide, should prove to be useful tools in dissecting the mechanism of PGHS turnover.

A continuing problem in the treatment of inflammatory diseases with NSAIDs is the gastrointestinal and renal damage observed among patients who receive chronic treatment with these drugs. The side effects are mechanism-based in that inhibition of PGHS not only blocks the symptoms of inflammation but also interferes with the maintenance of the protective mucosal linings of the digestive system. In the early 1990s it was discovered that humans and other mammals contain two isoforms of this enzyme: PGHS1, which is constitutively expressed in a wide variety of tissues, including gastrointestinal and renal tissue, and PGHS2, which is induced in response to inflammatory stimuli and is primarily localized to cells of the immune system and the brain. This discovery immediately suggested a mechanism for treating inflammatory disease without triggering the side effects of traditional NSAID therapy, namely, selective inhibition of the inducible isoform, PGHS2.

In the hope of developing new and safer anti-inflammatory drugs, a number of laboratories, including ours, set out to identify compounds that would inhibit PGHS2 selectively over PGHS1. One compound that seemed to fit this selectivity profile was DuP697, a methyl sulfonyl containing diaryl thiophene (Figure 9.13). Kinetic studies of DuP697 inhibition of PGHS1 and PGHS2 revealed an unusual basis for the isozyme selectivity of this compound. DuP697 appeared to bind weakly, but with equal affinity, to both isozymes (Copeland et al., 1994). For PGHS1, we demonstrated that DuP697 acted as a classic reversible competitive inhibitor (Copeland et al., 1995). For PGHS2, however, the binding of DuP697 induced an isomerization of the enzyme that led to much tighter association of the inhibitor–enzyme complex, according to Scheme C of Figure 9.1. This isomerization step in fact led to such tight binding that the inhibition could be treated as a two-step irreversible inactivation of the enzyme (Scheme D of Figure 9.1). Plots of k_{obs} as a function of DuP697 concentration showed the hyperbolic behavior expected for inactivation where k_6 was zero or near zero. From this we determined values of K_i and k_{inact} of

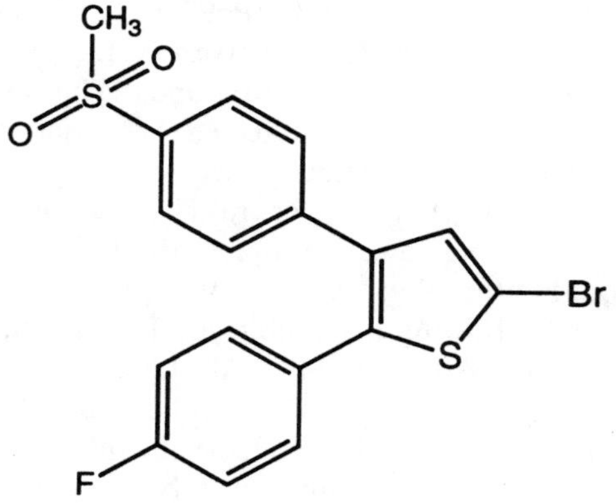

DuP697

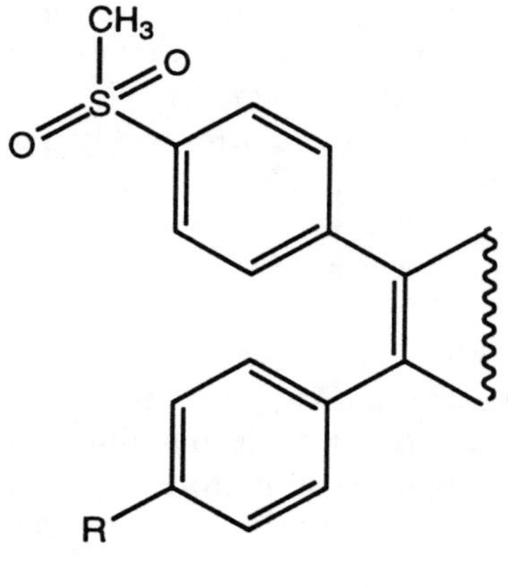

Generic PGHS2 Inhibitor

Figure 9.13 Chemical structures of DuP697 and the generic form of a PGHS2 selective inhibitor. [Based on the data from Copeland et al. (1995).]

2.19 μM and 0.017 s^{-1}, respectively, or a second-order rate constant k_{inact}/K_i of $7.76 \times 10^{-3}\ \mu M^{-1}, s^{-1}$. Thus the isozyme selectivity of this compound resulted from its ability to induce a conformational transition in one isozyme but not the other. The structural basis for this inhibitor-induced conformational transition remains to be fully elucidated.

We then explored analogues of DuP697 in an attempt to identify the minimal structural requirements for selective PGHS2 inhibition and to search for more potent and more selective compounds. The results of these studies identified the structural component labeled "generic PGHS2 inhibitor" in Figure 9.13 as the critical pharmacophore for selective PGHS2 inhibition (Copeland et al., 1995). Within this general class of compounds we were able to prepare inhibitors that showed complete discrimination between the two isozymes: that is, inhibitors that demonstrated potent, time-dependent inhibition of PGHS2, while showing no inhibition of PGHS1 at any concentration up to their solubility limits (Copeland et al., 1995). The information obtained from these studies, and similar studies from other laboratories, provides a clear direction for the development of PGHS2 specific inhibitors. These compounds may ultimately be useful in the design of new NSAIDs with significant benefits to patients suffering from inflammatory diseases.

9.6 Summary

In this chapter we have described the behavior of enzyme inhibitors that elicit their inhibitory effects slowly on the time scale of enzyme turnover. These slow binding, or time-dependent inhibitors can operate by any of several distinct mechanism of interaction with the enzyme. Some of these inhibitors bind reversibly to the enzyme, while others irreversibly inactivate the enzyme molecule. Irreversible enzyme inactivators that function as affinity labels or mechanism-based inactivators can provide useful mechanistic information concerning the types of amino acid residue that are critical for catalysis (see Silverman, 1988a; Lundblad, 1991).

We discussed kinetic methods for properly evaluating slow binding enzyme inhibitors, and data analysis methods for determining the relevant rate constants and dissociation constants for these inhibition processes. Finally, we presented examples of slow binding inhibitors of the serine proteases and prostaglandin G/H synthases to illustrate the importance of this class of inhibitors in enzymology.

References and Further Reading

Copeland, R. A. (1994) *Methods for Protein Analysis: A Practical Guide to Laboratory Protocols*, Chapman & Hall, New York, pp. 151–160.

Copeland, R. A., Williams, J. M., Giannaras, J., Nurnberg, S., Covington, M., Pinto, D., Pick, S., and Trzaskos, J. M. (1994) *Proc. Natl. Acad. Sci.* U.S.A. **91**, 11202.

Copeland, R. A., Williams, J. M., Rider, N. L., Van Dyk, D. E., Giannaras, J., Nurnberg, S., Covington, M., Pinto, D., Magolda, R. L., and Trzaskos, J. M. (1995) *Med. Chem. Res.* **5**, 384.

Kettner, C., and Shervi, A. (1984) *J. Biol. Chem.* **259**, 15106.

Kitz, R., Wilson, I. B. (1962) *J. Biol. Chem.* **237**, 3245.

Lundblad, R. (1991) *Chemical Reagents for Protein Modification*, CRC Press, Boca Raton, FL.

Morrison, J. F. (1982) *Trends Biochem. Sci.* **7**, 102.

Morrison, J. F., and Walsh, C. T. (1988) *Adv. Enzymol.* **61**, 201.

Picot, D., Loll, P. J., and Garavito, M. R. (1994) *Nature*, **367**, 243.

Rome, L. H., and Lands, W. E. M. (1975) *Proc. Natl. Acad. Sci.* U.S.A. **72**, 4863.

Silverman, R. B. (1988a) *Mechanism-Based Enzyme Inactivation: Chemistry and Enzymology*, Vols. I and II, CRC Press, Boca Raton, FL.

Silverman, R. B. (1988b) *J. Enzyme Inhib.* **2**, 73.

Tang, M. S., Askonas, L. J., and Penning, T. M. (1995) *Biochemistry*, **34**, 808.

Tian, W.-X., and Tsou, C.-L. (1982) *Biochemistry*, **21**, 1028.

Tipton, K. F. (1973) *Biochem. Pharmacol.* **22**, 2933.

Trzaskos, J. M., Fischer, R. T., Ko, S. S., Magolda, R. L., Stam, S., Johnson, P., and Gaylor, J. L. (1995) *Biochemistry*, **34**, 9677.

Vane, J. R. (1971) *Nature, New Biol.* **231**, 232.

Weissman, G. (1991) *Sci. Am.* January, p. 84.

CHAPTER

10

Enzyme Reactions with Multiple Substrates

Until now we have considered only the simplest of enzymatic reactions, those involving a single substrate being transformed into a single product. However, the vast majority of enzymatic reactions one is likely to encounter involve at least two substrates and result in the formation of more than one product. Let us look back at some of the enzymatic reactions we have used as examples. Many of them are multisubstrate and/or multiproduct reactions. For example, the serine proteases selected to illustrate different concepts in earlier chapters use two substrates to form two products. The first, and most obvious, substrate is the peptide that is hydrolyzed to form the two peptide fragment products. The second, less obvious, substrate is a water molecule that indirectly supplies the proton and hydroxyl groups required to complete the hydrolysis. Likewise, when we discussed the phosphorylation of proteins by kinases, we needed a source of phosphate for the reaction, and this phosphate source itself is a substrate of the enzyme. An ATP-dependent kinase, for example, requires the protein and ATP as its two substrates, and it yields the phosphoprotein and ADP as the two products. A bit of reflection will show that many of the enzymatic reactions in biochemistry proceed with the use of multiple substrates and/or produce multiple products. In this chapter we explicitly deal with the steady state kinetic approach to studying enzyme reactions of this type.

10.1 Reaction Nomenclature

A general nomenclature has been devised to describe the number of substrates and products involved in an enzymatic reaction, using the Latin prefixes uni,

Table 10.1 General Nomenclature for Enzymatic Reactions

Reaction	Name
$A \longrightarrow P$	Uni uni
$A + B \longrightarrow P$	Bi uni
$A + B \longrightarrow P_1 + P_2$	Bi bi
$A + B + C \longrightarrow P_1 + P_2$	Ter bi
⋮	⋮

bi, ter, and so on to refer to one, two, three, and more chemical entities. For example, a reaction that utilizes two substrates to produce two products is referred to as a *bi bi* reaction, a reaction using three substrates to form two products is as a *ter bi* reaction, and so on (Table 10.1).

Let us consider in some detail a group transfer reaction that proceeds as a bi bi reaction:

$$E + AX + B \rightleftharpoons E + A + BX$$

The reaction scheme as written leaves several important questions unanswered. Does one substrate bind and leave before the second substrate can bind? Is the order in which the substrates bind random, or must binding occur in a specific sequence? Does group X transfer directly from A to B when both are bound at the active site of the enzyme, or does the reaction proceed by transfer of the group from the donor molecule, A, to a site on the enzyme, whereupon there is a second transfer of the group from the enzyme site to the acceptor molecule B (i.e., a reaction that proceeds through formation of an E–X intermediate)?

These questions raise the potential for at least three distinct mechanisms for the generalized scheme; these are referred to as random ordered, compulsory ordered, and double-displacement or “Ping-Pong” bi bi mechanisms. Often a major goal of steady state kinetic measurements is to differentiate between these varied mechanisms. We shall therefore present a description of each and describe graphical methods for distinguishing among them.

In the treatments that follow we shall use the general steady state rate equations of Alberty (1953), which cast multisubstrate reactions in terms of the equilibrium constants that are familiar from our discussions of the Henri–Michaelis–Menten equation. This approach works well for enzymes that utilize one or two substrates and produce one or two products. For more complex reaction schemes, it is often more informative to view the enzymatic reactions instead in terms of the rate constants for individual steps (Dalziel, 1957). At the end of this chapter we shall briefly introduce the method of King and Altman (1956) by which relevant rate constants for complex reaction schemes can be determined diagrammatically.

10.2 Bi Bi Reaction Mechanisms

10.2.1 Random Ordered Bi Bi Reactions

In the random ordered bi bi mechanism, either substrate can bind first to the enzyme, and either product can leave first. Regardless of which substrate binds first, the reaction goes through an intermediate ternary complex (E · AX · B), as illustrated:

E + AX ⇌ E•AX
B
E•AX•B ⇌ E•A•BX
AX
E + B ⇌ E•B
BX
EA ⇌ E + A
A
E•BX ⇌ E + BX

Here the binding of AX to the free enzyme (E) is described by the dissociation constant K^{AX}, and the binding of B to E is likewise described by K^{B}. Note that the binding of one substrate may very well affect the affinity of the enzyme for the second substrate. Hence, we may find that the binding of AX to the preformed E · B complex is described by the constant αK^{AX}. Likewise, since the overall equilibrium between E · AX · B and E must be path independent, the binding of B to the preformed E · AX complex is described by αK^{B}. When B is saturating, the value of αK^{AX} is equal to the Michaelis constant for AX (i.e., K_m^{AX}). Likewise, when AX is saturating, $\alpha K^{B} = K_m^{B}$. The velocity of such an enzymatic reaction is given by Equation 10.1:

$$v = k_{cat}[E \cdot AX \cdot B] = \frac{k_{cat}[E_t][E \cdot AX \cdot B]}{[E] + [E \cdot AX] + [E \cdot B] + [E \cdot AX \cdot B]} \qquad (10.1)$$

If we express the concentrations of the various species in terms of the free enzyme concentration [E], we obtain:

$$v = \frac{V_{max}\left(\dfrac{[AX][B]}{\alpha K^{AX} K^{B}}\right)}{1 + \dfrac{[AX]}{K^{AX}} + \dfrac{[B]}{K^{B}} + \dfrac{[AX][B]}{\alpha K^{AX} K^{B}}} \qquad (10.2)$$

If we fix the concentration of one of the substrates, we can rearrange and simplify Equation 10.2 significantly. For example, when [B] is fixed and [AX] varies, we obtain:

$$v = \frac{V_{max}[AX]}{\alpha K^{AX}\left(1 + \dfrac{K^{B}}{[B]}\right) + [AX]\left(1 + \dfrac{\alpha K^{B}}{[B]}\right)} \qquad (10.3)$$

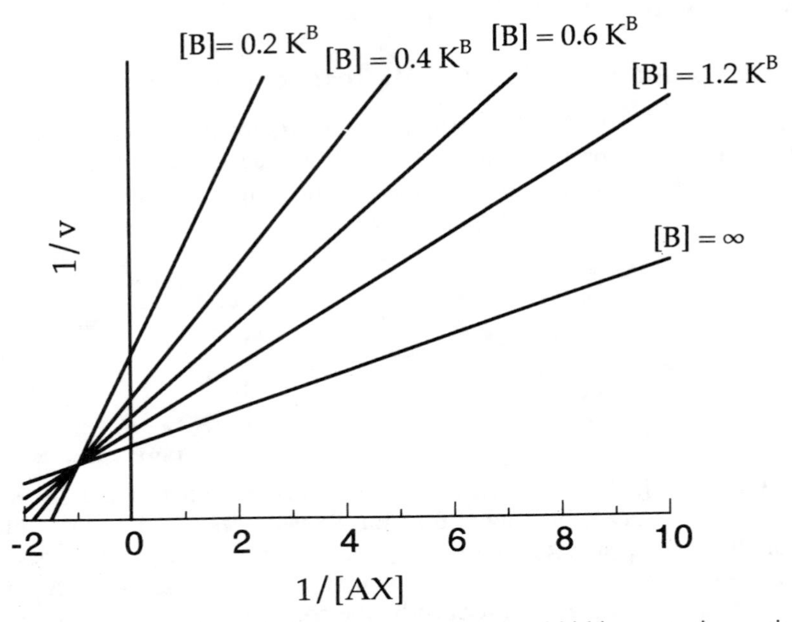

Figure 10.1 Double-reciprocal plot for a random ordered bi bi enzymatic reaction.

At high, fixed concentrations of B, the terms $K^{B}/[B]$ and $\alpha K^{B}/[B]$ go to zero. Thus, at saturating concentrations of B we find:

$$v = \frac{V_{max}^{app}[AX]}{K_{m}^{AX,app} + [AX]} \tag{10.4}$$

and likewise, at fixed, saturating [AX]:

$$v = \frac{V_{max}^{app}[B]}{K_{m}^{B,app} + [B]} \tag{10.5}$$

If we measure the reaction velocity over a range of AX concentrations at several, fixed concentrations of B, the reciprocal plots will display a nest of lines that converge to the left of the y axis, as illustrated in Figure 10.1. The data from Figure 10.1 can be replotted as the slopes of the lines as a function of $1/[B]$, and the y intercepts (i.e., $1/V_{max}^{app}$) as a function of $1/[B]$ (Figure 10.2). The y intercept of the plot of slope versus $1/[B]$ yields an estimate of $\alpha K^{AX}/V_{max}$, and the x intercept of this plot yields an estimate of $-1/K^{B}$. The y and x intercepts of the plot of $1/V_{max}^{app}$ versus $1/[B]$ yield estimates of $1/V_{max}$ and $-1/\alpha K^{B}$, respectively. Thus from the data contained in the two replots, one can calculate the values of K^{AX}, K^{B}, and V_{max} simultaneously.

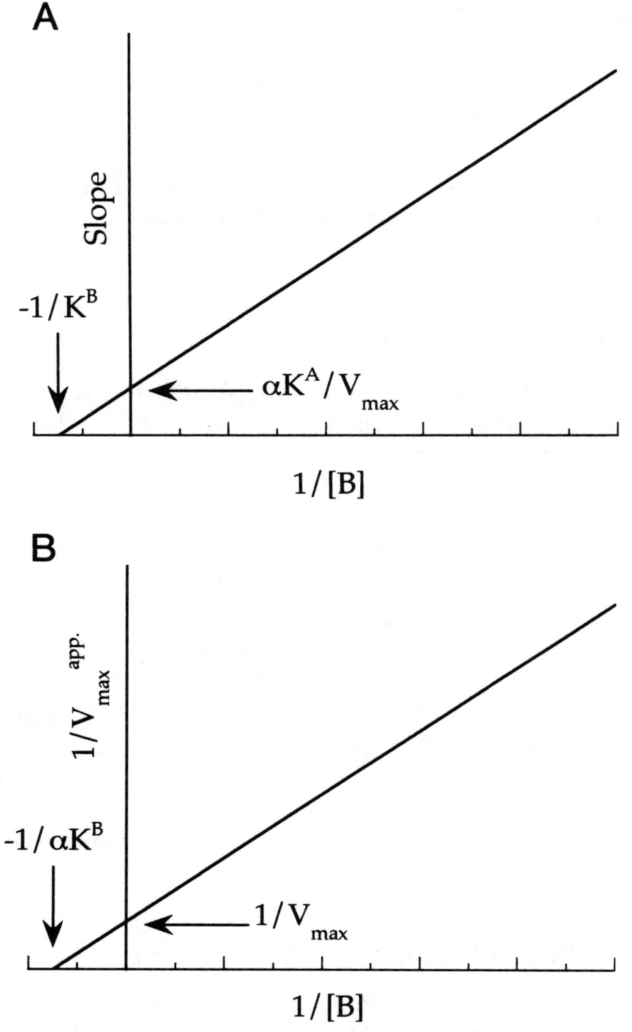

Figure 10.2 (A) Slope and (B) *y*-intercept replots of the data from Figure 10.1, illustrating the graphical determination of K^{AX}, K^{B}, and V_{max} for a random ordered bi bi enzymatic reaction.

10.2.2 Compulsory Ordered Bi Bi Reactions

In compulsory ordered bi bi reactions, one substrate, say AX, must bind to the enzyme before the other substrate (B) can bind. As with random ordered reactions, the mechanism proceeds through formation of a ternary intermediate. In this case the reaction scheme is as follows:

$$\mathrm{E + AX \rightleftharpoons E\cdot AX \overset{B}{\rightleftharpoons} E\cdot AX\cdot B \rightleftharpoons E\cdot A\cdot BX \overset{-BX}{\rightleftharpoons} E\cdot A \rightleftharpoons E + A}$$

If conversion of the E · AX · B complex to E · A · BX is the rate-limiting step in catalysis, then E, AX, B, and E · AX · B are all in equilibrium, and the velocity of the reaction will be given by:

$$v = \frac{V_{max}[\mathrm{AX}][\mathrm{B}]}{K^{\mathrm{AX}}K^{\mathrm{B}} + K^{\mathrm{B}}[\mathrm{AX}] + [\mathrm{AX}][\mathrm{B}]} \tag{10.6}$$

If, however, the conversion of E · AX · B to E · A · BX is as rapid as the other steps in catalysis, steady state assumptions must be used in the derivation of the velocity equation. For a compulsory ordered bi bi reaction, the steady state treatment yields Equation 10.7:

$$v = \frac{V_{max}[\mathrm{AX}][\mathrm{B}]}{K^{\mathrm{AX}}K_m^{\mathrm{B}} + K_m^{\mathrm{B}}[\mathrm{AX}] + K_m^{\mathrm{AX}}[\mathrm{B}] + [\mathrm{AX}][\mathrm{B}]} \tag{10.7}$$

As we have described before, the term K^{AX} in Equation 10.7 is the dissocation constant for the E · AX complex, and K_m^{AX} is the concentration of AX that yields a velocity of half V_{max} at fixed, saturating [B].

The pattern of reciprocal plots observed for varied [AX] at different fixed values of [B] is identical to that seen in Figure 10.1 for a random ordered bi bi reaction. Hence, *one cannot distinguish between random and compulsory ordered bi bi mechanisms on the basis of reciprocal plots alone.* It is necessary to resort to the use of isotope incorporation studies, or studies using product-based inhibitors.

10.2.3 Double Displacement or Ping-Pong Bi Bi Reactions

The double displacement, or Ping-Pong, reaction mechanism involves binding of AX to the enzyme and transfer of the group, X, to some site on the enzyme. The product, A, can then leave and the second substrate, B, binds to the E–X form of the enzyme (in this mechanism, B cannot bind to the free enzyme form). The group, X, is then transferred (i.e., the second displacement reaction) to the bound substrate, B, prior to the release from the enzyme of the final product, BX. This mechanism is diagrammed as follows:

$$\mathrm{E + AX \rightleftharpoons E\cdot AX \overset{-A}{\rightleftharpoons} EX\cdot A \rightleftharpoons EX \overset{B}{\rightleftharpoons} EX\cdot B \rightleftharpoons E\cdot BX \rightleftharpoons E + BX}$$

Using steady state assumptions, the velocity equation for a double-displace-

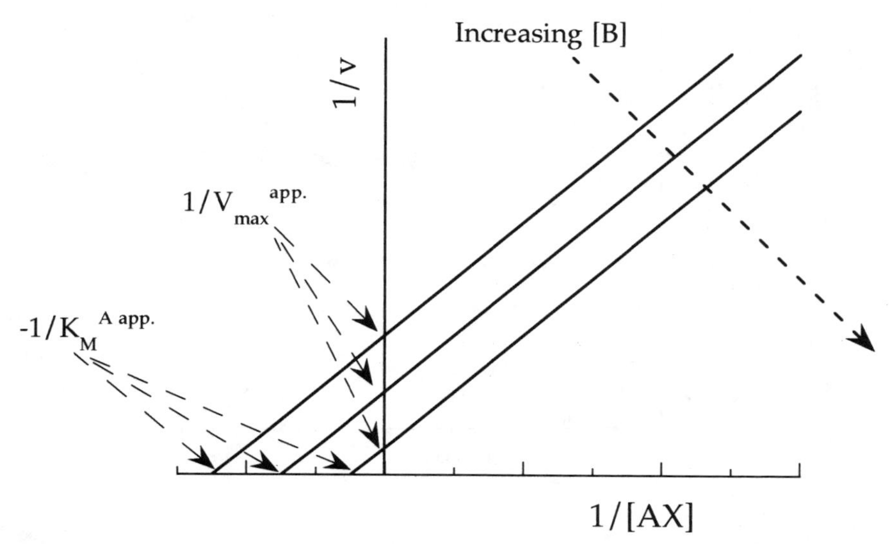

Figure 10.3 Double-reciprocal plot for a double-displacement (Ping-Pong) bi bi enzymatic reaction.

ment reaction can be obtained:

$$v = \frac{V_{\text{max}}[\text{AX}][\text{B}]}{K_{\text{m}}^{\text{B}}[\text{AX}] + K_{\text{m}}^{\text{AX}}[\text{B}] + [\text{AX}][\text{B}]} \tag{10.8}$$

If we fix the value of [B], then Equation 10.8 for variable [AX] becomes:

$$v = \frac{V_{\text{max}}[\text{AX}]}{K_{\text{m}}^{\text{AX}} + [\text{AX}]\left(1 + \frac{K_{\text{m}}^{\text{B}}}{[\text{B}]}\right)} \tag{10.9}$$

Reciprocal plots of a reaction that conforms to the double-displacement mechanism for varying concentrations of AX at several fixed concentrations of B will yield a nest of parallel lines, as seen in Figure 10.3. For each concentration of substrate B, the values of $1/V_{\text{max}}^{\text{app}}$ and $-1/K_{\text{m}}^{\text{AX,app}}$ can be determined from the y and x intercepts, respectively, of the double-reciprocal plot. The data contained in Figure 10.3 can be replotted in terms of $1/V_{\text{max}}^{\text{app}}$ and $1/K_{\text{m}}^{\text{AX,app}}$ as a functions of 1/[B], as illustrated in Figure 10.4. The value of $-1/K_{\text{m}}^{\text{B}}$ can be determined from the x intercepts of either replot in Figure 10.4. The y intercepts of the two replots yield estimates of $1/V_{\text{max}}$ (for the $1/V_{\text{max}}^{\text{app}}$ versus 1/[B] replot) and $1/K_{\text{m}}^{\text{AX}}$ (for the $1/K_{\text{m}}^{\text{app}}$ versus 1/[B] replot) for the reaction, as seen in Figure 10.4.

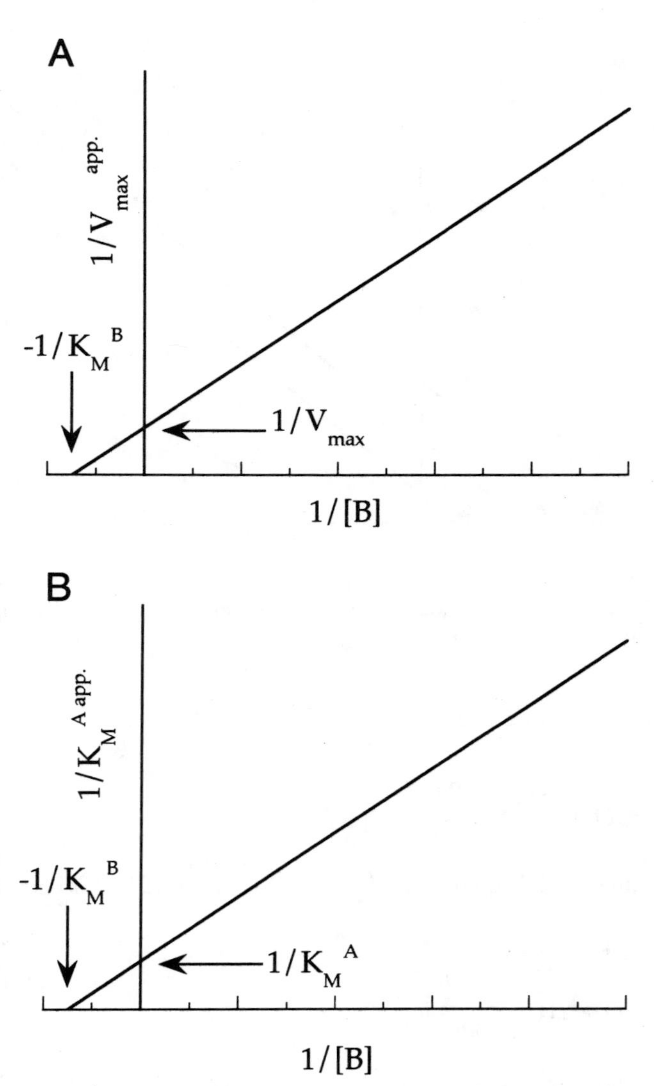

Figure 10.4 Replots of the data from Figure 10.3 as (A) $1/V_{max}^{app}$ versus 1/[B] and (B) $1/K_m^{AX,app}$ versus 1/[B], illustrating the graphical determination of K_m^{AX}, K_m^B, and V_{max} for a double-displacement (Ping-Pong) bi bi enzymatic reaction.

10.3 Distinguishing Between Random and Compulsory Ordered Mechanisms by Product Inhibition

It should be clear from Figures 10.1 and 10.3, and the foregoing discussion, that the qualitative form of the double-reciprocal plots makes it easy to distinguish between a double-displacement mechanism and a mechanism involving ternary complex formation. But again, it is not possible to further distinguish between random and compulsory ordered mechanism on the basis of reciprocal plots alone. If, however, there is available an inhibitor that binds to the same site on the enzyme as one of the substrates (i.e., is a competitive inhibitor with respect to one of the substrates), addition of this compound will slow the overall forward rate of the enzymatic reaction and can allow one to kinetically distinguish between random and compulsory ordered reaction mechanisms. Because of their structural relationship to the substrate, many times the product molecules of enzymatic reactions are themselves competitive inhibitors of the substrate binding site; this situation is referred to as *product inhibition.*

Recall from Chapter 7 that competitive inhibition is observed when the inhibitor binds to the same enzyme form as the substrate that is being varied in the experiment, or alternatively, binds to an enzyme form that is connected by reversible steps to the form that binds the varied substrate. The pattern of reciprocal lines observed with different inhibitor concentrations is a nest of lines that converge at the y intercept (see Chapter 7). For an enzyme that requires two substrates, a competitive inhibitor of one of the substrate binding sites will display the behavior of either a competitive or a mixed inhibitor, depending on which substrate is being varied, and the mechanism of substrate interaction with the enzyme. For a bi bi reaction, one observes specific inhibitor patterns for the different mechanisms we have discussed when a product of the reaction is used as an inhibitor while one of the substrates is varied and the second substrate is held at a fixed, nonsaturating concentration.

The relationships leading to these differing patterns of product inhibition for bi bi reactions have been derived elsewhere (see, e.g., Segel, 1976). Rather than rederiving these relationships, we present them as diagnostic tools for deter-

Table 10.2 Patterns of Product Inhibition Observed for the Bi Bi Reaction E + AX + B → E + A + BX for Differing Reaction Mechanisms

Mechanism	Product Used As Inhibitor	Inhibitor Pattern Observed	
		For Varied [AX]	For Varied [B]
Compulsory ordered with [AX] binding first	A	Competitive	Mixed
Compulsory ordered with [AX] binding first	BX	Uncompetitive	Competitive
Compulsory ordered with [B] binding first	A	Competitive	Uncompetitive
Compulsory ordered with [B] binding first	BX	Mixed	Competitive
Random Ordered	A	Competitive	Mixed
Random ordered	BX	Mixed	Competitive
Double displacement	A	Competitive	Uncompetitive
Double displacement	BX	Uncompetitive	Competitive

mining the mechanisms of reactions. The patterns are summarized in Table 10.2. By measuring the initial velocity of the reaction in the presence of several concentrations of inhibitor, and varying separately the concentrations of AX and B, one can identify the reaction mechanism from the pattern of double-reciprocal plots and reference to Table 10.2.

10.4 Isotope Exchange Studies for Distinguishing Reaction Mechanisms

An alternative means of distinguishing among reaction mechanisms is to look at the rate of exchange between a radiolabeled substrate and a product molecule under equilibrium conditions (Boyer, 1959; Segel, 1975).

The first, and simplest mechanistic test using isotope exchange is to ask whether exchange of label can occur between a substrate and product in the presence of enzyme, but in the absence of the second substrate. Looking over the various reaction schemes presented in this chapter, it became obvious that such an exchange could take place only for a double-displacement reaction:

$$E + A^*X \rightleftharpoons E{\cdot}A^*X \rightleftharpoons EX \cdot A^* \overset{-A^*}{\rightleftharpoons} EX \overset{B}{\rightleftharpoons} EX \cdot B \rightleftharpoons E \cdot BX \rightleftharpoons E + BX$$

For random or compulsory ordered reactions, the need to proceed through the ternary complex before initial product release would prevent the incorportion of radiolabel into one product in the absence of the second substrate.

Next, let us consider what happens when the rate of isotope exchange is measured under equilibrium conditions for a general group transfer reaction:

$$AX + B \rightleftharpoons A + BX$$

Under these conditions the forward and reverse reaction rates are equivalent, and the equilibrium constant is given by:

$$K_{eq} = \frac{[BX][A]}{[AX][B]} \tag{10.10}$$

If under these conditions radiolabeled substrate B is introduced in an amount so small that it is insufficient to significantly perturb the equilibrium, the rate of formation of labeled BX can be measured. The measurement is repeated at increasing concentrations of A and AX, to keep the ratio [A]/[AX] constant (i.e., to avoid a shift in the position of the equilibrium). As the amounts of A and AX are changed, the rate of radiolabel incorporation into BX will be affected.

Suppose that the reaction proceeds through a compulsory ordered mechanism in which B is the first substrate to bind to the enzyme and BX is the last product to be released. If this is the case, the rate of radiolabel incorporation into BX will initially increase as the concentrations of A and AX are increased. As the concentrations of A and AX increase further, however, the formation of the ternary complexes E·AX·B and E·A·BX will be favored, while dissociation of the EB and EBX complexes will be disfavored. This will have the effect

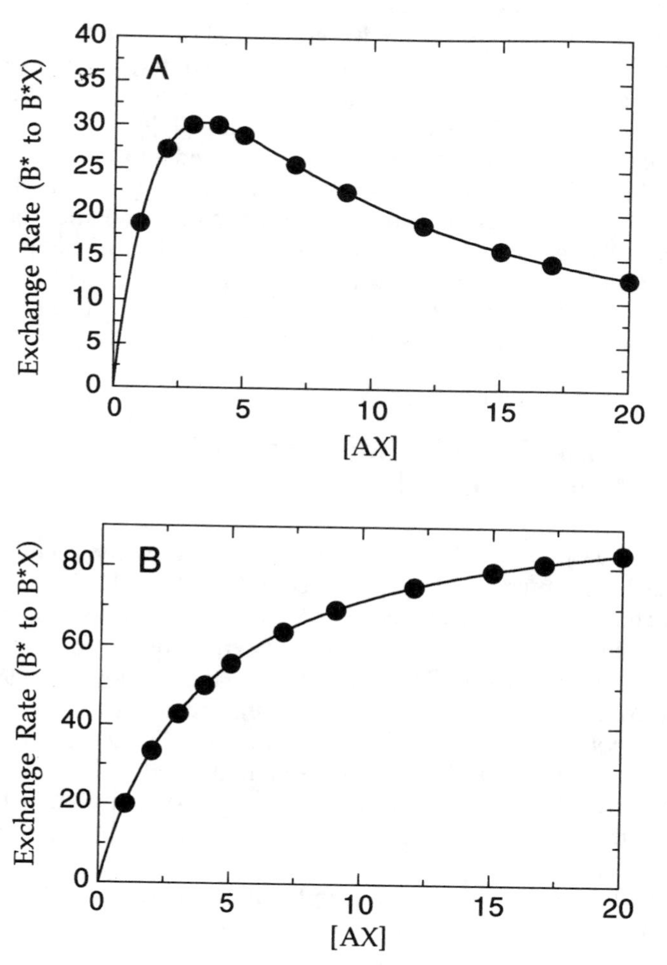

Figure 10.5 Plots of the equilibrium rate of radioisotope exchange between B and BX as a function of [AX] for (A) a compulsory ordered bi bi reaction in which B is the first substrate to bind to the enzyme and BX is the last product to be released, and (B) either a compulsory ordered bi bi reaction in which AX binds first or a random ordered bi bi reaction.

of lowering the rate of isotope exchange between B and BX. Hence, a plot of the rate of isotope exchange as a function of [AX] will display substrate inhibition at high [AX], as illustrated in Figure 10.5A.

The effect of increasing [AX] and [A] on the rate of radiolabel exchange between B and BX will be quite different, however, in a compulsory ordered reaction that requires initial binding of AX to the enzyme. In this case, increasing concentrations of AX and A will disfavor the free enzyme in favor of the EAX and EA forms. The EAX form will react with B, leading to formation of BX, while the EA form will not. Hence, the rate of radiolabel incorporation into BX will increase with increasing [AX] as a hyperbolic function (Figure 10.5B). The same hyperbolic relationship would also be observed for a reaction that proceeded through a random ordered mechanism. In this latter case, however, the hyperbolic relationship also would be seen for experiments performed with labeled AX and varying [B].

Thus isotope exchange in the absence of the second substrate is diagnostic of a double-displacement reaction, while compulsory ordered and random ordered reactions can be distinguished on the basis of the relation of the rate of radiolabel exchange between one substrate and product of the reaction to the concentration of the other substrate and product under equilibrium conditions. (See Segel, 1975, for a more comprehensive treatment of isotope exchange studies for multisubstrate enzymes.)

10.5 Determining Velocity Equations Using the King–Altman Method

The velocity equations for bi bi reactions can be easily related to the Henri–Michaelis–Menten equation described in Chapter 5. However, for more complex reaction schemes, such as those involving multiple intermediate species, it is often difficult to derive the velocity equation in simple terms. An alternative method, devised by King and Altman (1956), allows the derivation of a velocity equation for essentially any enzyme mechanism in terms of the individual rate constants of the various steps in catalysis. On the basis of the methods of matrix algebra, King and Altman derived empirical rules for writing down the functional forms of these rate constant relationships. We provide a couple of illustrative examples of their use and encourage interested readers to explore this method further.

To begin with, we shall consider a simple uni uni reaction as first encountered in Chapter 5:

$$E + S \rightleftharpoons ES \longrightarrow E + P$$

In the King and Altman approach we consider the reaction to be a cyclic process and illustrate it in a way that displays all the interconversions among the various enzyme forms involved:

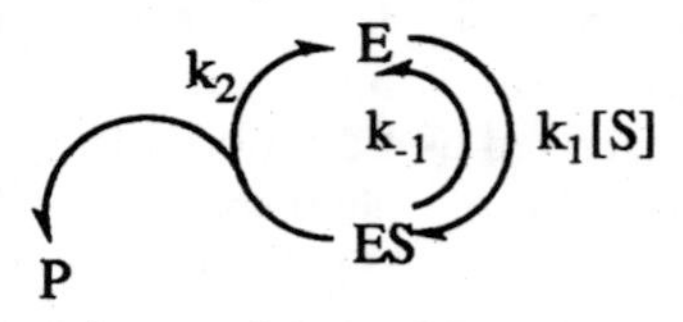

For each step in the reaction we can define a term κ (kappa) which is the product of the rate constant for that step and the concentration of free substrate involved in the step. Next, we determine every pathway by which a particular enzyme species might be formed in the reaction scheme. For the simple uni uni reaction under consideration we have:

Enzyme Form	Pathways to That Form	Σ of Kappa Products
E	$E \xleftarrow{k_{-1}}$ and $E \xleftarrow{k_2}$	$k_{-1} + k_2$
ES	$\xrightarrow{k_1[S]} ES$	$k_1[S]$

For any particular enzyme species, the following relationship holds:

$$\frac{[\text{form}]}{[E_t]} = \frac{\Sigma \kappa_{\text{form}}}{\Sigma \kappa} \tag{10.11}$$

where [form] is the concentration of the particular enzyme form under consideration, $\Sigma\kappa_{\text{form}}$ is the sum of the kappa products for that enzyme form, and $\Sigma\kappa$ is the sum of the kappa products for all species. Applying this to our uni uni reaction we obtain:

$$\frac{[E]}{[E_t]} = \frac{k_{-1} + k_2}{k_{-1} + k_2 + k_1[S]} \tag{10.12}$$

and

$$\frac{[ES]}{[E_t]} = \frac{k_1[S]}{k_{-1} + k_2 + k_1[S]} \tag{10.13}$$

The overall velocity equation can be written as follows:

$$v = k_2[ES] \tag{10.14}$$

Substituting the equalities given in Equations 10.12 and 10.13 into Equation 10.14, we obtain:

$$v = \frac{k_2 k_1[S][E_t]}{k_{-1} + k_2 + k_1[S]} = \frac{k_2[E_t][S]}{\left(\dfrac{k_{-1} + k_2}{k_1}\right) + [S]} \tag{10.15}$$

Inspecting Equation 10.15, we immediately see that k_2 is equivalent to k_{cat}, and $(k_{-1} + k_2/k_1)$ is equivalent to the Michaelis constant, K_m. If we invoke the further equality that $V_{max} = k_{cat}[E_t]$, we see that the King–Altman approach results in the same velocity equation we had derived as Equation 5.16.

Now let us consider the more complex case of a double-displacement bi bi reaction using the King–Altman approach. Note here that the initial concentrations of the two products A and BX are zero, and the release of these products from the enzyme is essentially irreversible. Hence, the cyclic form of the reaction scheme is:

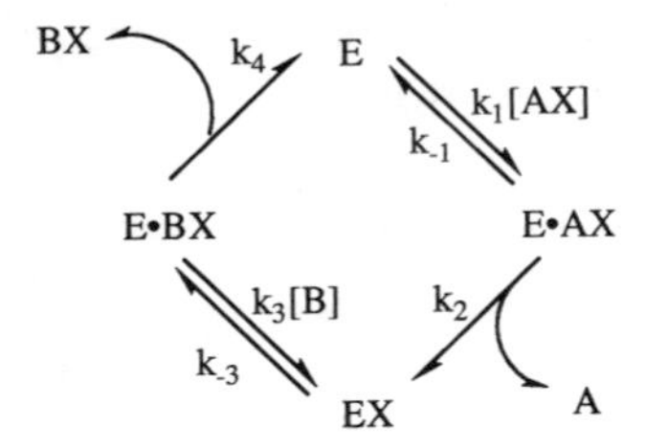

Consideration of this reaction yields the relationships given in Table 10.3. The overall rate equation for a double-displacement reaction is:

$$v = k_2[EAX] = k_4[EBX] \tag{10.16}$$

From the preceding relationships, we see that:

$$\frac{[EAX]}{[E_t]} = \frac{k_1k_3k_4[AX][B]}{k_1k_3[AX][B](k_2+k_4) + k_3k_4[B](k_{-1}+k_2) + k_1k_2[AX](k_{-3}+k_4)} \tag{10.17}$$

Combining Equations 10.16 and 10.17, and performing a few rearrangements we obtain:

$$v = \frac{\left(\frac{k_2k_4}{k_2+k_4}\right)[E_t][AX][B]}{\frac{k_2}{k_3}\left(\frac{k_{-3}+k_4}{k_2+k_4}\right)[AX] + \frac{k_4}{k_1}\left(\frac{k_{-1}+k_2}{k_2+k_4}\right)[B] + [AX][B]} \tag{10.18}$$

With the appropriate substitutions, Equation 10.18 can be recast, using the approach of Alberty, to yield the more familiar form first presented as Equation 10.8.

With similar considerations, the velocity equations for random ordered and compulsory ordered bi bi mechanisms can likewise be derived. With some practice, this seemingly cumbersome approach provides a clear and intuitive means of deriving the appropriate velocity equation for complex enzymatic systems. A more thorough treatment of the King–Altman approach can be found in the text by Segel (1975) as well as in the original contribution by King and Altman (1956).

Table 10.3 King–Altman Relationships for a Double Displacement Bi Bi Reaction

Enzyme Form	Pathways to Form	Σ of Kappa Products
E	k_{-1}, E, k_4, k_3[B] and k_2, E, k_4, k_3[B]	$k_{-1}k_3k_4[B] + k_2k_3k_4[B] = k_3k_4[B](k_{-1} + k_2)$
E·AX	k_1[AX], E•AX, k_4, k_3[B]	$k_1k_3k_4[AX][B]$
EX	k_1[AX], k_2, EX, k_4 and k_1[AX], k_2, EX, k_{-3}	$k_1k_2k_4[AX] + k_1k_2k_{-3}[AX] = k_1k_2[AX](k_{-3} + k_4)$
E·BX	k_1[AX], k_2, k_3[B], E•BX	$k_1k_2k_3[AX][B]$

10.6 Summary

In this chapter we have briefly introduced the concept of multisubstrate enzyme reactions and have presented steady state equations to describe the velocities for these reactions. We have seen that enzyme reactions involving two substrates and two products can proceed by three distinct mechanisms: random ordered, compulsory ordered, and double-displacement reactions. Experimental methods were presented to allow the investigator to distinguish among these methods on the basis of kinetic measurements, product inhibition studies, and radioisotope exchange studies. We briefly described the method of King and Altman for deriving the velocity equation of complex enzymatic reaction, such as those involving multiple substrates.

The importance of multisubstrate enzymatic reactions can hardly be overstated. In fact, the vast majority of enzymatic reactions in nature proceed through the utilization of more than one substrate to yield more than one product.

References and Further Reading

Alberty, R. A. (1953) *J. Am. Chem. Soc.* **75**, 1928.

Boyer, P. D. (1959) *Arch. Biochem. Biophys.* **82**, 387.

Cleland, W. W. (1963) *Biochim. Biophys. Acta,* **67**, 188.

Cornish-Bowden, A., and Wharton, C. W. (1988) *Enzyme Kinetics,* IRL Press, Oxford, pp. 25–33.

Dalziel, K. (1975) Kinetics and mechanism of nicotinamide-dinucleotide-linked dehydrogenases, in *The Enzymes,* 3rd ed. (P. D. Boyer, Ed., Academic Press, San Diego, CA, pp. 1–60.

King, E. L., and Altman, C. (1956) *J. Phys. Chem.* **60**, 1375.

Palmer, T. (1981) *Understanding Enzymes,* Wiley, New York, pp. 170–189.

Segel, I. H. (1975) *Enzyme Kinetics,* Wiley, New York, pp. 506–883.

CHAPTER

11

Cooperativity in Enzyme Catalysis

As we described in Chapter 3, some enzymes function as oligomeric complexes of multiple protein subunits, each subunit being composed of copies of the same or different polypeptide chains. In some oligomeric enzymes, each subunit contains an active site center for ligand binding and catalysis. In the simplest case, the active sites on these different subunits act independently, as if each represented a separate catalytic unit. In other cases, however, the binding of ligands at one active site of the enzyme can increase or decrease the affinity of the active sites on other subunits for ligand binding. When the ligand binding affinity of one active site is affected by ligand occupancy at another active site, the active sites are said to be acting *cooperatively*. In *positive cooperativity* ligand binding at one site *increases* the affinity of the other sites, and in *negative cooperativity* the affinity of other sites is *decreased* by ligand binding to the first site.

For cooperative interaction to occur between two active sites some distance apart (e.g., on separate subunits of the enzyme complex), ligand binding at one site must induce a structural change in the surrounding protein that is transmitted, via the polypeptide chain, to the distal active site(s). This concept of transmitted structural changes in the protein, resulting in long-distance communication between sites, has been termed "allostery," and enzymes that display these effects are known as *allosteric enzymes*. (The word "allosteric," which derives from two ancient Greek words—*allos* meaning different, and *stereos*, meaning structure or solid—was coined to emphasize that the structural *change* within the protein mediates the cooperative interactions among different sites.)

Allosteric effects can occur between separate binding sites for the *same* ligand within a given enzyme, as just discussed, in *homotropic cooperativity*. Also, ligand binding at the active site of the enzyme can be affected by binding of a structurally unrelated ligand at a distant separate site; this effect is known as *heterotropic cooperativity*. Thus small molecules can bind to sites other than the enzyme active site and, as a result of their binding, induce a conformational change in the enzyme that regulates the affinity of the active site for its substrate (or other ligands). Such molecules are referred to as *allosteric effectors*, and they can operate to enhance active site substrate affinity (i.e., serving as allosteric activators) or to diminish affinity (i.e., serving as allosteric repressors). Both types of allosteric effector are seen in biology, and they form the basis of metabolic control mechanisms, such as feedback loops.

In this chapter we shall describe some examples of proteins that display cooperativity and allostery that not only illustrate these concepts but also have historic significance in the development of the theoretical basis for understanding these effects. We shall then briefly describe two theoretical frameworks for describing the two effects. Finally, we shall discuss the experimental consequences of cooperativity and allostery, and appropriate methods for analyzing the kinetics of such enzymes.

The treatment to follow discusses the effects of cooperativity in terms of substrate binding to the enzyme. The reader should note, however, that ligands other than substrate also can display cooperativity in their binding. In fact, in some cases enzymes display cooperative inhibitor binding, but no cooperativity is observed for substrate binding to these enzymes. Such special cases are beyond the scope of the present text, but the reader should be aware of their existence. A relatively comprehensive treatment of such cases can be found in the text by Segel (1975).

11.1 Historic Examples of Cooperativity and Allostery in Proteins

The proteins hemoglobin and the Trp repressor provide good examples of the concepts of ligand cooperativity and allosteric regulation, respectively. Hemoglobin is often considered to be the paradigm for cooperative proteins. This primacy is in part due to the wealth of information on the structural determinants of cooperativity in this protein that is available as a result of detailed crystallographic studies on the ligand-replete and ligand-free states of hemoglobin. Likewise, much of our knowledge of the regulation of Trp repressor activity comes from detailed crystallographic studies.

Hemoglobin, as described in Chapter 3, is a heterotetramer composed of two copies of the α subunit and two copies of the β subunit. These subunits fold independently into similar tertiary structures that provide a binding site for a heme cofactor (i.e., an iron-containing porphyrin cofactor: see Figure

3.19). The heme in each subunit is associated with the protein by a coordinate bond between the nitrogen of a histidine residue and the central iron atom of the heme. Iron typically takes up an octahedral coordination geometry composed of six ligand coordination sites. In the heme groups of hemoglobin, four of these coordination sites are occupied by nitrogens of the porphyrin ring system and a fifth is occupied by the coordinating histidine, leaving the sixth coordination site open for ligand binding. This last coordination site forms the O_2 binding center for each subunit of hemoglobin.

A very similar pattern of tertiary structure and heme binding motif is observed in the structurally related monomeric protein myoglobin, which also binds and releases molecular oxygen at its heme iron center. Based on the similarities in structure, on would expect that each of the four hemes in the hemoglobin tetramer would bind oxygen independently, and with an affinity simlar to that of myoglobin. In fact, however, when O_2 binding curves for these two proteins are measured, the results are dramatically different, as illustrated in Figure 11.1. Myoglobin displays the type of hyperbolic saturation curve one would expect for a simple protein–ligand interaction. Hemoglobin, on the other hand, shows not a simple hyperbolic saturation curve but, instead, a sigmoidal dependence of O_2 binding to the protein as a function of O_2 concentration. This is the classic signature for cooperatively interacting binding sites. That is, the four heme groups in hemoglobin are not acting as independent oxygen binding sites, but instead display positive cooperativity in their binding affinities. The degree of cooperativity among these distant sites is such

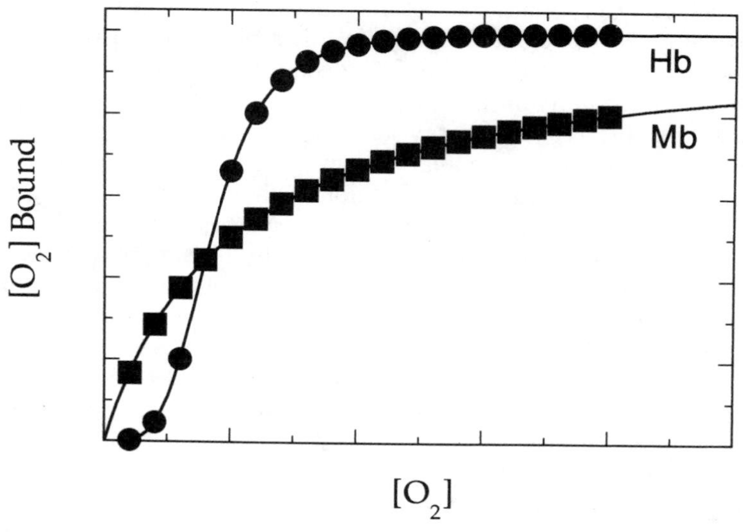

Figure 11.1 Plot of bound molecular oxygen as a function of oxygen concentration for the proteins hemoglobin (Hb) and myoglobin (Mb), illustrating the cooperativity of oxygen binding for hemoglobin.

that the data for oxygen binding to hemoglobin are best described by a two-state model in which all the molecules of hemoglobin contain either 4 or 0 moles of bound O_2; under equilibrium conditions, no significant population of hemoglobin molecules exist with intermediate (i.e., 2 or 3) stoichiometries of O_2 binding.

The crystal structure of oxy- (with four O_2 molecules bound) and deoxy- (with no O_2 bound) hemoglobin provide a clear structural basis for this cooperativity. We know from Chapter 3 that hemoglobin can adopt two distinct quaternary structures; these are referred to as the R (for relaxed) and T (for tense) states (see Section 11.2). The differences between the R and T quaternary structures are relative rotations of two of the subunits, as described in Figure 3.18. These changes in quaternary structure are mediated by changes in intersubunit hydrogen bonding at the subunit interfaces. The crystal structures of oxy- and deoxyhemoglobin reveal that loss of oxygen at the heme of one subunit induces a change in the strength of the iron–histidine bond that occupies the fifth coordination site on the heme iron. This change in bond strength results in a puckering of the porphyrin macrocycle and a displacement of position for the coordinated histidine (Figure 11.2). The coordinated histidine is located in a segment of α-helical secondary structure in the hemoglobin subunit, and the motion of the histidine in response to O_2 binding or release results in a propagated motion of the entire α helix. Ultimately, this propagated motion produces alterations of the intersubunit hydrogen-bonding pattern at the α_1/β_2 subunit interface that acts as a quaternary structure "switch." The accompanying movements of the other subunits leads to the alterations of the oxygen affinities for their associated heme cofactors.

The availability of detailed structural information for both the oxy and deoxy structures of hemoglobin has made this molecule the classic model of cooperativity in proteins, illustrating how distant binding sites can interact to control the overall affinity for a single ligand. Likewise, the structural information available for the Trp repressor protein has made this molecule an excellent example of allosteric regulation in biology. As its name implies, the Trp repressor protein acts to inhibit the function of the Trp operon, a segment of DNA that is ultimately responsible for the synthesis of the amino acid tryptophan. The protein accomplishes this task by binding within the major groove of the DNA in its tryptophan-bound form and, when not bound by tryptophan, releasing the DNA. The activity of the Trp repressor is an example of a negative feedback loop, in which the synthesis of an essential molecule of the cell is controlled by the concentration of that molecule itself. At low tryptophan concentrations, the synthesis of tryptophan is required by the cell. Under these conditions the Trp operon must be functional, and thus the Trp repressor must not bind to the DNA.

The crystal structure of the tryptophan-depleted protein shows that the α-helical segments of the protein are arranged in a way that precludes effective DNA binding (Figure 11.3A). Thus, when the tryptophan concentration is low, the protein is found in a conformation that does not allow for DNA binding,

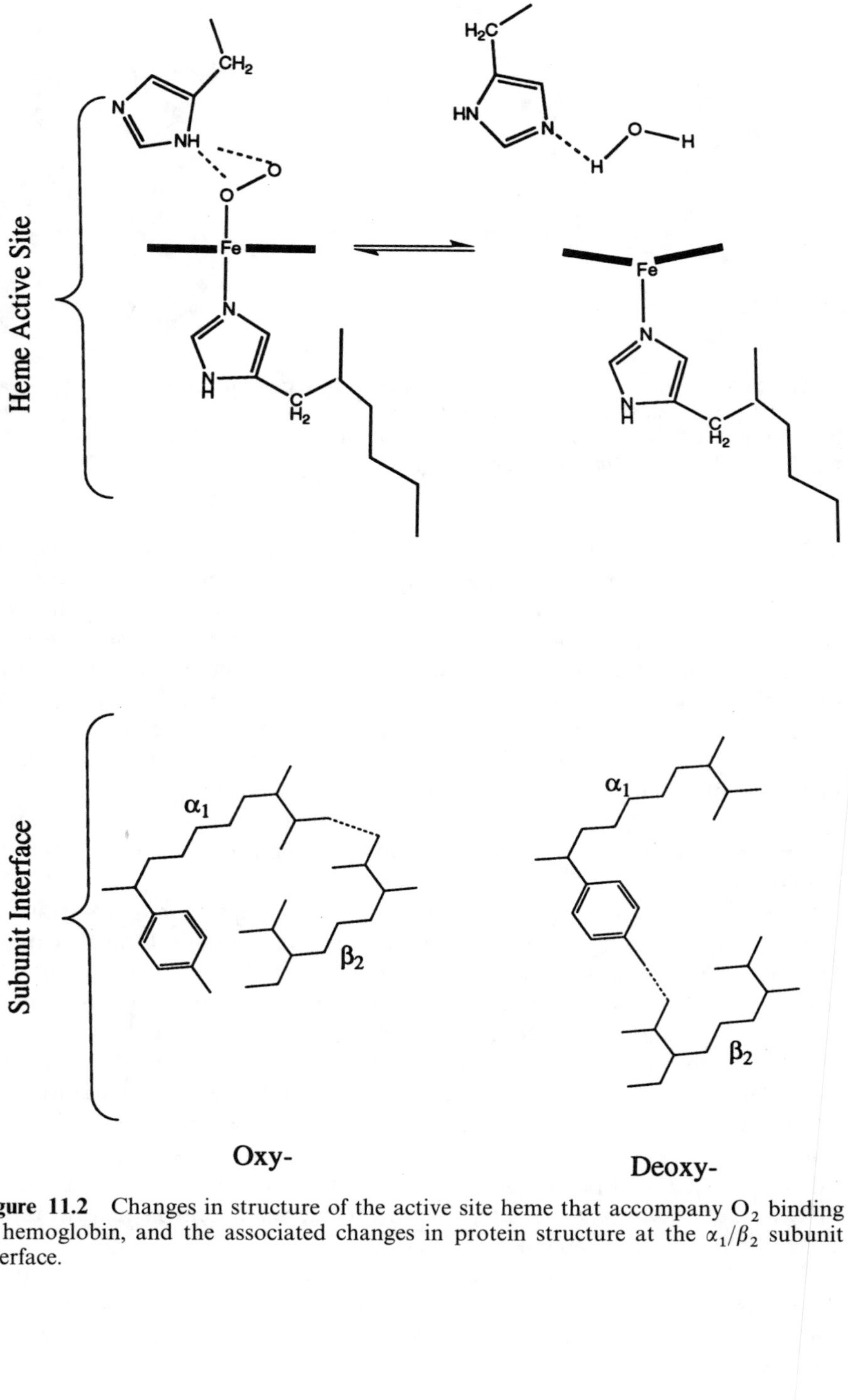

Figure 11.2 Changes in structure of the active site heme that accompany O_2 binding to hemoglobin, and the associated changes in protein structure at the α_1/β_2 subunit interface.

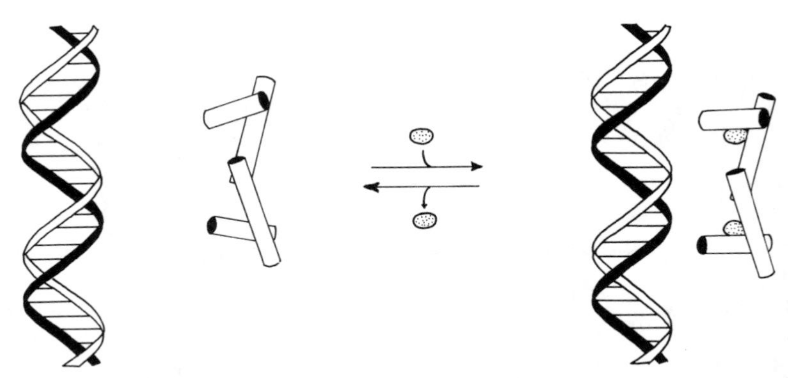

Figure 11.3 Cartoon of the interactions of the Trp repressor protein with Trp operon DNA in the absence (A) and presence (B) of bound tryptophan. This tryptophan-binding-induced conformational transition is the basis for the negative feedback regulation of tryptophan synthesis.

and the operon is functional, leading to tryptophan synthesis. When the tryptophan concentration in the cell exceeds some critical concentration, however, the Trp repressor binds tryptophan and, as a result, changes its conformation. The tryptophan-replete form of the protein now has an α-helical arrangement in which two helices are positioned for effective binding to the Trp operon, via interactions between the helices and the double-stranded DNA helical structure (Figure 11.3B). When the Trp repressor binds to the operon, it effectively shuts down the action of this DNA, thus leading to inhibition of further tryptophan synthesis. This simple method of conformationally controlling the activity of the Trp repressor, by binding of tryptophan, provides an elegant mechanism for the metabolic control of the production of an essential amino acid.

Again, we have used hemoglobin and the Trp repressor to illustrate the concepts of cooperativity and allosteric control in structural terms because of the wealth of structural information available for these two proteins. The reader should be aware, however, that the same mechanisms are common in enzymatic systems as well. Numerous examples of cooperativity and allosteric control of enzymatic activity can be found in biology, and these control mechanisms serve vital metabolic roles. For example, many enzymes involved in de novo biosynthetic cascades display the phenomenon of feedback inhibition. Here a metabolite that is the ultimate or penultimate produce of the cascade will act as a heterotropic inhibitor of one of the enzymes that occurs early in the biosynthetic cascade, much as tryptophan controls its own rate of synthesis by binding to the Trp repressor.

One of the first examples of this phenomenon came from studies of threonine deaminase from the bacterium *E. coli.* Abelson (1954) observed that addition of isoleucine to cultures of the bacterium inhibited the further biosynthesis of isoleucine. Later workers showed that this effect is due to specific inhibition by isoleucine of threonine deaminase, the first enzyme in the biosynthetic route to isoleucine. Further studies with purified threonine deaminase revealed that the substrate threonine and the inhibitor isoleucine bind to the enzyme at different, nonidentical sites; thus isoleucine is an example of a heterotropic allosteric inhibitor of threonine deaminase.

Another classic example of feedback inhibition comes from aspartate carbamoyltransferase. This enzyme catalyzes the formation of carbamoyl-aspartate from aspartate and carbamoylphosphate, which is the first step in the de novo biosynthesis of pyrimidines. The enzyme shows a sigmoidal dependence of reaction rate on the concentration of the substrate, aspartate, demonstrating cooperativity between the active sites of this oligomeric enzyme. Additionally, the enzyme is inhibited by the pyrimidine analogues cytidine, cytidine-5-phosphate, and cytosine triphosphate (CTP), and is activated by adenosine triphosphate (ATP). Structural studies also have revealed that the substrate binding sites and the CTP inhibitory binding sites are separate and distinct; binding of CTP at its exclusive site, however, influences the affinity of the active site for aspartate, via heterotropic allosteric regulation.

Threonine deaminase and aspartate carbamoyltransferase are examples of what is now known to be a ubiquitous means of metabolic control, namely, feedback inhibition. To understand fully this important biological control mechanism requires a theoretical framework in which to describe how distant sites within an enzyme can interact to affect one another's affinity for similar or different ligands. We now turn to a brief description of two such theortical frameworks that have proved useful in the study of allosteric enzymes.

11.2 Models of Allosteric Behavior

When an oligomeric enzyme contains multiple substrate binding sites that all behave identically (i.e., display the same K_m and k_{cat} for substrate) and independently, the velocity equation can be shown to be identical to that for a single active site enzyme (Segel, 1975). Regardless of whether the binding sites interact, an oligomeric enzyme will have different distributions of ligand occupancy among its different subunits at different levels of substrate saturation. For instance, six possible combinations of subunit occupancies can be envisaged for a tetrameric enzyme with two substrate molecules bound. This is illustrated for the example of a tetrameric enzyme in Figure 11.4.

When the ligand binding sites of an oligomeric enzyme interact cooperatively, we need to modify the existing kinetic equations to account for this intersite interaction. Two theoretical models have been put forth to explain allostery in enzymes and other ligand binding proteins. The first, the *simple sequential interaction* model, proposed by Koshland and coworkers (Koshland et al., 1966), is in some ways an extension of the induced-fit model introduced

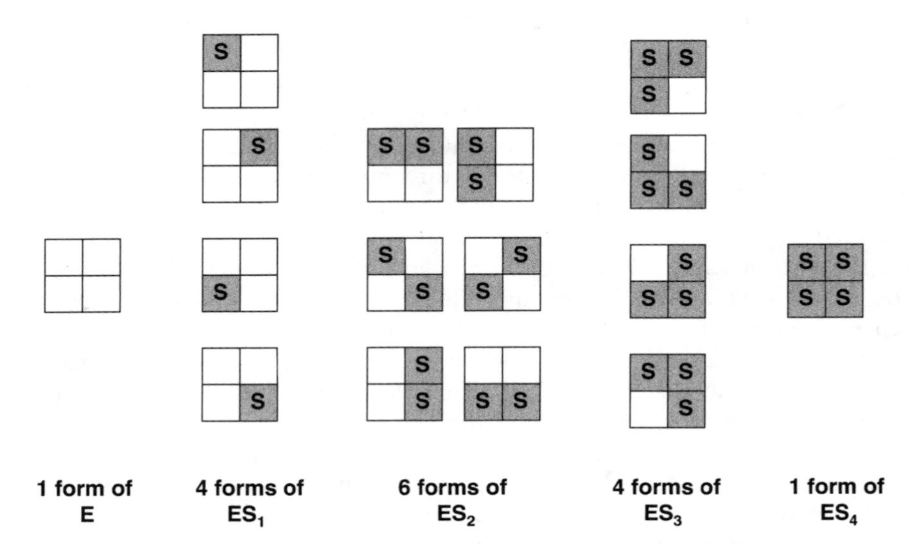

Figure 11.4 Schematic illustration of the number of possible forms of ligand binding to an enzyme that is a homotetramer: open squares, are subunits with empty binding site; shaded squares with S, are subunits to which a molecule of substrate has bound to the active site.

in Chapter 4. The second, the *concerted transition* or *symmetry* model, is based on the work of Monod, Wyman, and Changeux (1965) and has been widely applied to proteins such as hemoglobin, to explain ligand binding cooperativity.

The simple sequential interaction model assumes that a large conformational change attends each ligand binding event at one of the enzyme active sites. It is this conformational transition that affects the affinity of the enzyme for the next ligand molecule. Let us consider the simplest case of an allosteric enzyme with two substrate binding sites that display positive cooperativity. The equilibria involved in substrate binding and their associated equilibrium constants are schematically illustrated in Figure 11.5 (Segel, 1975). The dissociation constant for the first substrate molecule is give by K_S. When one of the substrate binding sites is occupied, however, the dissociation constant for the second site is modified by the factor a, which for positive cooperativity has a value less than 1. The overall velocity equation for such an enzyme is given by:

$$v = \frac{V_{\max}\left(\frac{[S]}{K_S} + \frac{[S]^2}{aK_S^2}\right)}{1 + \frac{2[S]}{K_S} + \frac{[S]^2}{aK_S^2}} \tag{11.1}$$

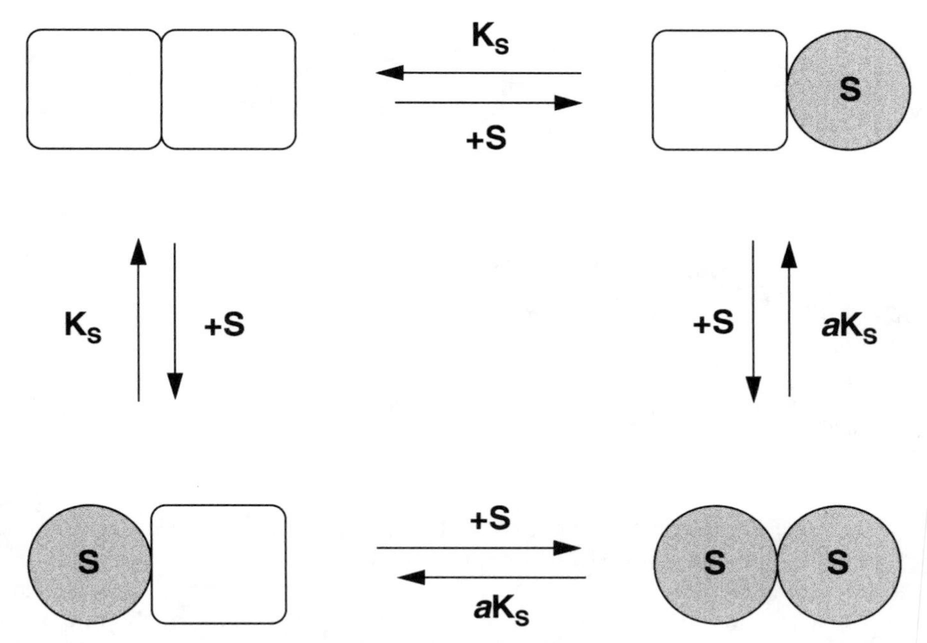

Figure 11.5 Schematic representation of the equilibria involved in substrate binding to a homodimeric enzyme where substrate binding is accompanied by a conformational transition of the subunit to which it binds, according to the model of Koshland et al. (1966).

Now let us extend the model to a tetrameric enzyme (Figure 11.6). In this case the binding of the first substrate molecule modifies the dissociation constant of all three other binding sites by the factor a. If a second substrate molecule now binds, the two remaining vacant binding sites will have their dissociation constants modified further by the factor b, and their dissociation constants will be abK_S. When a third substrate molecule binds, the final empty binding site will be modified still further by the factor c, and the dissociation constant will be $abcK_S$. Taking into account all these factors, and the occupancy weighing factors from Figure 11.4, we can write overall velocity of the enzymatic reaction as follows:

$$v = \frac{V_{\max}\left(\frac{[S]}{K_S} + \frac{3[S]^2}{aK_S^2} + \frac{3[S]^3}{a^2bK_S^3} + \frac{[S]^4}{a^3b^2cK_S^4}\right)}{1 + \frac{4[S]}{K_S} + \frac{6[S]^2}{aK_S^2} + \frac{4[S]^3}{a^2bK_S^3} + \frac{[S]^4}{a^3b^2cK_S^4}} \tag{11.2}$$

Equation 11.2 provides a velocity equation that can account for either positive

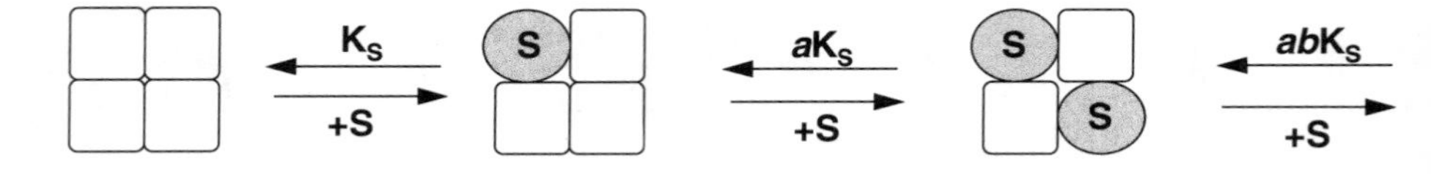

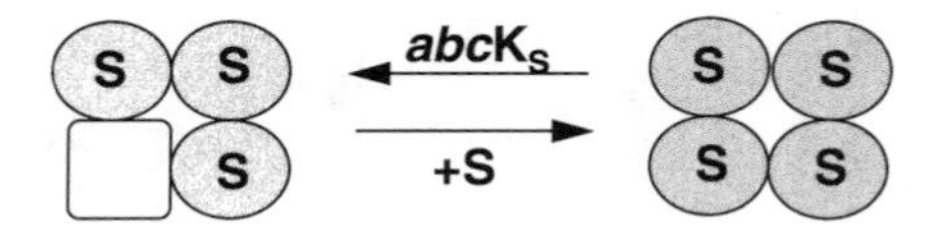

Figure 11.6 Extension of the Koshland model, from Figure 11.5, to a tetrameric enzyme.

or negative homotropic cooperativity, depending on the numerical values of the coefficients *a*, *b*, and *c*. In this model each binding event is associated with a separate conformational readjustment of the enzyme. Since, however, the effects are additive and progressive, once a single ligand has bound, the subsequent steps are strongly favored.

The second model for homotrophic cooperativity is the concerted transition or symmetry model, which is also known as the MWC model in honor of its original proponents: Monod, Wyman, and Changeux. This model assumes that allosteric enzymes are oligomers made up of identical minimal units (subunits or "protomers") arranged symmetrically with respect to one another and that each unit contains a single ligand binding site. The overall oligomer can exist in either of two conformational states, reflecting either a change in quaternary structure or tertiary structure changes within the individual protomer units, and these two conformations are in equilibrium. Another feature of the MWC model is that the transition between the two conformational states occurs with a retention of symmetry. For this to be so, all the protomer units must change in concert—one cannot have an oligomer in a mixed conformational state (i.e., some protomers in one conformation and some in the other).

Hence, in contrast to the Koshland model, in the MWC model the transition between the two conformational states is highly concerted, and there are no hybrid or intermediate states involved. The affinity of the ligand binding site on a protomer depends on the conformational state of that protomer unit. In other words, the ligand of interest will bind preferentially to one of the two conformational states of the protomer. Thus, binding of a ligand to one binding site will shift the equilibrium between the conformational states in favor of the preferred ligand binding conformation. Since the protomeric units of the oligomer shift conformation in concert, ligand binding to one site has the effect of switching all the ligand binding sites to the higher affinity form. Thus the

MWC model explains strong positive cooperativity in terms of the observation that occupancy at a single ligand binding site induces all the other binding sites of the oligomeric protein to take on their high affinity conformation.

The original MWC model assumes that the conformational state with low ligand affinity is a strained structure and that the strain is relieved by ligand binding and the associated conformational transition. For this reason, the state of low binding affinity is often referred to as the "T" state (for tense), and the high affinity conformation is referred to as the "R" state (for relaxed). While the transitions between these states are concerted, as described, for bookkeeping purposes diagrams of the MWC model designate different ligand occupancy states of the two conformations as R_x and T_x, where x indicates the number of ligand bound to the oligomer. Therefore, a tetrameric enzyme could in principle occur in states R_0 through R_4, and T_0 through T_4. The states R_0 and T_0 thus refer to the two conformational states with no ligands bound to the enzyme. The equilibrium constant between these two "empty" states, the *allosteric constant*, is symbolized by L:

$$L = \frac{[T_0]}{[R_0]} \tag{11.3}$$

This dissociation constant of a binding site for ligand, S, on a protomer in the T state is termed K_{ST}, and for the protomer in the R state this dissociation constant is K_{SR}. The ratio K_{SR}/K_{ST} is referred to as the *nonexclusive binding coefficient* and is symbolized by c. Both L and c influence the degree of cooperativity displayed by the enzyme. As L becomes larger, the velocity curve for the enzymatic reaction displays greater sigmoidal character, because the R_0–T_0 equilibrium favors the T_0 state more. As c increases, the affinity of the T state for ligand increases relative to the R state. Hence, high cooperativity is associated with small values of c.

The simplest example of the MWC model is that for a dimeric enzyme in which the T state is assumed to have no affinity at all for the substrate (i.e., $c = 0$). Figure 11.7 schematically represents the equilibria involved in such a system, where substrate binds exclusively to the R state of the dimer. (Since only the R state has a noninfinite substrate dissociation constant here, we shall use the symbol K_S in place of K_{SR} for this system.) The velocity equation for such a system, which can be derived from a rapid equilibrium set of assumptions, yields the following functional form:

$$v = \frac{V_{max}\dfrac{[S]}{K_S}\left(1 + \dfrac{[S]}{K_S}\right)}{L + \left(1 + \dfrac{[S]}{K_S}\right)^2} \tag{11.4}$$

For oligomers larger than dimers, a generalized form of Equation 11.4 can be

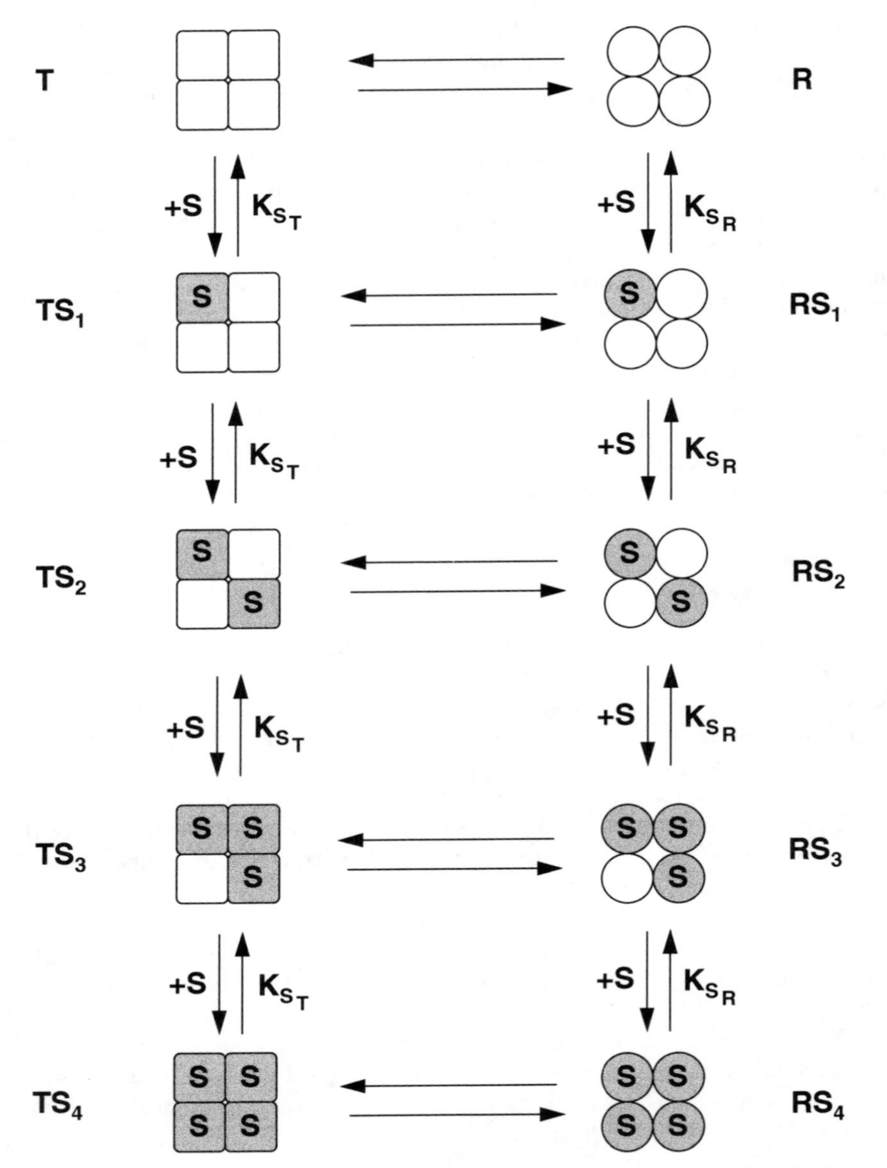

Figure 11.7 Schematic representation of the equilibria involved in the binding of substrates to a tetrameric enzyme according to the model of Monod, Wyman, and Changeux (1965). See text for further details.

derived (Segel, 1975):

$$v = \frac{V_{max} \dfrac{[S]}{K_S}\left(1 + \dfrac{[S]}{K_S}\right)^{h-1}}{L + \left(1 + \dfrac{[S]}{K_S}\right)^{h}} \qquad (11.5)$$

where h is the total number of ligand binding sites on the oligomeric enzyme.

The MWC model provides a useful framework for understanding positive homotropic cooperativity, and it can be modified to account for heterotropic cooperativity as well (Segel, 1975; Perutz, 1990). This model cannot, however, account for the phenomenon of negative homotrophic cooperativity. When negative cooperativity is encountered, it is usually explained in terms of the Koshland sequential interaction model.

11.3 Effects of Cooperativity on Velocity Curves

Referring back to the Koshland simple sequential interaction model, we can state that if the cooperativity is large, the concentrations of species with at least one substrate binding site unsaturated will be very small at any concentration of substrate greater than K_S. In the case of a tetrameric enzyme, for example, under these conditions Equation 11.2 reduces to the much simpler equation:

$$v = \frac{V_{max}[S]^4}{K' + [S]^4} \qquad (11.6)$$

where $K' = a^3b^2cK_S^4$. Equation 11.6 is a specific case (i.e., for tetrameric enzymes) of the more general simple equation:

$$v = \frac{V_{max}[S]^h}{K' + [S]^h} \qquad (11.7)$$

in which h is the total number of substrate binding sites on the oligomeric enzyme molecule and K' is a constant that relates to the individual interaction coefficients a through h, and the intrinsic dissociation constant K_S. Note that in the absence of cooperativity, and when $h = 1$, Equation 11.6 reduces to an equation reminiscent of the Henri–Michaelis–Menten equation from Chapter 5. When cooperativity occurs, however, the constant K' no longer relates to the concentration of substrate required for the attainment of half maximal velocity.

Equation 11.7 is known as the Hill equation, and the coefficient h is referred to as the Hill constant. This simple form of this equation can be readily used to fit experimental data to enzyme velocity curves, as introduced in Chapter 5 (see Figure 5.14). When the degree of cooperativity is moderate, however, contributions from intermediate occupancy species (i.e., number of bound substrate molecules $< h$) may contribute to the overall reaction. In these cases, the experimental data are often still well modeled by Equation 11.7, although the empirically determined value of h will no longer reflect the total number of binding sites on the enzyme, and may not in fact be an integer. In this situation the experimentally determined coefficient is referred to as an *apparent h* value (sometimes given the symbol h_H). The next highest integer value above the apparent h value is considered to represent the *minimum* number of binding sites on the oligomeric enzyme.

For example, suppose that the experimentally determined value of h is 1.65. This could be viewed as representing and enzyme with 1.65 highly cooperative substrate binding site, but of course this make no physical sense. Instead we might say that the minimum number of binding sites on this enzyme is 2 and that the sites display a more moderate level of cooperativity. However, there is no compelling evidence from this experiment that the enzyme has only two binding sites. It could have three or four or more binding sites with weaker intersite cooperativity. This is why the value of 2 in this example is said to represent the minimum number of possible binding sites.

As we saw in Chapter 5, the Hill equation can be linearized by taking the logarithm of both sides and rearranging to yield:

$$\log\left(\frac{v}{V_{max} - v}\right) = h\log[\mathrm{S}] - \log(K') \tag{11.8}$$

This equation can be used to construct linearized plots from which the values of h and K' can be determined graphically. An example of a linearized Hill plot was given in Chapter 5 (Figure 5.15). Despite the form of Equation 11.8, the experimental graphs usually deviate from linearity in the low substrate region, where species with fewer than h substrate molecules bound can contribute to the overall velocity. Typically, the data conform well to a linear function between values of [S] yielding 10–90% saturation (i.e., V_{max}). The slope of the best fit line between these limits is commonly taken as the average value of h_H.

The degree of sigmoidicity of the direct velocity plot is a measure of the strength of cooperativity between sites in an oligomeric enzyme. This is best measured by taking the ratio of substrate concentrations required to reach two velocities representing different fractions of V_{max}. Most commonly this is done using the substrate concentrations for which $v = 0.9V_{max}$, known as $[\mathrm{S}]_{0.9}$, and for which $v = 0.1V_{max}$, known as $[\mathrm{S}]_{0.1}$. The ratio $[\mathrm{S}]_{0.9}/[\mathrm{S}]_{0.1}$, the *cooperativity index*, is an inverse measure of cooperative interactions; in other words, the larger the difference in substrate concentration required to span the range of $v = 0.1V_{max}$ to $v = 0.9V_{max}$, the larger the value of $[\mathrm{S}]_{0.9}/[\mathrm{S}]_{0.1}$ and the weaker the degree of cooperativity between sites. The value of the cooperativity index is related to the Hill coefficient h, and K' as follows:

When $v = 0.9V_{max}$

$$v = 0.9V_{max} = \frac{V_{max}[\mathrm{S}]_{0.9}^{h}}{K' + [\mathrm{S}]_{0.9}^{h}}$$

$$\therefore [\mathrm{S}]_{0.9} = (9K')^{1/h} \tag{11.9}$$

and when $v = 0.1V_{max}$,

$$v = 0.1V_{max} = \frac{V_{max}[S]_{0.1}^h}{K' + [S]_{0.1}^h}$$

$$\therefore [S]_{0.1} = \left(\frac{K'}{9}\right)^{1/h} \tag{11.10}$$

Therefore:

$$\frac{[S]_{0.9}}{[S]_{0.1}} = \frac{(9K')^{1/h}}{(K'/9)^{1/h}} = (81)^{1/h} \tag{11.11}$$

or

$$h = \frac{\log(81)}{\log\left(\frac{[S]_{0.9}}{[S]_{0.1}}\right)} \tag{11.12}$$

Thus the Hill coefficient and the cooperativity index for an oligomeric enzyme can be related to each other, and together they provide a measure of the degree of cooperativity between binding sites on the enzyme and the minimum number of these interacting sites.

While the Hill coefficient is a convenient and commonly used index of cooperativity, it is not a direct measure of the change in free energy of binding ($\Delta\Delta G$) that must exist in cooperative systems. Recently Forsen and Linse (1995) presented a thermodynamic treatment of cooperativity for a two-site system that discusses the changes in binding affinities in terms of changes in binding free energies. This alternative treatment offers a straightforward means of describing the phenomenon of cooperativity in more familiar thermodynamic terms. It is well worth reading.

Another useful method for diagnosing the presence of cooperativity in enzyme kinetics is to plot the velocity curves in semilog form (velocity as a function of log[S]), as presented in Chapter 7 for dose–response plots of enzyme inhibitors. Such plots always yield a sigmoidal curve, regardless of whether cooperativity is involved. The steepness of the curve, however, is related to the degree of positive or negative cooperativity. When the enzyme displays positive cooperativity, the curves reach saturation with a much steeper slope than in the absence of cooperativity. Likewise, when negative cooperativity is in place, the saturation curves displays a much shallower slope (Neet, 1980). The data in these semilog plots is still well described by Equation 11.7, as illustrated in Figure 11.8 for examples of positive cooperativity. The steepness of the curves in these plots is directly related to the value of h that appears in Equation 11.7.

These plots are useful because the presence of cooperativity is very readily apparent in them. The effect of positive or negative cooperativity on the steepness of the curves is much more clearly observed in the semilog plot as

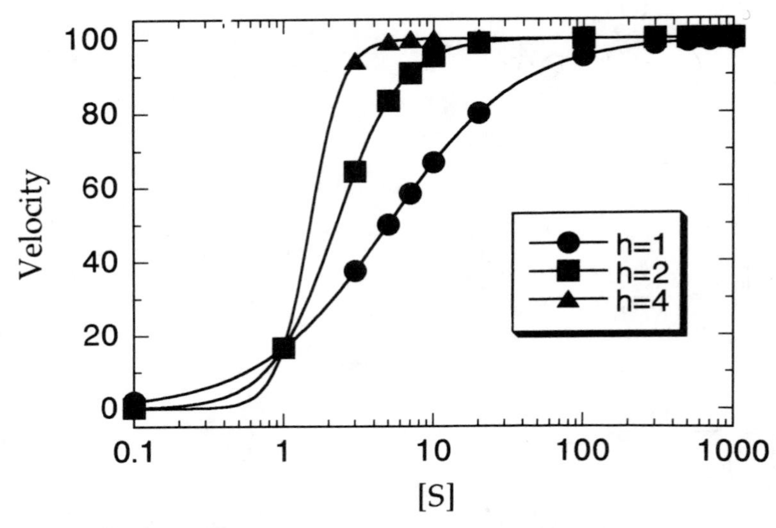

Figure 11.8 Velocity as a function of substrate concentration plotted in semilog fashion. The circles represent the data for a noncooperative enzyme, for which the apparent Hill coefficient is 1. The squares and triangles represent data for enzymes displaying positive cooperativity, with Hill coefficients of 2 and 4, respectively. Each solid line through the data represents the best fit of an individual data set to Equation 11.7.

opposed to the linear plot, especially in the case of small degrees of cooperativity. The steepness of the curves in such semilog plots is also diagnostic of cooperative effects in ligands other than substrate. Thus, for example, the IC_{50} equation introduced in Chapter 7 (Equation 7.21) can be modified to include a term to account for cooperative effects in inhibitor binding to enzymes as well.

11.4 Sigmoidal Kinetics for Nonallosteric Enzymes

The appearance of sigmoidal kinetics in enzyme velocity curves for allosteric enzymes is a reflection of the cooperativity of the substrate binding events that precede the catalytic steps at the enzyme active sites. The same cooperativity should be realized in direct studies of ligand binding by the enzyme, which can be performed by equilibrium dialysis, certain spectroscopic methods, and so on. If true allostery is involved, the cooperativity of ligand binding should be measurable in the enzyme velocity curves and in the separate binding experiments as well. In some cases, however, the direct ligand binding experiments fail to display the same cooperativity observed in the velocity measurements.

One must assume that such ligand binding events are not cooperative, which means that some other explanation must be sought to account for the sigmoidal velocity curve.

One way of observing sigmoidal kinetics in the absence of true cooperativity entails an enzyme preparation containing a mixture of enzyme isoforms that have different K_m values for the substrate (Palmer, 1985). In such cases the velocity curve will be the superposition of the individual curves for the varied isoforms. If two or more isoforms differ significantly in K_m for the substrate, a nonhyperbolic curve, resembling the sigmoidal behavior of cooperative enzymes, may result.

Also, it has been noted that a two-substrate enzyme that follows a random ordered mechanism can display sigmoidal kinetics without true cooperativity. This occurs when one of the two ordered reactions proceeds faster than the competing ordered reaction; for example, formation of E·AX then E·AX·B and subsequent product release is faster than formation of E·B then E·B·AX and product release. In the case of two ordered reactions of unequal speed, the affinity of the free enzyme for substrate B is less than the affinity of the E·AX complex for B. If $[E_t]$ and [B] are held constant while [AX] is varied, at low concentrations of AX the enzyme will react mainly with substrate B first, and thus will proceed through the slower of the two pathways to product. As the concentration of AX increases, there will be a greater probability of the enzyme first binding this substrate and proceeding via the faster pathway. The observed result of this pathway "switching" with increasing substrate concentration is a sigmoidal plot of velocity as a function of [AX].

Finally, sigmoidal kinetics can be observed even for a monomeric single binding site enzyme if substrate binding induces a catalytically required conformational transiton of the enzyme. If the isomerization step after substrate binding is rate limiting, the relative populations of the two isomers, E and E′, can influence the overall reaction velocity. If the equilibrium between E and E′ is perturbed by substrate, the relative populations of these two forms of the enzyme will vary with increasing substrate concentration. Again, the end result is the appearance of a sigmoidal curve when velocity is plotted as a function of substrate concentration.

11.5 Summary

In this chapter we presented the concept of cooperative interactions between distal binding sites on oligomeric enzymes, which communicate through conformational transitions of the polypeptide chain. These allosteric enzymes display deviations from the normal Henri–Michaelis–Menten behavior that is seen with single substrate binding enzymes, as introduced in Chapter 5. Examples of allosteric proteins and enzymes were described that provide some structural rationale for allosteric interactions in specific cases, and two theoretical models of cooperativity were described. The classic signature of cooperativity in enzyme kinetics is a sigmoidal shape to the curve of velocity versus [S]. The appearance of such sigmoidicity in the enzyme kinetics is not sufficient, however, to permit us to conclude that the substrate binding sites interact cooperatively. Direct measurements of ligand binding must be used to confirm the cooperativity of ligand binding. We saw that in some cases sigmoidal enzyme kinetics exist in the absence of true cooperativity—when, for example, multisubstrate enzymes proceed by different rates depending on the order of substrate addition, and when rate-limiting enzyme isomerization occurs after substrate binding.

The understanding of allostery and cooperativity in structural terms is an active area of research today. This fascinating subject was recently reviewed by one of the leading experts in the field of allostery, Max Perutz, who has spent most of his career studying the structural determinants of cooperativity in hemoglobin. The text by Perutz (1990) is highly recommended for those interested in delving deeper into this subject.

References and Further Reading

Abelson, P. H. (1954) *J. Biol. Chem.* **206**, 335.

Forsen, S., and Linse, S. (1995) *Trends Biochem. Sci.* **20**, 495.

Koshland, D. E., Nemethy, G., and Filmer, D. (1966) *Biochemistry*, **5**, 365.

Monod, J., Wyman, J., and Changeux, J. P. (1965) *J. Mol. Biol.* **12**, 88.

Neet, K. E. (1980) *Methods Enzymol.* **64**, 139.

Palmer, T. (1985) *Understanding Enzymes*, Wiley, New York, pp. 257–274.

Perutz, M. (1990) *Mechanisms of Cooperativity and Allosteric Regulation in Proteins*, Cambridge University Press, New York.

Segel, I. H. (1975) *Enzyme Kinetics*, Wiley, New York, pp. 346–464.

APPENDIX I

Suppliers of Reagents and Equipment for Enzyme Studies

Some of the commercial suppliers of reagents and equipment that are useful for enzyme studies are given here. A more comprehensive listing can be found in the ACS Biotech Buyers Guide, which is published annually. The Buyers Guide can be obtained from the American Chemical Society, 1155 16th Street N.W., Washington, DC 20036. Telephone (202) 872-4600.

Aldrich Chemical Company, Inc.
940 West Saint Paul Avenue
Milwaukee, WI 53233
(800) 558-9160

Amersham Corporation
2636 South Clearbrook Drive
Arlington Heights, IL 60005
(800) 323-9750

Amicon
24 Cherry Hill Drive
Danvers, MA 01923
(800) 343-1397

Bachem Bioscience, Inc.
3700 Horizon Drive
Renaissance at Gulph Mills
King of Prussia, PA 19406
(800) 634-3183

Beckman Instruments, Inc.
P.O. Box 3100
Fullerton, CA 92634-3100
(800) 742-2345

Bio-Rad Laboratories
1414 Harbour Way South
Richmond, CA 94804
(800) 426-6723

Biozymes Laboratories International Limited
9939 Hilbert Street, Suite 101
San Diego, CA 92131-1029
(800) 423-8199

Boehringer-Mannheim Corporation
Biochemical Products
9115 Hague Road
P.O. Box 50414
Indianapolis, IN 46250-0414
(800) 262-1640

Calbiochem
P.O. Box 12087
San Diego, CA 92112
(800) 854-9256

Eastman Kodak Company
343 State Street
Building 701
Rochester, NY 14650
(800) 225-5352

Enzyme Systems Products
6497 Sierra Lane
Dublin, CA 94568
(510) 828-6618

Hamilton Instruments
P.O. Box 100030
Reno, NV 89520
(702) 786-7077

Hoefer Scientific Instruments
P.O. Box 77387
654 Minnesota Street
San Francisco, CA 94107
(800) 227-4750

Millipore Corporation
80 Ashby Road
Bedford, MA 01730
(800) 225-1380

Novex
4202 Sorrento Valley Boulevard
San Diego, CA 92121
(800) 456-6839

Pharmacia LKB Biotechnology AB
800 Centennial Avenue
Piscataway, NJ 08854
(800) 526-3618

Pierce Chemical Company
P.O. Box 117
Rockford, IL 61105
(800) 874-3723

Schleicher & Schuell, Inc.
10 Optical Avenue
Keene, NH 03431
(800) 245-4024

Sigma Chemical Company
P.O. Box 14508
St. Louis, MO 63178
(800) 325-3010

Spectrum Medical Industries, Inc.
1100 Rankin Road
Houston, TX 77073-4716
(800) 634-3300

United States Biochemical Corporation
P.O. Box 22400
Cleveland, OH 44122
(800) 321-9322

Upstate Biotechnology, Inc.
199 Saranac Avenue
Lake Placid, NY 12946
(800) 233-3991

Worthington Biochemical Corporation
Halls Mill Road
Freehold, NJ 07728
(800) 445-9603

APPENDIX

II

Useful Computer Software for Enzyme Studies

There is available a large and growing number of commercial software packages that are useful for enzyme kinetic data analysis. Also, several authors have published the source code for computer programs they have specifically written for enzyme kinetic analysis and other aspects of enzymology. I have listed some of the programs I have found useful in the analysis of enzyme data, together with the source of further information about them. This list is by no means comprehensive, but rather gives a sampling of what is available. The material is provided for the convenience of the reader; I make no claims as to the quality or accuracy of the programs.

Another useful source of programs and information about enzymes and protein biochemistry in general is the Internet. A recent search of the term "Enzyme" on the Internet yielded more than 600 addresses with useful information and analysis packages that can be accessed electronically. It is well worth the reader's time to do a little "surfing" on this new and exciting medium.

Cleland's Package of Kinetic Analysis Programs. This is a suite of FORTRAN programs written and distributed by the famous enzymologist W. W. Cleland. The programs include methods for simultaneous analysis of multiple data for determination of inhibitor type and relevant kinetic constants, as well as statistical analyses of ones data. Reference: W. W. Cleland, *Methods Enzymol.* **63**, 103 (1979).

Enzfitter. A commercial package for data management and graphic displays of enzyme kinetic data. [See **Ultrafit** for similar version, compatible with

Macintosh hardware.] Available from Biosoft, P.O. Box 10938, Ferguson, MO. Telephone: (314) 524-8029. Internet address: ab47@cityscape.co.uk.

Enzyme Kinetics. A commercial package for data management and graphic displays of enzyme kinetic data. Distributed by ACS Software, Distribution Office, P.O. Box 57136, West End Station, Washington, DC 20037. Telephone: (800) 227-5558.

EZ-FIT. A practical curve-fitting program for the analysis of enzyme kinetic data. Reference: F. W. Perrella, *Anal. Biochem.* **174**, 437 (1988).

k·cat. A commercial package for data management and graphic displays of enzyme kinetic and receptor ligand binding data. Available from BioMetallics, Inc., P.O. Box 2251, Princeton, NJ 08543. Telephone: (800) 999-1961.

Kinlsq. A program for fitting kinetic data with numerically integrated rate equations. Provides data analysis routines for tight binding inhibitors as well as classical inhibitors. Reference: W. G. Gutheil, C. A. Kettner, and W. W. Bachovchin, *Anal. Biochem.* **223**, 13 (1994).

MPA. A program for analyzing enzyme rate data obtained from a microplate reader. Provides a convenient means of transforming and analyzing data directly from 96-well formatted data arrays. Reference: S. P. J. Brooks, *BioTechniques*, **17**, 1154 (1994).

Ultrafit. Similar to Enzfitter software, but designed for Apple Macintosh computers. Available from Biosoft, P.O. Box 10938, Ferguson, MO. Telephone: (314) 524-8029. Internet address: ab47@cityscape.co.uk.

Index